The Subjectivities an Occupational Risk

MW00682250

The Subjectivities and Politics of Occupational Risk links restructuring in three industries to shifts in risk subjectivities and politics, both within workplaces and within the safety management and regulative spheres, and examines the conflict and changes in law, political discourses and management approaches that often result. The state and corporate governance emphasis on worker participation and worker rights, internal responsibility, and self-regulative technologies are understood as corporate and state efforts to reconstruct control and responsibility for Occupational Health and Safety (OHS) risks within the context of a globalized neoliberal economy. Part 1 presents a conceptual framework for understanding the subjective bases of worker responses to health and safety hazards using Bourdieu's concept of habitus and the sociology of risk concepts of trust and uncertainty. Part 2 demonstrates the restructuring arguments using three different industry case studies of multiple mines, farms and auto parts plants.

The final chapter draws out the implications of the evidence and theory for social change and presents several recommendations for a more worker-centred politics of health and safety. The book will appeal to social scientists interested in health and safety, work, employment relations and labour law, as well as worker advocates and activists.

Alan Hall recently retired from the Department of Sociology at Memorial University where he now holds an Honorary Research Professor position. His most recent 2020 publications are a co-authored book on employment standards published by University of Toronto Press, and an article in *Economic and Industrial Democracy* on vulnerable workers.

Routledge Advances in Sociology

For more information about this series, please visit: https://www.routledge.com/Routledge-Advances-in-Sociology/book-series/SE0511

The Subjectivities and Politics of Occupational Risk

Mines, Farms and Auto Factories

Alan Hall

Routledge
Taylor & Francis Group

LONDON AND NEW YORK

First published 2021
by Routledge
2 Park Square, Milton Park, Abingdon, Oxon OX14 4RN

and by Routledge
52 Vanderbilt Avenue, New York, NY 10017

Routledge is an imprint of the Taylor & Francis Group, an informa business

British Library Cataloguing-in-Publication Data
A catalogue record for this book is available from the British Library

Library of Congress Cataloging-in-Publication Data
A catalog record has been requested for this book

ISBN: 978-0-367-46995-5 (hbk)
ISBN: 978-1-003-03268-7 (ebk)

Typeset in Times New Roman
by codeMantra

This book is dedicated to all workers in Canada who have suffered injury and death in their workplaces, and to all those worker representatives and activists who have struggled diligently to improve their workplaces.

Contents

Graphs

Tables

Figures

Acknowledgements

I thank my colleagues for their crucial participation in the research data collection, including Dr Anne Forrest (University of Windsor) and Dr Alan Sears (Ryerson University) for the auto study research and Dr Veronica Mogyrody for the organic farm study research. For the accident and injury reporting study, thanks to Dr Tanya Basok (University of Windsor), Dr Jamey Essex (University of Windsor) and Dr Urvashi Soni Sinha (University of Windsor) for their work. For their invaluable contributions to the second worker representative study, thanks to Andy King (USWA), Dr Wayne Lewchuk (McMaster University), Dr John Oudyk (OHCOW) and Dr Syed Naqvi (OHCOW). A special thanks as well to several research assistants who worked on these studies, especially Dr Susan Sverdup-Phillips and Niki Carlan. I also wish to acknowledge the funding received from the Social Science and Research Council Canada, Auto21, the University of Windsor Environmental Centre and the Ontario Workplace Safety and Insurance Board. A major expression of gratitude to all the workers, farmers, union officials, managers and others who participated in these studies. I have always felt that I failed to fully present your voices in my previous work and I hope this presentation rectifies that failure. Finally, a special thanks to my partner, Lynne Phillips, who has cheerfully endured my frustration and grumpiness while encouraging the completion of this work.

Abbreviations

CAW	Canadian Auto Workers Union
CF	Cut and Fill Mining Method
CFIB	Canadian Federation of Independent Business
CWS	Classification Wage System (INCO)
HRM	Human Resources Management
IAPA	Industrial Accident Prevention Association
IRS	Internal Responsibility System
ISO	International Safety Organization
KA	Knowledge Activist Reps.
LOARC	Labour, OHCOW and Academics Research Collaboration
LSI	Lost Time Injuries
MA	Medical Aid Injuries without Lost Time
MAPAO	Mining Accident Prevention Association of Ontario
MSD	Musculo-Skeletal Disorder
OFSA	Ontario Farm Safety Association
OHCOW	Occupational Health Clinics for Ontario Workers
OHS	Occupational Health and Safety
OHSA	Occupational Health and Safety Act
OMAF	Ontario Ministry of Agriculture and Food
OMOL/MOL	Ontario Ministry of Labour
ONA	Ontario Nurses Association
OPSEU	Ontario Public Service Employees Union
PA	Political Activist Reps.
QWL	Quality of Work Life
Reps. OHS	Representatives
RSI	Repetitive Strain Injury
RT	Risk Theory
SF	Safety First
SFT	Social Field Theory
SM	Scientific Management
TL	Technical-Legal Reps.
TLV	Threshold Limit Values

UCF	Undercut and Fill Mining Method
UFCW	United Food and Commercial Workers International Union
USWA	United Steelworkers of America
VRM	Vertical Retreat Mining
WAC	Workers Action Centre (Toronto)
WWAC	Windsor Workers' Action Center
WHSC	Workers Health and Safety Centre
WOHIS	Windor Occupational Health Information Service
WOSH	Windsor Occupational Safety and Health Group

Chapter 1

Introduction and research methods

Though they rarely garner the headlines of a terrorist bombing, workplace injuries can be just as horrific. The young summer student who is torn apart in a large mixing machine, the middle-aged worker whose scalp is ripped from her head when her hair is caught in a machine or the older worker who is burnt to ashes in a blast furnace, each event unspeakable in terms of the pain and horror for the persons involved, to speak nothing of the victims' families, co-workers and friends (Dudgeon, 2016). Workplace-contracted diseases such as asbestosis or silicosis are no less disturbing (Leyton, 1975: Storey and Lewchuk, 2000). The slow painful deterioration and death that many workers and their families are forced to endure make one wonder if the people who suffer a quick accidental death are the lucky ones. Even those who live through the excruciating pain of non-fatal chronic ailments from musculoskeletal degeneration (MSDs) to repetitive injuries (RSIs) making life a living hell for many may well pause to wonder if the living workplace victims are the ones who drew the short straws (Beardwood, Kirsh and Clark, 2005; Tarasuk and Eakin, 1995). Of course, thinking about which is worse – the quick fatal insult or the agonizing deterioration of the body – is not what is on the minds of most injured workers or their families. Along with getting through each day, for them the nagging question is why did this happen?

In any standard occupational health management textbook, the usual answer to this question is that some responsible person or persons did not take the correct steps to prevent the accident or exposure from happening (Dyck, 2015; Fingnet and Smith, 2013; Kelloway, Francis and Gatien, 2016). While always insisting that prevention requires committed and knowledgeable management and expert leadership to ensure that the safety systems, proper equipment, training and procedures are in place, the worker is often cast in this narrative as the weakest link in the health and safety chain (Crawford, 1977; Geller, 2000). Accordingly, although occupational health and safety (OHS) management theory (as well as OHS law) place primary responsibility on managers to ensure and communicate clear procedures and goals, and train and motivate workers to follow those procedures, once those boxes are ticked, it is ultimately the workers who carry the final responsibility for following the rules (Hakkinen, 2015). To ensure that these boxes are ticked, management textbooks and management safety experts advise employers to set up and implement *OHS management* systems (Elefterie, 2012; Frick, 2000, 2011; Saksvik and Quinlan, 2003), which establish organizational structures, procedures, rules, standards and auditing mechanisms aimed at ensuring a *systematic* approach to OHS risk assessment and management. The widely acknowledged need for occupational health and safety law, regulations and enforcement, which have increasingly included regulative requirements for safety management systems, recognize that not all firms or managers have seen the light offered by the OHS management literature and must be persuaded or induced in some way to embrace safety management principles and technologies. Despite the lack of strong evidence on effectiveness (Breslin et al., 2006b; Frick, 2011; Robson et al., 2007; Walters, 2003), advocates and authorities frequently claim that declines in the official rates of workplace injuries reflect the widening adoption of these principles and systems by employers and managers (CCOHS, 2020).

With a few notable exceptions (e.g. Boyd, 2003), the assumption underlying this management literature and government policy is that there is a natural correspondence of labour and employer interests surrounding health and safety which can be realized through the adoption of OHS management principles. This discourse denies underlying contradictions or conflicts surrounding profit-making and OHS prevention. If managers of a particular firm are not committed to health and safety, it must be because they fail to understand that health and safety prevention, as constructed within the OHS management framework, works to increase efficiency, productivity and profitability. If workers are not following health and safety rules and procedures, then it is likely because their attitudes have been distorted unintentionally by poor management and/or a poor safety culture (Cunningham and Sinclair, 2014; Guldenmend, 2018). While corrective changes such as the introduction of safety teams, safety culture audits and safety incentives are often proposed, education is ultimately understood as the principal means of preventing future accidents and disease. Managers must be educated to create and abide by OHS management systems and principles, and workers must be educated and conditioned to recognize that it is in their interests to abide by the rules and standards of the OHS management system (Frick, 1997). As will be shown later, this way of thinking is also prevalent in government approaches to regulation and enforcement, expressed by a long-standing policy that education on best management practices is the primary means of achieving widespread compliance with the law (Tucker, 1990, 2003; Walters, 2003).

This book begins from a different starting position. Grounded first in labour process theory (LPT), structural conflicts between capital accumulation and worker health and safety are seen as central to under-standing the production and reproduction of workplace illness and disease (Burawoy, 1985; Navarro, 1982; Thompson, 2010; Tucker, 1990). The crea-tion and implementation of OHS management programmes or systems are understood as being shaped by these conflicts, representing to varying de-grees, substantive management efforts to limit the political and economic effects of close calls, injuries, illnesses and disruptive high-risk conditions. While control over these effects is achieved partly by managing and seek-ing to reduce the risks themselves, control is also gained by managing the risk attitudes, knowledge and responses of workers. As this argument also implies, safety management systems can serve as much to obscure hazards and their production as prevent them. More broadly, injuries, diseases and their prevention are understood first and foremost as *political* outcomes grounded in the relative *power* exercised by workers, worker representa-tives, supervisors and managers in their day-to-day production relations. By conceptualizing health and safety activities and relations as political, including what happens within and through safety management systems, the perception and understanding of risk and the means of identifying, controlling or taking risks become the central focus of analysis. The re-search and theoretical objectives accordingly centre around explaining why managers and workers accept and take risks as integral aspects of their employment relationships and roles, despite and, sometimes because, elaborate safety management systems have been established (Boyd, 2003; Frick, 2011).

There is no easy or single explanation for the many workers who have been killed, injured or made sick, and why workplaces continue to take their terrible toll despite decades of regulative, technological and management system "improvements" (AWCBC, 2018; WSIB, 2018). However, this book is premised on the argument that more effective prevention requires a critical examination of how and why workplace actors – that is, workers, worker rep-resentatives, supervisors, safety engineers and managers – come to perceive and understand the origins and risks of workplace accidents and diseases within their workplaces, and how those perceptions and understandings in turn shape their risk-related practices. Grounded in this premise, the ob-jective of this project is to examine the structural and subjective bases of workplace decisions and actions regarding workplace hazards, injuries and risks. By comparing worker and management actions within specific pro-duction contexts or, more precisely, three different kinds of workplaces – underground mines, farms and auto part manufacturing plants – this book seeks to shed light on the origins and development of workplace risk sub-jectivities and the social expression of those subjectivities in risk-related politics, practices, systems and technologies. Although nested in a work-place-level analysis of *production politics* (Burawoy, 1985), this project also employs a political economy approach consistent with LPT which locates and compares specific workplaces and corporate entities within a broader industrial and political economic context, emphasizing in particular labour and commodity market conditions, laws, government policies and enforcement, and broader cultural discourses on governance, risk, safety and health (Jessop, 1995; Tucker, 1990; Vosko et al., 2020).

Building answers: risk, control and responsibility

A large body of research shows that workers often blame themselves or other workers at least in part for their workplace injuries or diseases (Basok, Hall and Rivas, 2014; Breslin et al., 2007; Crawford, 1977; Haas, 1977; Hall, 1999, 2016; Wilde, 1999). We also know, however, that many injured workers feel just as strongly that the conditions of their job or their workplace are the primary causes of the accident or disease, and will often point to specific actions or inactions of their employer and/or specific managers and supervisors as causing what happened to them (Glasbeek and Tucker, 1999; Storey, 2006). While locating blame in different places, these narratives share a common concern with fixing responsibility grounded in perceptions of where control and power lie. As shown in this book, few workers fit neatly into internalizing or externalizing all blame for the OHS events or hazards in their workplaces. However, the contrast between the two extremes points to two closely related questions which form the central focus of this book – who (or what) is constructed as responsible for preventing injuries and diseases in a workplace, and underlying that construction of responsibility, who (or what) is understood as identifying, assessing and controlling the risk associated with working conditions and practices?

These questions are identified as central to this work for three reasons. First, the history of Canadian OHS law, state policy, management theory and worker advocacy over the last hundred years or so has revolved substantially around a struggle over employer and worker responsibility, blame and control over work and working conditions (Bennett, 2011; MacDowell, 2012; Tucker, 1990; Storey, 2005, 2006), a struggle which, I will argue, parallels historical transformations in labour processes and forms of state and management governance over work and employment. Second, those struggles have also been pacified to different degrees and in diverse ways by, on the one hand, nesting and internalizing responsibility and self-regulation *in* workers, and on the other, encouraging workers to defer to management-controlled knowledge and technology (Hall, 1999; Gray, 2009). Finally, responsibility and control are the essential concepts which link the different theoretical perspectives employed in this analysis, labour process theory, social field theory, risk theory and governance theory (Bourdieu, 1977, 1990; Burawoy, 1985; Dean, 1999, 2007; Mollering, 2006; O'Malley, 2004; Thompson, 2010; Snider, 2009; Zinn, 2004).

Fixing responsibility: a short history of occupational health and safety in Canada

The debate about who is responsible for workplace injuries and disease is not a new one. As Tucker and others (Aldrich, 1997; Figlio, 1985; Fudge and Tucker, 2001; Tucker, 1990) have shown, historical evidence on this point comes from the records of early civil court cases of the late 1800s and early 1900s as workers sought some form of compensation for their injuries. Workers were rarely successful in their early efforts to locate responsibility with employers, often because the courts measured worker and employer responsibility in a less than balanced manner. While early health and safety legislation in Canada such as the 1884 Ontario Factories Act represented a new development in workplace safety within Canada (Fudge and Tucker, 2001; Tucker, 1990), vague regulations, weak enforcement and low fines for infractions did relatively little to alter the hazardous nature of most workplaces. As limited as these early statute laws were, they introduced the idea that *employers* had legal duties to *prevent* some of the conditions seen as causing workplaces accidents and illness. This implied a level of employer control over some conditions with at least limited responsibility to prevent injuries. The move from private to public law in governing workplace safety also signalled that the state was taking external responsibility for commanding and controlling employers who failed to take reasonable preventive steps, implying further that the balance of power in private market relations and private law were no longer seen by the public authorities as sufficient in all cases (Fudge and Tucker, 2001).

Yet, even as the carnage of early industrialization continued and, the economic and political costs of workplace injuries and diseases mounted, Canadian governments by and large did not respond by expanding and altering their OHS regulative regimes (Fudge and Glasbeek, 1992; Fudge and Tucker, 2001; Tucker, 1990). Instead, the legislative developments through much of the first half of the 20th century in Canada, as elsewhere, were restricted to minor reforms of weakly enforced command-and-control industry-specific laws (e.g. Mining Act, Factories Act), and the introduction of injury compensation law (Figlio, 1985; Gunderson and Hyatt, 2000; Ison, 2015; Tucker, 2003). Significantly, most compensation laws in Canada and other western countries were constructed as *no fault* insurance programmes (e.g. Meredith Report, 1913; Ontario Workmen's Compensation Act, 1914) providing workers (and their dependents) with the right to fixed compensation without reference to blame or fault.

Although the no-fault framing of Ontario compensation law had removed the right of workers to sue their employers, responsibility was not entirely externalized in as much as employers were required to fund the compensation system as a cost of doing business. Yet as Ison (2015) notes, the incentive to boost prevention was limited by employers' capacity to transfer these costs to workers in the form of reduced wages. Nevertheless, as Fordism and post-depression Keynesian policies pushed increased wages and health costs, industrial employers especially in the US were encouraged to pay more attention to prevention (Heron and Storey, 1986; Taksa, 2009). This emerging idea that "management" and industrial engineering needed to attend to accident prevention was reflected most notably by the development of what is known as the "safety first" movement (SFM), which began in the early 1900s, led initially by large US corporations such as US Steel and then adopted over time across America, Canada Britain, Japan and elsewhere (Aldrich, 1997; Horiguchi, 2011; Taksa, 2009). For some analysts such as Aldrich (1997), the SFM was a much-needed effort to engineer safety into workplaces through the standardization of production steps or processes using technology, supervision and tight rule-based controls over workers. Although aspects of "safety first" programmes likely led to some reductions in injuries and fatalities, more critical analysts have pointed out that "safety first" programmes were grounded in the assumption that worker control over the labour process was a significant source of high injury rates – and that the solution was accordingly to remove as much worker control as possible through whatever means possible (Sass and Grook, 1981; Storey, 2004; Taksa, 2009). In the "safety first" (SF) context, then, injuries were understood ultimately to be the result of the workers' failure to follow specified procedures, standards and rules. Companies thus exercised their OHS responsibilities by imposing strict engineering and supervisory controls.

Less than coincidently, these prescriptions jived nicely with emerging Taylorist or scientific management (SM) approaches which were grounded in the imposition of strict procedural requirements, standards and control technologies as the means to greater productivity (Braverman, 1974; Taksa, 2009). Consistent with a key premise in SM was the SF idea that injuries, like low productivity, were caused by *too much* worker control and autonomy. While an SM approach to engineering involved the introduction of new machinery and/or altered labour processes, the emerging disciplines of industrial engineering and SF management largely took as given that these changes would simultaneously improve productivity and safety to the extent they directly limited worker control – that is as long as workers were properly trained and conditioned to follow management directions (Dwyer, 1992). As such, accident prevention within the SF movement was often defined principally by management efforts to control the individual "habits" of workers through training, reward systems, close supervision and disciplinary actions. It was in this context of heightened responsibilization of workers that we start to see concepts such as "accident proneness" and "operator error" gaining popularity among managers and engineers to explain the persistence of workplace injuries in the modern industrial workplace (Burnham, 2009; Crawford, 1977; Greenwood and Woods, 1919; Rawson, 1944; Sass and Grook, 1981).

Although labour process theorists traditionally see SM as reducing worker decision-making and control over the labour process (Braverman, 1974; Burawoy, 1985; Edwards, 1979; Heron and Storey, 1986), adoptions of SM and SF management often included experiments with the use of worker committees (Taksa, 2009). Some management historians saw the creation of worker committees as further evidence that SM and SF were part of a liberal progressive period in the US aimed at improving the lives and health of workers (Savino, 2016; Schacter, 2010). A more critical interpretation, as Taksa (2009, p. 19) argued, was that the committees likely reflected not only the demonstrated capacity of workers to organize and resist SM collectively, but also the recognition that even within the context of rapid industrialization, employers were still often dependent on worker knowledge and discretion to achieve their productivity goals (see also Ramsay, 1977).

The 1940s to the 1950s were periods of major change in labour relations law in Canada, notably the introduction of provincial and federal "Labour Relations Acts" which provided labour unions with their first substantive state protection around unionization and strike powers (Fudge and Tucker, 2001). However, very little changed in workplace safety legislation or enforcement during this time period. The same limited command-and-control laws in health and safety that had been established in the late 1800s and 1900s were still in place, along with their long-standing pattern of weak enforcement (Carson, 1982; Doern, 1977; Tucker, 1990). It was not until the 1970s that governments in Saskatchewan, Ontario and elsewhere in Canada moved away from the industry-specific command-and-control acts (e.g. Factory Acts, Mining Acts) and introduced legislation grounded in a self-regulative participative model of OHS legislation. The new laws revolved substantially around general duties of care and performance and process requirements while mandating worker representation and joint health and safety committees as core features of a cooperative self-regulative system (Doern, 1977; Tucker, 1995, 2003). While different in some ways, similar self-regulative forms of legislation were introduced around the same time in Australia, New Zealand and Britain (Sirrs, 2015, 2016; Walters, 2003; Williams, 1995). In Ontario, Canada's most industrialized province, this self-regulative model was framed substantially by a Royal Commission on Health and Safety in Mining under the direction of Dr James Ham. The Commission was established after a 1974 wildcat uranium mine strike in Elliot Lake, Ontario, over high rates of cancer deaths among the miners (MacDowell, 2012; Ontario, 1976).

Central to this legislative development was an enforcement philosophy, known in Canada as the "internal *responsibility* system" (IRS), a term coined in the Commission's Report (Ontario, 1976, p. 250). The IRS continues to this day as the core OHS governance policy in Ontario and much of Canada (OMOL, 2010, 2019). Backed by new worker participation rights including the right to joint advisory labour/management health and safety committees, the right to hazard information and the right to refuse unsafe work, this approach to OHS governance pushed the idea that workers and employers were workplace partners with *common* interests and *separate* spheres of influence and responsibilities in prevention (Tucker, 2003). Enforcement of the safety standards was achieved through internal self-regulation by the two cooperating parties, workers and management. *Both* workers and employers were given explicit responsibilities to prevent workplace injuries and disease. Although the threat of more significant penalties including prison sentences was meant to signal the new seriousness with which the government was approaching occupational health and safety, the primary emphasis was on internal responsibility and self-regulation, which meant ultimately a relatively weak external enforcement regime that emphasized compliance through education rather than deterrence (Bittle and Snider, 2015; Storey and Tucker, 2006; Tucker, 2003; Walters, 1985). Given that the legislation also moved away from command and control regulations, the responsibility for defining workplace OHS hazards and acceptable levels of risk fell increasingly to the managers and workers of workplace specific IRSs.

Within this IRS discourse, workers' interests in preventing accidents and disease were largely assumed as given while considerable government attention was directed at identifying and highlighting employer interests. The core insight is captured by the simple and often repeated mantra that "safety pays." While this was initially cast mainly in terms of compensation savings (Walters, 1985), the "safety pays" discourse was transformed into the much broader idea that attention to safety automatically leads to better productivity, quality and profitability. As noted, this mantra was also central to current OHS management theory and government policy (Kelloway, Francis and Gatien, 2016; OMOL, 2010; Sirrs, 2015). As a government's appointed "Expert Occupational Health and Safety Panel" stated in its 2010 report, "sophisticated business owners and CEOs understand that health and wellness in the workplace go hand in hand with productivity, quality and profitability" (OMOL, 2010, p. 6). The government belief that some employers would act upon this "truth" in their approach to health and safety management was critical to the promotion of the original proposal of internal responsibility, in as much as it underscored the basic faith that the IRS works as long as employers come to the realization that they can "partner" with workers to achieve cost savings, quality and productivity improvements at the same time as they improve health and safety. Again, as Walters (2003) and others have noted, similar developments occurred in the UK, Australia and some Nordic countries (Frick, 1997; Robson et al., 2005; Sirrs, 2016; Tombs and Whyte, 2007).

While the Ontario government continued to insist that employers have common interests with labour in preventing injuries and disease, the province's workers and unions soon discovered that their new rights left much to be desired. As worker and union protests mounted, several studies including a cross-province survey done by the tripartite Ontario Advisory Council on Occupational Health and Safety (OACOHS, 1986) confirmed that many workers were unable to exercise their rights to safety information, participate in safety decisions or refuse unsafe work (see also Burkett, Riggin and Rothney, 1981; Hall, 1992; NDP, 1983, 1986; OACOHS, 1986; Storey, 2004; Storey and Tucker, 2006; Tucker, 1995; Walters and Haines, 1988). In the Advisory Council survey, for example, 78% of the Ontario employers surveyed were not in full compliance with the committee and representation elements of the Occupational Health and Safety Act.

The early controversies surrounding the Ontario Health and Safety Act (OHSA) (e.g. see McKenzie and Laskin, 1987; NDP, 1986) ultimately led to Bill 208 in 1990, the first major reform of the Act (OHSA, 1990; OMOL, 1989a,b). While the reform contained some new provisions strengthening institutional procedures and requirements for joint committees, the major proposal from the Labour side at the time, the right of worker representatives to unilaterally shut down unsafe work sites, was not introduced in the final version of the law after vigorous opposition, much of it from small businesses (CFIB, 1990; Hall, 1992; Storey and Tucker, 2006, p. 165). The requirement for management policies and programmes were also defined more clearly and linked explicitly to joint health and safety committees in terms of policy development, implementation and monitoring.

Relatively few changes were introduced to OHSA over the next 20 plus years, despite several turnovers in the ruling political parties and despite continuing evidence of weak enforcement and weaknesses in the IRS (Basok, Hall and Rivas, 2016; Eakin and MacEachen, 1998; Gray, 2002; Lewchuk, Robb and Walters, 1996; MacEachen, 2000; O'Grady, 2000). External enforcement levels have gone up and down over the years but throughout, Ontario governments have continued to anchor their approach to health and safety to the internal responsibility system, including the idea of worker participation in all its forms (Bittle and Snider, 2015; Tucker, 2003). As the most recent government-appointed Expert Panel on Health and Safety (2010) put it, "there is the widespread view that Dr. Ham [Ontario, 1976] got it right: government can set standards, monitor performance and enforce regulations, but it can't be in every workplace" (p. 6). As the Report also concludes, employers *and* workers must take primary responsibility within the context of the IRS. The Panel's Report was then followed by several reforms to OHSA, arguably the most significant since 1990. However, while concerns about the effects of precarious employment on health and safety were described by the government as a major motivation for the review, the resulting legislation (OHSA, 2011) continued the tradition of tinkering with the internal responsibility system without making substantial changes in the power relations between workers and employers. Recent reviews of those protections reveal major flaws in the protections that they were supposed to offer workers (King et al., 2019; Ontario, 2019).

Empowerment and the power to participate in the neoliberal era

This abridged history suggests that while worker participation and responsibility have been front and centre in the development of health and safety governance in Ontario since the 1970s, the questions of work power and control over the production process and the employment relationship have been kept to the margins in most OHS policy statements and government narratives. For many social scientists interested in health and safety, this history also demonstrates that management control over hazards is not something that employers have been willing to concede. The bulk of the evidence suggests that successive governments in most Western countries, including Ontario and Canada at large, have also been careful to leave that control largely intact in their policy and legislative reforms (Bennett, 2011; Bittle and Snider, 2015; Storey and Tucker, 2006; Tombs and Whyte, 2007; Tucker, 2003; Walters, 2006).

While some historians have suggested that the OHS policy emphasis on worker participation and self-regulation was prompted largely by developments within the field of health and safety, notably the perceived ineffectiveness of command-and-control laws and progressive developments in safety engineering (Sirrs, 2016), others have argued that the development of health and safety law and management since the 1970s cannot be understood in isolation from the broader governance shifts taking place at the same time, most notably neoliberalism (Bittle and Snider, 2015; Gray, 2009; MacEachen, 2000; Mogensen, 2006; Quinlan et al., 2016; Tucker, 2003). From this latter critical perspective, the dominance of neoliberal economic theory over postwar Keynesian policies beginning in the 1970s and stretching into the 2000s has meant trade liberalization and competition, deregulation and reduced public spending, while reflecting and fuelling the ideological emphasis on voluntarism and individual responsibility over collectivism and state-based protection and support (Birch and Springer, 2019; Harvey, 2005). Along with shaping the self-regulative orientations of OHS governance and management (Gray, 2009), neoliberalism is understood as gravely affecting working and employment conditions and relations. Critics within the OHS literature argue that increased global competition in deregulated environments have increased the intensification of work especially evident in new production systems such as lean production and high performance, while undermining labour union protection and worker security making it harder for workers to challenge unsafe conditions and report injuries (Aronsson, 1999; Clayson and Halpern, 1983; Hutter, 2001; King et al., 2019; Lewchuk, Clarke and de

Wolff, 2011; Mogenson, 2006; Quinlan, 1999; Sass, 1986; Storey and Tucker, 2006; Tombs & Whyte, 2007; Tucker, 1995).

From a labour process perspective, neither of these developments, the emergence of neoliberalism nor the focus on IRS for health and safety regulation, can be understood in isolation from the 1970s "Fordist crisis" (Jessop, 1990, 1995; Russell, 1999). In Canada, this meant high unemployment, inflation and recessionary pressures, which undermined Keynesian policies and led to growing labour control problems in industrial workplaces, reflected by major increases in both legal and illegal strikes as well as worker absenteeism in the 1960s (Wells, 1986). Conflicts over health and safety, including the 1974 wildcat strike in northern Ontario that led to the Ham Royal Commission (see above), can be understood as specific expressions of these larger developments. These labour problems also underpinned new developments in management theories of workplace governance. A good example of this in the late 1970s was state and corporate support for quality of work life (QWL) programmes which encouraged the acceptability of worker participation as a way of improving worker satisfaction and productivity. In Ontario, the Ministry of Labour established the Ontario Quality of Working Life Centre along with a bipartite industry/labour QWL Advisory Committee, to promote wider employer and worker adoption of QWL programmes across the economy (Wells, 1986). The activities of the Centre and the committee included training seminars, consultation and the publication of a quarterly journal called *QWL Focus* (e.g. Ontario QWL Centre, 1984; Ontario QWL Centre, Annotated Bibliography, 1974–80). The federal US government established a similar organization in the 1970s (National Center for Productivity and Quality of Working Life, 1976, 1978) as did European countries (Wilpert and Quintanilla, 1984).

Lean production, new public management and high performance management models were also emerging developments around this time period which further pushed support for participative forms of management and governance, with their advocates claiming that a primary objective was to move responsibility *and* control from middle management *to* workers at the same time that the hierarchy was being flattened, implying a promise of more control, responsibility and power for workers (Adler, Goldoftas and Levine, 1997; Womack, Jones and Roos, 1990). Participative management approaches were also being increasingly presented in management classes and journals, boardrooms and industry think tanks as the best way to manage labour and production within neoliberal and post-Fordist knowledge economies characterized by increased global competition, automation, communication and flexible production technologies (Piore and Sabel, 1984).

The question of whether these management systems and technologies yielded actual or perceived increases in worker skill, control and power spawned an enormous work literature, much of it on the auto industry where lean production first came to prominence (Babson, 1995; Graham, 1995; Lewchuk, Stewart and Yates, 2001; Rinehart, Huxley and Robertson, 1997; Russell, 1999; Thompson, 1983; Yates, Lewchuk and Stewart, 2001). While most of this literature, especially those researchers operating within an LPT framework, were highly sceptical of management claims of worker empowerment (Lewchuk and Roberston, 1997; Lewchuk, Stewart and Yates, 2001; Rinehart, Huxley and Robertson, 1997; Thompson, 2010), many analysts acknowledged that the form that these changes took and their effects were often quite mixed in different workplaces and over time (Belanger, Edwards and Wright, 2003; Russell, 1999; Thompson, 2010). A subset of this literature was focused on the specific impacts of participative management systems, including quality circles, management/worker committees and teams, and again the consensus, to the extent that there was one, was that workplace committees and teams had somewhat diverse impacts on autonomy and decision-making power, with often mixed results (Applebaum, 2002; Belanger, Edwards and Wright, 2003; Edwards, Geary, & Sisson, 2002; Hodson, 2002; Rinehart, Huxley and Roberston, 1997). For example, several studies found that while work teams (and the supporting technologies) were often perceived as offering workers more decision-making autonomy from management, they often simultaneously intensified their work and put more stress on them to perform as employers reduced the workforce, undermining a key aspect of their sense of control; and, with direct relevance to their health and safety, increased the chances of error-, speed- and exhaustion-related incidents and exposures to risk (Lewchuk, Stewart and Yates, 2001; Rinehart, Huxley and Roberston, 1997; Thompson, 2010; Thompson and Smith, 2007). As noted, many critical researchers in the occupational health area have sought to establish links between work intensification within lean production and other production models to higher levels of accidents and injuries, as well as increased stress and health-related symptoms of stress, including sleep disorders, heart disease, and behavioural disorders (Kalleberg, 2012; Landbergis, Cahill and Schnall, 1999; MacEachen, 2000; Novek, Yassi and Spiegel, 1991).

Within the OHS literature on participation, the major focus has been on the impact or effectiveness of joint workplace health and safety committees, with relatively little attention to management's strategic use of joint committees or other participative mechanisms to exercise more control over workers (Boyd, 2003; Eaton and Nocerino, 2000; Facey et al., 2017; Geldart et al. 2010; Gunningham, 2008; Hall et al., 2006; 2016; Lewchuk, Robb and Walters, 1996; Mendandez, Benach and Vogel, 2007; O'Grady, 2000; Olle-Espluga et al., 2014; Walters, 2016). While finding that joint health and safety committees can often have beneficial effects on workplace health and safety, those effects were often recognized as limited and short term, contingent on certain conditions such as management cooperation and/

or often achieved only after the exemplary political efforts by individual worker representatives and the collective action of workers (Boyd, 2003; Eaton and Nocerino, 2000; Facey et al., 2017; Hall, 1993; Hall et al., 2006, 2016; Lewchuk, Robb and Walters, 1996; Olle-Espluga et al., 2014; Storey and Lewchuk, 2000; Walters, 2006; Walters et al., 2016; Walters and Haines, 1988). Less emphasized explicitly in this literature is the concern that OHS committees frequently serve to locate more responsibility for prevention, and more blame for the failures of prevention *in* workers, without providing substantial or sustained gains in control and power over the conditions and practices that lead to workplace injuries and disease (Gray, 2009; Hall, 1992, 1993; Storey and Tucker, 2006; Walters, 1983), Again, for these critical scholars, the weaknesses of the IRS have been greatly accentuated by the neoliberal transformations in work and employment relationships (Dorman, 2006; Lewchuk, Clarke and de Wolfe, 2009; Storey and Tucker, 2006).

Much of what has been discussed thus far would seem to suggest that a broad answer to my starting question about the persistence of workplace accidents and diseases is that the changes we've seen in state and management OHS governance have not spoken to the persistent conflicting forces, interests and power imbalances underlying the production and taking of risks in the workplace; and indeed, have tended to obscure them from view. While there are good grounds to argue that power imbalances between employers and employees have gotten worse in the last four decades, most notably because of increased worker insecurity (Vosko, 2010; Vosko et al., 2020; Quinlan, 1999; Walters, 2006), there is enough evidence to say with some confidence that workers and worker representatives are still out there contesting their conditions, sometimes with positive outcomes, whether through formal participative mechanisms or through other means (e.g. Hall, 2016; Hall et al., 2016; Lewchuk and Dassinger, 2016; Storey, Walters et al., 2016). It is also demonstrable that government reforms and enforcement are not always one-sided and oriented entirely towards self-regulation in part *because* of worker and labour union resistance (e.g. see Tucker, 2003). However, I would go further and acknowledge that safety technologies and management OHS systems coupled with joint worker/management committees are often beneficial at some level in preventing injuries and disease exposures (Hall et al., 2016; Walters et al., 2016). I would also suggest that worker and management interests in preventing particular injuries and disease *are* sometimes aligned in some situations (Boyd, 2003).

However, this outline of the broad parameters of OHS success and failure is not sufficient to answer our core question regarding how we prevent the large numbers of workers who are still being injured, killed or made sick (WSIB, 2018). Along with a critique of neoliberalism and OHS legislation and management (Dekker, 2020; Gray, 2009) and their negative consequences, we need to explain when, how and why positive changes in OHS and in worker power and control over OHS are achieved and, if those changes are not sustained, why not. As this also implies, it is not enough to argue that the claims of empowerment through worker participation are simply a management fiction or ploy. Instead, we need to employ the view largely prevailing in the current labour process literature that worker control (and responsibility) is changing in diverse and contradictory ways rather than a simple trajectory of one-way losses for workers (Belanger, Edwards and Wright, 2001; Boyd, 2003; Guaimet, 2019; Thompson, 2010). This means an effort to identify and understand when and why workers sometimes make individual and collective gains in control within the context of labour process and management changes, whether those gains in control over their work translate into power which they can use to protect themselves, and finally, whether those gains can be sustained through conscious political organizing and action.

Building and integrating theory

The core concern in this book is understanding control over workplace hazards – that is, who produces and controls worker exposures to hazardous conditions, how is that control exercised and understood, and what shapes the limits and forms of that control. Connected to these questions is the concern with how responsibility is distributed and constructed by management and workers within these governance structures, and how this in turn translates into more or less hazard production and risk-taking. Reflecting the interest in control and responsibility, I draw on four theoretical approaches: labour process theory (LPT), governmentality theory (GT), Bourdieu's social field theory (BSF) and risk theory (Bourdieu, 1977, 1990; Brown and Calnan, 2012; Dean, 1999, 2007; Delbridge and Ezzamel, 2005; Kosla, 2015; Mollering, 2006; O'Malley, 2004; Snider, 2009; Thompson, 2010; Zinn, 2004, 2008a,b).

Labour process theory

Historically for labour process theorists, the defining concern has been to understand the reproduction of class-based control, especially at the points of production, given the contradictions of capitalist development and labour/capital relations (Burawoy, 1985; Friedmann, 1977; Herman, 1982; Thompson, 2010; Thompson and Smith, 2009). Two core aspects of Marxist political economic theory are critical to most LPT analyses (Mandell, 1977; Marx, 1977; Thompson, 2010). One, the central dynamic underlying the problem of control in a capitalist workplace is the challenge of realizing productive labour from workers' labour power (i.e. how do employers get workers to produce in ways and at levels that yield surplus value or profit?). Two, competitive pressures within capitalism produce a constant restructuring of the labour process (i.e. a productive activity of human beings using tools and instruments to produce a product or service), and/or the products being produced and marketed, which more often than not means changes in how the work and workers are organized and managed. Much of LPT and research based on it is accordingly about understanding how the challenges of employer/management control, including worker resistance, fuel and interact with the competitive demands for restructuring and the restructuring itself. Following this line of thinking, this book first nests the examination of control in an analysis of labour process and management restructuring within the three industries and the firm case studies.

While focusing on the OHS effects and origins of restructuring, the LPT that I'm employing here is one that has moved some distance from the early arguments of Braverman (1974) that control is entirely nested in the organization of the labour process governed by laws of capitalist labour process development. According to Braverman, a widening separation of intellectual and manual labour and increasing hierarchical control and fragmentation and deskilling of work were universal tendencies occurring across all industries and occupations which would inexorably undermine worker control and power. Based on a large body of research since then, current LPT analysts recognized that different industries, firms and occupations have moved in diverse directions with management control taking different forms both within the organization of the labour process and through other governance mechanisms, operating both from within (including, of course, participative programmes) and from outside the workplace (the state and the law) (Burawoy, 1985; Lewchuk, Stewart and Yates, 2001; Thompson, 1983, 2010). While some firms and industries were documented as moving to more coercive systems grounded in close surveillance and the use of contract and temporary workers, others were found to focus on nesting more control and responsibility in workers, emphasizing self-regulation over direct management or supervisory control. However, as other LPT analysts such as Michael Burawoy (1985) have argued, underlying and in concert with the shift to self-regulation were broader shifts in the control regimes of Western industrial countries, what he called "hegemonic despotism," which created and exploited worker insecurity (achieved principally through deregulation, offshoring production, de-unionization strategies and expanded use of non-standard forms of employment). In contrast to the standard employment relationship of the Fordist post-war period, where workers were understood as trading wage improvements and lifetime security for productivity growth, workers within post-Fordism took on more responsibility and worked in exchange for short-term security (Burawoy, 1979, 1985).

Although most labour process theorists agreed that worker power had been significantly undermined by increased employment insecurity since the 1970s, workers were also widely seen as continuing to exhibit capacities to resist in different ways (Burawoy, 1979, 1985; Edwards, 1979; Lewchuk and Dassinger, 2016; Littler and Salaman, 1982; Thompson, 1983, 2010). The key explanations for this persistence were twofold: one, the control effects of management's strategies were often revealed as contradictory, sometimes undermining worker consent while also offering opportunities for workers to gain and use substantive power to protect themselves; two, competitive pressures were constantly pushing the restructuring of industries and businesses, altering and often undermining existing management control regimes and labour consent (Thompson, 2010).

As already indicated, the LPT theory employed here also pays central attention to worker subjectivity in trying to understand both management control and worker resistance (O'Doherty and Willmott, 2001; 2009), When talking about risk subjectivity, control and power relations in this book, two concepts drawn principally from LPT will be central to the analysis – consent and compliance (Burawoy, 1979, 1985). Applied here to the question of worker responses to risks, these concepts speak to whether and on what basis workers are accepting or challenging their working conditions on health and safety grounds, or on other grounds, which, in turn, is critical to our understanding of the capacity of workers to exercise greater control over their health and safety. Within LPT, the distinction between consent and compliance is derived in part at least from Gramsci's (1971) concept of hegemony – that control can be understood by actors as being consensual "freely given" to others, or as coerced compliance due to some overt or assumed loss, sanction or threat. With respect to the acceptance or rejection of risk, this distinction recognizes that workers in different contexts and workplaces experience some of their risk-taking in political terms as unwanted actions dictated by their insufficient power and/or control over their work, while experiencing other risks which they willingly take because they thought they were appropriate, necessary or in their interests. It was in these latter contexts that management control was being achieved through self-regulative mechanisms and ideologies, which as discussed above, were distinguishable from other management approaches which relied more heavily on direct control, threats and reprisals. However, as I will also show in my analysis, the distinctions between consent and compliance, and consensual and coercive controls, are not simple matters of either/or, nor is consent or compliance determined in any simple way by any one form of management action or strategy. Accordingly, I adopt the view often made in the LPT literature that coercion and consent are best viewed as "two sides of the same coin" in the sense that overall management control or hegemony is always grounded in elements of coercion and consent which can both support and undermine the extent to which workers self-regulate (Burawoy, 1979; Dick and Hyde, 2006; Thompson, 1983).

I also acknowledge the spirited LPT debate that took place in the late 1990s and 2000s between the so-called structuralists and post-structuralists (Friedmann, 2004; Knights and Willmott, 1999; O'Doherty and Willmott, 2001, 2009; Thompson, 2010), principally around the question of whether the labour process matters as much as a source of control in the post-Fordist neoliberal workplace. In this book, I favour the position of those who still see the control exercised through technology and the organization of the labour process as critical to our understanding of workers' responses to hazards, while also accepting the argument that there has been an important shift towards subjective controls in many workplaces and industries as capital seeks to better valorize worker knowledge and energy (Thompson, 2010). Moreover, while still insisting that it is important to recognize the demands of capital accumulation as the principal pressures underlying change in labour processes and management strategies, as well as the structural tensions in labour–capital relations, I adopt the post-structural position within LPT that subjectivity is not simply a resource for resistance or domination, but rather is a critical dynamic on its own underlying change in worker and management behaviour. As O'Doherty and Willmott (2009, p. 937) put it, "the indeterminate productivity of workers' labour power is but one aspect, or articulation, of the indeterminacy of subjectivity, albeit one that is indeed of key importance for analysing the dynamics of the political economy of capitalism." From this perspective, worker and management interests in risk-taking and prevention are not simply given by their structural relationship but are also being socially constructed.

Governmentality (GT)

Governmentality theory (GT) offers some further insights into how we can understand the subjectivity of control as it relates to OHS risk. In GT, the defining concern is to identify and understand the strengths and limitations of control discourses, mechanisms and technologies without the narrow focus on class power and objective interests characteristic of LPT, while also encouraging a recognition of non-state institutions in mediating power and interests. When trying to theorize questions of worker and management control, governmentality theorists have been instrumental in connecting lean production and flexible management ideas to neoliberal government, cultural and media discourses, which stress the importance of competition, knowledge, individual initiative, freedom and responsibility as the hallmarks of personal and social success (Dean, 1999, 2007; O'Malley, 2004).

This recognition of the broader cultural shifts contained in neoliberalism as a system of ruling is perhaps where a "governmentality" approach offers its greatest value to this study. From this perspective, we can see more clearly why the shift in power over health and safety is not fully nested in the changing character of the labour process, and better recognize how neoliberalism or, what Rose (1993) calls "advanced liberalism," brings with it new ideas and technologies of ruling, which transform what it means to say that workers have gained more or less agency within this new regime. For example, when LP analysts ask whether workers perceive more or less control over their work or over health or safety hazards within a restructured workplace (Burawoy, 1985; Jessop, 1990; Littler and Salaman, 1984), they generally assume that workers are judging this change from a common and persistent framework and experience of personal agency. The worker is understood as having more or less direct control through his/her personal skills and organizational and decision-making capacities as permitted or allowed within the organization of the labour process and the jobs within that process (Edwards, 1979; Friedmann, 1977). Yet one of the core arguments from the governance literature is that neoliberalism and neoliberal management brings with it a new intensified sense of individual and self-regulating responsibility and agency. As limited as these impulses may be, corporate and state policies which emphasize worker participation and internal responsibility are from this perspective much more than simple ideological smokescreens, but represent part of what Dean (1999: 171) calls a broader rethinking of ourselves "as *self-managing* individuals and communities, enterprising persons, and active citizens," constituting a wider pluralism that breaks down previous social and political identities and communities "coincident with the national state." This argument is particularly relevant to the governance of health and safety in as much as the management and responsibilization of "risk" have been central to what Dean, Rose and others understand as neoliberal governance, whether it flows from government, corporations or other organizations.

The recognition of audits and the so-called audit culture are also central elements to an analysis of neoliberal governance which can be readily applied to a discussion of health and safety governance and especially safety management systems which are often saturated in audit procedures and discourse (Dyck, 2015; Kelloway, Francis and Gatien, 2016). If these "technologies," as Dean (1999) calls them, engage workers as informed responsible free agents capable of taking control of their own risks, or what O'Malley called the "new prudentialism," then we have another conceptual basis for understanding worker perceptions of control within the context of workplace, corporate and state restructuring, and, indeed more broadly, an understanding of the politics of workplace safety and health.

Bourdieu and social field theory (BSF)

The weakness of a governmentality theory approach, a weakness which post-structuralism shares more generally, is the failure to locate these politics in a material context. In the case of occupational health and safety in particular, the politics of health and safety, whether at the workplace or the state level, cannot be divorced from the continuing demands on capital to accumulate through surplus value extraction or, for that matter, from the often competing and sometimes intensifying demands on workers to sustain themselves and their families Indeed, by concentrating mainly on the micropolitics of change, we can lose sight of the macroeconomics underlying corporate and state restructuring. The neoliberal focus on flexibility, efficiency and marketization are not just ways of ruling, they are ways of ruling *so that* the capitalist class can maximize profitability and reproduce their class power over the dominated class. One early point made by LPT analysts such as Littler and Salaman (1984) and others, is that management is not just blindly implementing some managerial imperative to restructure or reorganize for the sake of control, but rather acting strategically to maximize shareholder return on investment (Friedmann, 1977; Thompson, 1983). This is, of course, a central proposition underlying classical as well as post-structuralist LPT – that ruling in a *capitalist society* is principally about sustaining capitalism.

However, as the debate within LPT suggests, the tension between structuralist and post-structuralist positions has not been entirely resolved (Thompson, 2010, p. 8), and governmentality theory does not entirely resolve that tension. In this book, I suggest that Bourdieu's (1990) social field theory is especially helpful in bridging structuralist and post-structuralist theory; first, because he emphasizes the importance of agency and actors in shaping and being shaped by structured outcomes (i.e. "structuring and structured"); and second, because he focuses more attention on the tensions and linkages between capitalist-based economic power and cultural power at the level of practice (Bourdieu and Wacquant, 1992, p. 121). Accordingly, we are encouraged to understand that control, consent and change in the productive spheres are tied in important ways to the (embodied and institutional) cultural, social and symbolic capital that workers, worker representatives and managers are able to build, access and mobilize within the workplace "field," that is, "a network or configuration of objective [work] relations between [worker, supervisor, management and related] positions" (Bourdieu and Wacquant, 1992, p. 97). This provides a more complex and comprehensive concept of power which does not rely on economic relations alone, while at the same time, an analysis of strategic agency *within* economically structured relations is preserved. Bourdieu thus helps us to avoid the reductionist tendency of labour process theory to explain motivation, power and action largely in terms of capitalist interests and forces, while at the same time, continuing to place significant weight on class relations without reducing them to economic-based interests and powers.

Bourdieu's concept of habitus or discursive rationale is also helpful in understanding the links between coercive and consensual power relations, and consent and compliance, such as the complex ways that employment insecurity and worker identities can be connected (Bourdieu, 1990, p. 53). For example, applying Bourdieu's theory, risk subjectivities and actions grounded in firm commitment and pride in doing one's job can be understood as aspects of what I call a work "risk habitus" which workers have developed overtime within a context of structured field relations involving elements of coercion, cooperation and asymmetric power relations. Some of these meanings may have been encouraged by management actions including the selective use of praise and rewards, but it is just as possible that pride and commitment are the historical outcomes of workers' efforts to use risk-taking to gain power and achieve some security within contexts of limited worker power and coercive management controls. Moreover, as Bourdieu also points out, many aspects of gender- and class-based habitus are well established through family and formal education systems (Bourdieu, 1990, p. 54). As such, we are reminded that ways of thinking about work, responsibility and occupational risk are often set in motion earlier in their family and early employment experiences. Some of those early experiences may have included learning the rules of the employment game, including that working hard offers a level of security and that any challenges to supervisors have to be grounded in various sources of security and leverage. Responsibility for one's health can also be a significant disposition developed at an early age. More broadly, workers learn through everyday experience where their interests lie in different fields and where they can gain power and rewards, and in the process, they learn the limits and risks to that power given their positions in those fields (i.e. employees and occupations within a workplace). As such, workers may be disposed to work hard and take certain risks as part of what this means to them, often without a second thought, and indeed working hard may have become a part of their identity so that they express pride in taking risks. However, underlying those constructions are series of assumptions and understandings about what is possible and reasonable to expect from working hard, and what would happen to them if they did not work hard, grounded in family, educational and work experiences, all of which have been structured in important ways by their class position as well as their gender, race and ethnicity.

While helping to explain how power relations shape patterns of risk behaviour which become almost second nature, the concept of habitus understood within organizational or workplace fields is also very helpful in better explaining how changes in power relations and the "rules of the game" within the workplace can alter those behaviours as workers discover that the old rules and assumptions no longer apply – that is, aspects of the risk habitus are "disrupted" (e.g. working hard and taking risks are challenged by safety managers or worker representatives as reckless behaviour).

Risk theory

Along with the "risk society" argument that neoliberalism and post-modernity have focused greater attention on risk, simultaneously pushing both an increase in fatalism and insecurity, and a greater demand for scientific and risk governance solutions (Beck, 1992; Giddens, 1991; O'Malley, 2004), the risk literature draws our attention to the less rational ways in which people deal with risk, insecurity and uncertainty. This line of analysis points to the reliance on trust, intuition, faith and even blind hope when people are dealing with the uncertainty of many risks (Brown, 2009; Hall, 2016; Kosla, 2015; Luria, 2010; Zinn, 2004), while also questions the dichotomy of rational versus non-rational decision-making which tends to be assumed in LPT. Similar to Bourdieu, these "less than rational" strategies are understood as reflecting not only a lack of knowledge and the time to engage in rational decisions but, more importantly, as reflecting their lack of power and control. Several risk theorists argue that everyday practices reveal a number of different reasonable strategies – "in between" the ideal types of "rational" and "non-rational" – for dealing with risk and uncertainty such as intuition, trust and emotions (Brown and Calnan, 2012; Luria, 2010). Even the most ostensibly "irrational" approaches to uncertainty, such as hope or faith, can be deemed reasonable when facing risk, uncertainty or pending disaster (Zinn, 2004). It has also become clear that, far from recourse to any one of these strategies, social actors adopt varied strategies across different circumstances and often combine multiple such strategies simultaneously (Mollering, 2006).

At the same time, and again like Bourdieu (1990), risk theory emphasizes the importance of personal life histories, their cultural and class upbringing and the particular social and organizational contexts in which they are operating, as shaping the often varied and complex mix of strategies that people use (Fredericksen, 2014; Zinn, 2015). Risk theory thus retains our view of individuals as active agents who are engaged in a localized politics of risk, while underlying this conception of politics is an understanding that the strategies people use are forms of coping within the structural limitations of their positions and relations. This reference to risk responses as forms of coping also leaves room for a recognition of psychological mechanisms such as fatalism and cognitive dissonance in explaining how workers deal with the contradictions and stress of competing risks (Dorman, 2006).

In sum, while power is understood in different ways within these four perspectives, I seek to show that when applied together, they provide a better understanding of how and why workers understand and respond to their work in the ways they do, while relating those responses to the actions, relations and discourses of others, in particular, owners, managers, supervisors, technical and medical experts, union and government officials.

Worker control and the governance of health and safety: a theory

Consistent with LPT, I argue that we can understand the politics and subjectivities of occupational risk within workplaces *and* within the broader OHS political and legal spheres by examining two core aspects of labour/capital relations: (1) the relative distribution of worker, management and

technical control over work and working conditions, both actual and perceived, and the associated power and responsibility attached to workers,
managers and technology for controlling and preventing any resulting
risks; and, (2) the relative dependency of labour and capital on their specific
employment relationship, and the resulting variations in employment insecurity among workers. To put it in simple terms, when workers have and/or
perceive relatively more control and power over their work and employment
relationship, they are more likely to construct the control of risk as their
personal or collective responsibility underpinning dispositions towards the
acceptance of those risks. While workers may use that control and power
to demand or negotiate improvements in their working and employment
conditions, certain elements of risk-taking are also understood by workers
and employers as an integral aspect of the employment relationship, the
extent to which varies with the workers' dependency and insecurity. Although these assumptions, understandings (i.e. habitus), arrangements and
exchanges can reproduce labour consent to risk for a period of time, they
also often introduce constraints on productivity and efficiency. As per LPT,
private firms are understood as being pushed by competition and worker
resistance to restructure the labour process or intensify the work in some
way, altering and often reducing aspects of worker control and power; or
in Bourdieu (1990) terms, changing the power relations and the rules of
the games. I show that conflicts about health and safety in a workplace
or industry tend to emerge in these dynamics of labour process and employment restructuring, often contributing to changes in OHS law, political
discourses and OHS management approaches, which further alter work/
risk habitus and the political and subjective bases of worker consent and
resistance to workplace risks.

It is important to stress that while labour process transformations and
new technologies are understood as introducing new hazards and aggra-
vating existing ones, these changes also often eliminate hazards and reduce
others. At the same time, employers may have heightened production interests within certain emerging production regimes such as just-in-time management in minimizing injuries which leads to more substantive prevention
efforts by management. While the objective production of hazards, risks
and injuries is part of the analysis, as in most LP-based theories the principal concern is to understand the transformation effects on risk subjectivities and politics which stem from the objective shifts in worker control and
power.

Given this LP theory of occupational risk construction and politics, I theorize further that we can understand several key developments in health and
safety management and state regulation discourse, notably the emphasis on
worker participation and worker rights, internal joint management/worker
responsibility, and self-regulative technologies such as audits and risk monitoring programmes, as efforts to reconstruct worker and management

control and responsibility for OHS risks within the contexts of a globalized neoliberal economy and post-Fordist work arrangements. The analysis of each industry and workplace(s) examined in this book seeks to show how global market pressures pushed firms to restructure their labour processes and/or employment relations, how those changes altered OHS hazards and producer control in the workplace, and how producers, managers and regulators responded to those changes. While demonstrating that management and regulators are often successful in reproducing labour consent within their new participative and risk governance regimes, I argue further that corporations and managers frequently fail to anticipate the full range of physical, social and political effects of labour process changes. Management, and indeed individual worksite managers and supervisors engage in a number of reactive and proactive strategies aimed at limiting the negative effects on production, accidents and incidents, and worker motivation and consent. However, as production problems and incidents occur, and the risk habitus is disrupted, some workers begin to contest their new conditions often pushing managers to make further modifications to those strategies.

As will also be recognized, individual managers and supervisors are not always in agreement reflecting their own work risk habitus and, as such, there may be conflicting perceptions, strategies and interests in play at any given point in time. For their part, although similar in many important ways given their common class and career experiences, workers also differ in the responses reflecting their different positions within the workplace and their distinct work and life histories tied to ethnicity, gender, age and geography (Bourdieu, 1977, 1990). As active agents, a view again taken from both LPT and Bourdieu's social field theory, workers [and managers] will alter and change their strategies and approaches to controlling work and working conditions over time, depending substantially on shifts in power resources or capital within and outside the workplace.

Although stressing the dynamic and localized aspects of change, the overarching argument is that the general forms or orientations that these management, corporate and state strategies have taken in recent history, specifically from the 1980s to now, have been grounded in neoliberal logics, ideas and technologies of rule which have come to increasingly define dominant approaches to all areas of risk, including workplace accidents/diseases (Dean, 1999; Gray, 2009; O'Malley, 2004). While some aspects of neoliberal governance have been weakening in recent years (Harvey, 2005), most notably the idea that economic growth is maximized through minimal government regulation, I suggest that the concepts of new prudentialism, technologies of agency and technologies of performance continue to offer a framework for understanding the dominant reconstruction and restructuring of OHS governance which is central to the reproduction of labour consent and class control.

At the same time, I argue that neoliberal state, corporate and management policies continue to restructure power relations in health and safety in important ways by altering employment security and dependence (Quinlan et al., 2016; Vosko et al., 2020). The key thesis in this respect is that neoliberal health and safety governance including the promotion of OHS management systems, as presented and encouraged by the state, non-state agencies, OHS professionals and management scholars, has sought to reconstitute workers as free individual risk-takers united by a common interest in and dependence on security. However, since security in this context is understood both as security of the body (i.e. health) and economic/employment security, these two dimensions of security are understood as being objectively in tension given production relations which continuously force workers to trade one for the other.

Finally, while making the above arguments, I demonstrate that workers, worker health and safety representatives, and worker organizations continue to carve out areas of limited contestation and control within restructuring contexts, which in turn push both corporations and governments to expand, temper or alter their reliance on neoliberal ideals and other forms of governance. I argue that the impact of post-Fordist restructuring and neoliberal forms of rule have been contradictory, offering up glimpses of management failures and uncontrolled risks and sometime new spaces in which workers can contest and challenge corporate and state claims. While focusing principally on workplace politics, I also try to show that it is these contradictory effects at the level of production which ultimately explain the variations not only in corporate applications of neoliberal governance but also in the actions of the state (Tucker, 2003; Vosko et al. 2020).

To fully address the multilevel aspects of the theory, the analysis hinges around an understanding of the ways in which businesses in different industries restructure their labour processes, labour markets and management, and how these changes shape and reshape labour consent and resistance at different points and places of production; and finally, how these responses in turn shape changes in the management of health and safety. One complexity that the analysis also seeks to address is the variety of strategies and responses that employers and managers employ within the same industry when dealing with market pressures, technological developments, state interventions and labour relations. With respect to health and safety, this type of theorizing moves us away from the simple question of whether restructuring yields improving or worsening levels of risk, and concentrates instead on recognizing the different forms that restructuring can take, on understanding how those various forms relate to government and industrial contexts and, on showing how they are linked to different OHS politics, subjectivities and outcomes within those contexts. This means looking at various aspects of restructuring including technological; organizational;

management; labour process and labour market reforms; local, national and global market shifts; changes in legislation, regulation and enforcement; changes in unionization; ideological, habitus and cultural shifts; and, finally, specific developments and applications in OHS discourses and programmes at the level of production.

Research challenges

Several research challenges are evident. First, to understand *how* all these changes in their various forms are worked out and reflected in actual day-to-day health and safety discourses and practices, implies the need for research at the level of the workplace, aimed in particular at tracing out the ways in which managers, union representatives, state officials and workers enact and react to these changes both contextually and over time. Second, if I also want to tease out the commonalities as well as differences linking restructuring and health and safety practices, it is not enough to focus on a single workplace or occupation – the research requires comparative features which will better allow me to make and demonstrate those links across industries and workplaces. Third, if production transformations and politics are to be tied to larger political-economic and cultural developments, the comparisons need to draw in historical comparisons which include these levels of developments.

This then is the somewhat ambitious task of this book. By relying on case studies of firms and different workplaces within three distinct industries – auto, nickel mining and agriculture, each examined at distinct points in time, I seek to identify and theorize the links between OHS discourses and practices at the points of production and the various kinds of macro-structural developments (markets, government, law, media, management theory, science, cultural and public discourse) occurring at the time which were shaping these industries and the particular firms and workplaces within those industries. Through comparative case studies of concrete workplace sites, I seek insights into what Burawoy (1985) calls the "politics of production," or what I call here "OHS politics," understood as relations, expressions and practices of power among worker and managers as they produce and respond to OHS risks. A focus on specific political processes allows me to better explain the similar *and* dissimilar changes and impacts which emerge in different contexts. The use of multiple studies conducted in different industries at different points in time also permits more coherent arguments regarding postfordist and neoliberal restructuring directions and patterns which acknowledge and explain distinct trajectories, and even contradictions within those commonalities, by theorizing and examining these connections more directly. Finally, the comparison of a large multinational mining firm with small- to moderate-sized auto parts firms and family farms allows a consideration of the significance of employer size and self-employment in shaping the development and application of OHS governance ideas and systems (Eakin, 1992; Walters, 2004).

A review and discussion of research methods

This book relies on data collected in 8 distinct studies conducted in Ontario by the author over a 30 year period, 4 of which were conducted in collaboration with different colleagues (see Table 1.1). Two are case studies of specific mining and auto parts firms, two are case studies of multiple family farm operations in Ontario, and two are cross-sectional survey and interview studies of union and health and safety representation – one in the auto industry and the second in several public and private sector industries including construction, mining, manufacturing, health, public administration and education (Stake, 2003). One additional cross-sectional survey was done of workers in public administration, construction and manufacturing which included some autoworkers with a specific focus on injury and hazard reporting. Multiple methods were used in all the case studies including archival research, formal interviews using both closed- and open-question formats, and participant and other observation techniques, while survey and interview data were collected in other studies (Creswell and Clark, 2011). Although the designs and objectives of each study differed in some significant ways, as will be outlined next, a central focus in all these studies was to understand the effects of different kinds of restructuring on health and safety.

Study 1: The first study was conducted in Sudbury from 1985 to 1987 within a single large multinational nickel mining firm. The research was aimed specifically at understanding the production politics of health and safety within the context of organizational and technological change. Data collection involved a full year of onsite fieldwork observing activities at the union hall and in mining operations; reviewing and collecting union documents and reports; attending safety training programmes; and interviewing managers (N = 19), union officials, stewards and worker health and safety representatives (N = 51) and government health and safety inspectors (N = 4). A face-to-face interview survey of 171 miners across the firm was conducted using a mixed sampling procedure pulled from five different mines (see Hall, 1989, p. 636 for more details). Ten miners were selected from the survey sample as individual case studies based on their job positions and survey answers. Lengthy and multiple taped interviews were conducted with each worker and their spouse, along with worksite observations of them over a four-month period in three mines. Case studies of three mines were also conducted which involved taped face-to-face interviews with all supervisors

Table 1.1 Main studies employed in the book

Study	Site	Main Methods
Study 1: Mine Study 1985–87	Single Company, Multiple Underground Mines	Mixed Including Observations, Survey, Qualitative Interviews, Archival, Case Studies
Study 2: First Farm Study 1993–94	Multiple Family Farms	Case Studies – Mixed Including Observation, Qualitative Interviews with Self-Employers Farmers and Workers, Archival
Study 3: Second Farm Study 1999–2001 Co-Researcher: Dr. Veronica Mogorody	Multiple Family Farms	Case Studies – Mixed Including Observation, Qualitative Interviews with Farmers and Workers, Archival
Study 4: 1st Auto Study 2003–4 Co-Researchers: Dr. Anne Forrest Dr. Alan Sears	Multiple Firms (Parts)	Survey and Interview Union/OH Reps
Study 5: 2nd Auto Part Study 2004–5 Co-Researchers: Dr. Anne Forrest Dr. Alan Sears	Single Parts Firm	Case Study – Mixed Method Observations, Interviews with Workers and Managers, Archival
Study 6: Mixed Industry Study 2012–13 Co-Researchers: Mr. Andrew King Dr. Wayne Lewchuk Dr. John Oudyk Dr. Syed Naqvi	Multiple Firms/OH Reps.	Survey, Interviews with Union/Non-Union Worker Reps.
Study 7: Mixed Industry Study 2011–12 Co-Researchers Dr. Tanya Basok Dr. Jamey Essex Dr. Urvashi Soni-Sinha	Multiple Firms	Survey, Interviews with Union/Non-Union Workers
Study 8: Archival/ Historical Interviews	Multiple Unions, Archival, Interviews Bill 208, 1992	Ontario Government

and managers in each mine; all the health and safety and union representatives; the collection of OHS records, memos, accident investigation reports and statistics; and on-the-job observations over a six-month period (see Hall, 1989 for Survey and Interview Instruments). Observations of four joint health and safety committees were also conducted over the year and committee minutes were collected covering 1980–86. While most of the data collection was completed by the summer of 1986, follow-up interviews were conducted with union and company health and safety representatives (N = 12) in the summer of 1987 aimed at determining the outcomes of certain health and safety issues that had developed over the course of the field work. Archival research focusing on both union and company documents and history-focused interviews with retired miners and union activists (N=18) were also conducted. One of the strengths of this study in terms of examining the impact of restructuring is that the company was in the process of major changes in its mining labour process and was experimenting with different forms of management structures and policies in different mines. This meant that I was able to compare miners in very different kinds of mining processes and mine operations, representing conventional and restructured labour process contexts. It is also worth noting that I worked for this company in two of its mines at a different location in a western province in 1975 and in 1976, employed in several different jobs – long-hole driller, tram operator, powder man and blaster. I operated most of the machinery discussed in the study including ore trains, scooptrams, stoper drills, long hole drills and jack legs.

Study 2: The second study involved intensive case studies of 12 grain and mixed grain/livestock farming operations in Southwest Ontario in 1993–94. The operations ranged in size from 200 to 2,000 acres and were selected purposely to represent three distinct farming methods – conventional tillage farming, no-till farming and organic farming. The original aim of this study was to examine and link both environmental and occupational health ideologies and practices to production and market restructuring (Hall, 1998a,b). Data collection involved multiple interviews with farm owners, family members and workers, and participant observations over a single growing season (April to late October). Participant observation included some operation of farm-machinery plowing; planting and harvesting; physical labour of various sorts; and the preparation of chemical sprays. Observations of farm organization meetings including the Ontario the Federation of Agriculture (OFA), Ontario Farm Safety Association (OFSA), Canadian Organic Growers (COG) and Ontario Soil and Crop Improvement Association (OSCIA) were conducted over the course of a year leading up to and during the case studies. These observation activities also involved informal discussions and interviews with 100+ farmers which then became the basis for selecting and connecting with the 12 case studies. Extensive archival data in the form of government documents, agribusiness and farm media articles on farm safety, and conventional, conservation and organic farming were also collected (Hall, 1998a,b, 2007).

Study 3: The third study was focused specifically on organic farmers in Ontario, and involved phone interviews conducted in 1999 with 240 farmers, and follow-up case studies of 20 farm operations in 2000, again entailing participant observation (two weeks at each farm at two different points over a single growing season) and qualitative interviews of owners and workers. An initial contact list was developed through contacts with several organic farm organizations and names were randomly selected for the survey from a compiled list of 400 farms. Observations of organic farm meetings and interviews with officials from farm organizations and the Ministry of Agriculture representatives were also conducted over a two-year period (1999–2001). While the focus of this study was principally on environmental issues, there was a component on health and safety, principally within the case study interviews and observations. For comparison purposes, the analysis in this book is based on 8 of 31 case studies that were grain and livestock farms that had moved from conventional to organic farming. These were selected because they mirrored the grain production systems and sizes of the non-organic farms studied in the second study (Hall and Mogyorody, 2002, 2003).

The farm analysis presented in Chapters 6 and 7 relies mainly on open-ended interviews (N = 40) and participant observation of the combined 20 farm case studies from the 2 farm studies. The OHS questions were similar to those used in the mine interviews, focusing on both assessing safety and health knowledge and actions with particular reference to tractor and combine accidents, road dangers, dust and fumes, and pesticide exposures. In those farms with employed workers and multiple operators (spouses, parents/children), both the "workers" and operators were interviewed and observed. The two-week observation periods per farm were also used to further interview farm owners and workers and detailed field notes were taken. As with the mine study, there was also a set of preliminary interviews (N=30) aimed at identifying key environmental and OHS issues and industry accident and disease prevention practices and standards, which were conducted with various farm organization leaders and representatives, Ministry of Agriculture officials and a small group of farmers.

Study 4 and 5: Two studies on the auto parts industry in Ontario were conducted from 2003–6. First, health and safety committee worker co-chairs and local union chairs from 36 auto parts firms across Ontario were surveyed with a written questionnaire and then 21 of the health and safety representatives were interviewed at some length with open-ended questions (1.5–2 hours) The survey and interview instruments were aimed at assessing the links between evidence of lean production and health and safety participation and other outcomes, as well as documenting the way in which health and safety representatives achieved change (Hall et al., 2006).

The union co-chairs provided most of the information on their firm's production, management and labour relations systems while the OHS co-chairs addressed health and safety management and issues.

The second auto parts study (Study 5) involved a case study of a three-plant non-union parts manufacturing firm. The research involved limited observations of the plant operation (two visits), observations of the two plants' company health and safety committees over a one-year period, qualitative interviews with workers (N=25), worker OHS committee representatives (N=6), managers (N=8), plant floor supervisors (N=4) and the owner, and a detailed review of company documents and records. Interviews were focused on gauging their perceptions of the main hazards in their workplace, their reactions to those hazards, and their explanations for their responses. Supervisors, managers and health and safety committee representatives were also interviewed regarding the history of production and management changes and their perceptions of hazards and the company's efforts to prevent injuries and disease exposures.

A wide range of documentary evidence regarding the firms and the industries was collected in all the case studies. This includes current and archived newspaper and industry magazine articles, government and Royal Commission Inquiry reports, corporate safety and injury reports, company safety manuals, audit materials and reports, minutes of joint health and safety committees, union and company submissions to government, statistical reports and communications by agencies such as WSIB and relevant provincial Accident Prevention Associations, grievance records, internal company and union memos, and other formal communications and newsletters on OHS issues. Field notes from observations of joint health and safety committees and the union health and safety office over the course of year in both the mining and the auto case studies were also important data sources. While there were no joint OHS committees in the farm context, numerous local farm organization meetings and provincial farm conventions dealing with safety and other issues were observed over several years.

Study 6: The sixth data source was a study done of worker health and safety representatives in a variety of different public and private sector industries in Ontario in 2012–13 (Hall et al., 2013, 2016). The study involved an initial non-randomized survey of 888 worker representatives who were contacted through hard-copy invitation cards, online notices and email appeals distributed via a range of participating unions (CAW/UNIFOR, OPSEU, SEIU, CUPE, USWA) and worker centres. Recruiting was also done at OHS conferences and training/certification programmes. Forty follow-up qualitative interviews were also conducted, sampled to reflect three clusters of representatives identified from the quantitative data (Hall et al., 2013, 2016). Both the survey and the interviews were focused on identifying different forms of worker representation based on the type and scope of issues they raised and their political strategies for getting management attention to those issues. Although this study was not specific to the auto industry, many of the participants in this study were working in the auto or manufacturing sector (N=126), offering some additional application to the discussion of this industry. A number were also working in the underground mining industry (N=40).

Study 7: Although used more sparingly in this book, a seventh source of data was a study of injury and hazard reporting (see Basok, Hall and Rivas, 2014; Hall et al., 2012 for more details). This study began with a 2011 survey of 1,100 unionized workers conducted with the recruitment assistance of four unions (CAW [now UNIFOR], OPSEU, LIUNA, and IBEW) on injury and hazard reporting. Follow-up qualitative interviews were also conducted with 56 of these workers. In addition, a sample of 65 non-unionized workers were recruited and interviewed using purposive sampling and snowballing with the assistance of a Workers' Center in Ontario. This study was particularly valuable in Chapters 2 and 3 where a framework for understanding the risk subjectivities of workers is elaborated. Again, while not confined to the auto or mining industry, there were workers from both of these industries in the sample which allowed me to draw them into the industry case study analysis.

Study 8: Finally, some of the historical analyses presented in the book rely on archival research and interviews conducted with activists, Ministry of Labour bureaucrats and provincial politicians (N=31). These were all people involved in labour and health and safety politics during the 1980s and 1990s (Study #8), focusing in particular on the political events leading to Bill 208, the first major reform of the Ontario Occupational Safety and Health Act in 1990 (Hall, 1991). Combined with historical data collected in Study #1 (see above), I use this evidence to link the workplace and industry levels of analysis to the macro developments in law and public discourse at the provincial level taking place at the time.

Data Coding and Analysis

The qualitative observation and interview data were coded and analysed thematically focusing on the distinct ways in which workers and managers described and explained their responses to working activities and actions (Fontana and Frey, 2003; Lofland and Lofland, 1995). In the mine study (Study 1) and the first farm study (Study 2), the thematic coding was done manually using a larger chart which graphed themes and relevant quotes for each case study, while the qualitative programme NVIVO was used for the auto research (Study 5 and 6) and the injury/hazard reporting studies (Study 7). Field notes and archival documents were not formally coded but thematic-based indices were constructed identifying the locations of quotes and descriptors within the field notes. In the mining context, this also allowed me to cross-reference varying accounts of particular workplace events and issues. Quantitative data from the mine survey, the two auto studies, and the injury/hazard reporting study were coded and analysed using SPSS. Where sample sizes warranted (i.e. the mining and auto studies), multiple regression and cluster analysis were employed to test certain predictive models (see Hall, 1989; Hall et al., 2016) but, for the purposes of this book, the focus is confined to simple frequency comparisons, pearson correlations and significant differences between worker groups and work locations.

Although conducted at different points in time, the first seven studies share similar mixed methodologies and analyses. What most of the studies also share is a focus on labour process changes within the context of broader social and economic changes. From a comparative perspective, what the case studies also offer is a way of looking at how restructuring and health and safety subjectivities and consent relate to producer control in three quite distinct power relational contexts – in family farms where the direct producer is self-employed and working in his own land, and as such, has in theory complete independence and control; in hard rock nickel mines where wage workers have historically had a fair amount of independence and control with relatively little SM; and in auto factory work, where wage workers under more Taylorist and Fordist production principles have historically had relatively little independence or control. The health and safety representation studies add something as well in that they were conceptualized within a framework that sought to tie different forms of representation to neoliberal regulation and industrial restructuring.

As a final note, I want to acknowledge that my understanding of health and safety as presented in this book is also informed by my experiences as a union health and safety activist and joint committee representative (and co-chair) in university settings and as a community activist advocating in both industrial and rural work contexts. While my impact was woefully inadequate, those experiences helped me to recognize the complexity and variety of management and worker responses to workplace hazards and appreciate the challenges involved in gaining sustained improvements in OHS conditions.

Organization of the book

Along with the Introduction and Conclusion chapters, I have organized the chapters into two parts, with two chapters in Part 1 and six chapters in Part 2. Part 1 and the chapters therein elaborate a framework for categorizing and analyzing the different kinds of risk subjectivities underlying consent and resistance. The introduction to Part 1 outlines four main types of understandings or dispositions underlying worker responses to conditions which are then illustrated with evidence from the three main work settings (mining, farming and auto manufacturing) in the subsequent two chapters. Chapter 2 first examines consent grounded in the exclusion of specific conditions (e.g. oil mist. dust) or activities (e.g. working with limited light) as representing risks to health or safety. The critical questions here are which conditions of work are defined by the authorities and workers as involving some risk and which are not, and how do these definitions contribute to management control and worker consent and resistance. Chapter 2 then moves on to examine how workers judge defined and recognized hazards as acceptable or unacceptable risks, with reference to their perceived probability and seriousness. Again, the focus is on the different ways that consent is gained through the construction and judgement of limited risk.

Chapter 3 examines workers' explanations for accepting or resisting risks which they recognize as more serious. The first part of the chapter focuses on the different explanations that workers give for taking recognized risks, while the second part examines when workers contest the hazards and their explanations for doing so. In both chapters, similarities and differences within and between different firms and workplaces are identified, with some preliminary analysis suggesting that these patterns can be linked to commonalities and variations in labour process changes, management discourses and technologies, economic conditions, and state and union politics.

Part 2 contains six chapters, with two chapters dedicated to the discussion of each industry. The first of each pair of chapters documents the major restructuring pressures and changes and the second relies mainly on case study data to show how these changes have shaped health and safety subjectivities and politics. However, the focus of the analysis also shifts to some extent with each industry. The emphasis in the mine study is to link labour process and OHS management changes to shifting subjectivities of risk among workers and management. The farm study is focused on comparing the impact of neoliberal and alternative economic development logics on risk subjectivities, while the auto industry analysis seeks to examine the impact of lean production and participative management on OHS politics to risks within unionized and non-unionized contexts.

The conclusion chapter integrates the case studies with the other research data on legislative reform and health and safety representation in an effort to provide comparative-based responses to the book's core questions about the origins of workers' consent and resistance, focusing in detail on the significance of restructuring and neoliberal governance. The chapter and the book end with a discussion of how worker advocates and organizations might achieve greater and more sustained gains in injury and disease prevention.

Part I

Risk subjectivities and practices

In this introduction to the first part of the book, I sketch out a framework for conceptualizing and categorizing the different subjectivities underlying worker consent and resistance to hazards, leading to two chapters which use the framework to map out the subjective bases of consent and resistance among the workers in the three types of workplaces, identifying both variations and similarities. Subsequent chapters in Part 2 will then look at each of the three industry cases in sequence to consider how these subjectivities and the associated politics of consent and resistance are related to restructuring and developments in work organization, management discourses and orientations, state intervention, and union involvement.

Two key contrasting patterns of worker subjectivity are explored. One is the manner and extent to which responsibility for identifying, judging and controlling risk is nested in the workers as opposed to being deferred to management and management systems, experts and technologies. This will feed into a main argument in later chapters that restructuring and the corporate discourses surrounding restructuring reconstitute the parameters, capacities and the appearances of worker versus management identification, judgement and control of risk, in ways which expand the scope of management control while continuing to reconstruct worker responsibility. The second pattern of worker subjectivity is the manner and extent to which workers understand OHS risk-taking and risk avoidance as a function of different forms of insecurity as opposed to avenues for enhancing security. While overall employment security is a key focus here, the perceived security of particular job positions, financial security and security of the body are all important aspects (Lewchuk, Clarke and de Wolff, 2011; Vosko, 2010). The core relevant argument as presented in later chapters is that although restructuring increases worker insecurity in various ways, neoliberal corporate and state discourses regarding risk offer up new avenues through which workers can reconstruct different levels and understandings of security and the means of achieving them (Dean, 1999; Rose, 1993).

Understanding worker subjectivity: hazard identification, risk judgement and control, and risk-taking

After detailed reviews of archival materials on safety issues in each type of workplace, and talking informally with workers, farmers and OHS activists in each study, sets of survey and/or interview questions were developed which directly asked the direct producers in the three different work settings to describe and explain their responses to specific working conditions or work processes such as dust, fumes, tool use or machine operations. After looking at the coded responses for all three work groups, four types of explanations were identified as capturing their acceptance or rejection of the condition or activity.

First, some denied that there was a *health and safety* issue, either because the condition or activity was not perceived as present in their particular workplace (e.g. Miner: "there is no radiation in this mine"), or, alternatively, because they didn't believe that the condition or activity presented *any* appreciable risk of injury or health effects (e.g. Autoworker: "it gets pretty hot in here in the summer but it's not really a health issue"). In terms of worker practices and OHS politics, workers accordingly worked within these conditions or performed these activities in question without attending in any way to personal or management assessment or control of risk, which also meant that these were not raised *as* health or safety issues within the context of their relations with other workers, union or worker OHS representatives, government inspectors, supervisors, managers, engineers, or safety experts. In other words, labour acceptance of these particular conditions was grounded in their *exclusion* from health and safety definitions and discourse.

When workers (including self-employed farmers) defined the condition or practice as a hazard or potential hazard, the second type of explanation offered for their acceptance revolved around their judgement or understanding that the risk of significant injury or disease was not high enough to warrant concern. Risk levels were typically judged along two dimensions (see Figure P1.1): either (1) the health or injury effects were seen as being too minor or too temporary to warrant concern, as in a small cut or muscle pain (Farmer: "I often get cut when fixing the planter or the combine but nothing serious"); and/or, (2) the likelihood of a more serious injury or health effect including death was seen as too low to warrant concern (Autoworker: You would have to be pretty stupid to get hurt badly in this job"). As will be demonstrated, the judgement of "serious" injuries and health effects was linked to their continued capacity to do their job and to the workers' construction of what is "normal" in terms of "healthy" body reactions (e.g. temporary coughs or headaches) as well as their construction of non-work causes of any "abnormalities" (e.g. "getting old," "smoking," "runs in the family"). The likelihood of a more serious injury was understood in a number of ways but a core distinction was between risks which were constructed as being

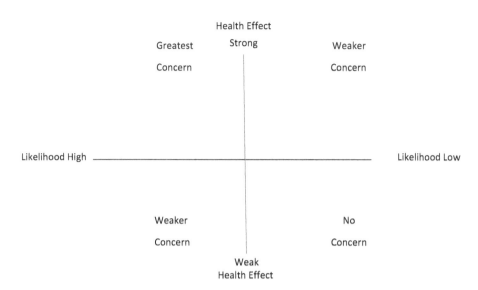

Figure P1.1 Dimensions of risk judgement.

inherently low (e.g. Miner: "The radiation levels are low in this mine") versus risks which were constructed as being intentionally controlled in some way to "acceptable" levels (e.g. Farm worker: "I wear a mask and ear protection when I'm on the tractor"). Often central to consent was the workers' capacity, or at least their self-perceived capacity to engage in actions which allowed them to detect, assess and control the risk at the outset and/or as the work proceeded. It is here where we see most clearly the internalization of responsibility for controlling risks. However, the control and monitoring of some risks were often understood as an externalized responsibility, with consent then being grounded at least partially in the knowledge, trust and faith that workers placed in other workers, managers, engineers, work processes, safety standards, technologies and the machines which they were operating (Hall, 2016; Jeffcott et al., 2006; Kosla, 2015).

When workers judged a condition or activity as involving significant risk of serious injury or disease, they might still situationally or routinely consent to the risk, explaining their risk-taking as necessary, appropriate and/or in their interests (see Figure P1.2). Risk-taking was understood here as "freely given" consent (as opposed to coerced compliance). The more typical explanations were that taking the risk was an accepted feature of their job or occupation, either routinely or in certain circumstances (e.g. Miner: "Every UG job is dangerous, it's up to you to make it safe"; Self-employed Farmer: "Sometimes you can't get the job done any other way"), and/or

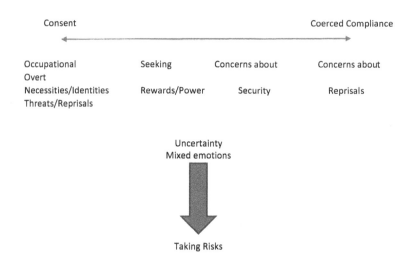

Figure PI.2 Taking acknowledged risks: acceptance continuum.

risk-taking was understood as a strategic means of gaining or retaining certain benefits, including their current job or employment (e.g. Miner: "To earn [production] bonus, you have to take some risks"; Auto Worker: "The company needs to get the parts out on time and my job depends on the company being successful").

The subjective content underlying consent in these various contexts was quite diverse in as much as workers gave a range of explanations for consenting to risky conditions, some of which were constructed as a personal choice and others constructed as more forced by their particular circumstances (albeit circumstances which were not seen as grounded in employer coercion). Occupational identity and commitment to firm, personal relations at work and family needs were particularly important rationales for taking risks, as was the perceived need to preserve their employment or current job situation (i.e. Miner: "If we can't reduce compensation costs, management tells they will have to close the mine"). As a *consent-based* response, the "need" to take risks was not understood as being forced by a coercive employer or manager, but rather, was accepted as a "fact of life," fair exchanges or the function of circumstances beyond the employers' control (Hynes and Prasad, 1999; Weber et al., 2018). While another identified pattern, the capacity of the employer or manager/supervisor to confer specific favours or benefits to workers in exchange for risk-taking, was recognized by workers as reflecting management's power to confer such favours, it was still experienced as more consensual than coercive in the sense that the power was constructed by the worker as legitimate or reasonable (Miner: "He scratches my back, I scratch his"). Some mixed emotions or resentment were evident depending, for example, on whether the manager or supervisor was seen as being fair with his/her favours. Risk-taking in this context was often understood as a political strategy which workers used to gain capital (Bourdieu, 1977) and which they used to secure or improve their position and their security (Miner: "There is a lot of skill to work in those conditions and the shift boss understands that"). The capacity to take and manage risks was sometimes a source of pride which underpinned a worker's occupational identity (Miner: "It takes a certain kind of person to work underground. Lots of guys can't cope with the danger") (Haas, 1977; Lawson, 2009/10; Paap, 2006).

However, reflecting the coercive side of the coin, workers often explained their acceptance of high risks in starker negative terms – that is, their *compliance* was constructed as being forced by their lack of power and their vulnerability to management/supervision reprisals if they contested the conditions (Auto Worker: "I know if I complain, I'll be out of a job"; Miner: "I've tried to do something about it but they put me in a ditch so I don't say anything anymore"). Even self-employed farmers sometimes referred to contractors as forcing them to take certain risks under threat of contract loss and blackballing. While some workers seemed to assume management resistance and reprisals without having had any direct experiences, the conviction that they were powerless or would be subject to reprisals if they took any action was often based on explicit current or past threats and/or reprisals, whether personal or observed (Autoworker: "I haven't said anything but I've heard from other people [workers] not to say anything"; Miner: "I saw what they did to the crew on my cross-shift when they complained. They were transferred off bonus to some shit hole job"). In contrast to the consensual situations cited above where workers took risks because they assumed or believed it was legitimate, necessary or in their interests, here workers were responding to fears, direct threats and past punishment.

In Labour Process theory (LPT) terms, this distinction is important conceptually in that this latter form of acceptance was experienced by workers as coerced *compliance* forced by a perceived lack of power *and* as reflecting a perceived direct conflict of interest between themselves and the employer or manager (Burawoy, 1979; Gramsci, 1971). However, as will be shown, whether the worker constructed their acceptance as more consensual or as more coercive, employment and other sources of insecurity were central to explaining both kinds of responses.

The distinction between consensual and coerced risk-taking was not always clear-cut in workers' minds. Just as workers were often not certain about their judgement of risk, they were also often uncertain about whether their risk-taking was justifiable or necessary (Kosny et al., 2012). Workers frequently expressed mixed emotions deciding to accept what they were being asked to do while still being wary and suspicious, especially as they began to realize more fully the seriousness of the risks they were taking. Moreover, what started off as consensual may eventually be experienced as more coercive as workers began to think that their commitment to take certain risks, such as working harder or faster, was being unfairly forced or exploited by management (Autoworker: "I don't mind when they rush us from time to time but lately it is happening all the time"). As this implies, risk responses are perhaps best understood as interactive ongoing processes which vary along an acceptance continuum from unambiguous consent to clearly coerced compliance, with degrees of uncertainty, commitment and conflicting emotion in between (see Figure P1.2).

Resistance

When workers reported that they had contested or resisted a perceived hazard in some way, and this again took different forms and degrees, it was usually because they defined the condition as a serious threat to their body which they understood could or should be controlled better or removed entirely from the workplace. Workers, in these contexts, might still be concerned about risking their employment and reprisals, as well as other negative consequences, and they might be sceptical or uncertain about their ability to get the risk reduced or eliminated. However, as will be shown, more often than not, workers complained or challenged management if: (1) they believed they had the political capacity to gain changes (Autoworker: "It was not easy but once we (the workers) got together, they had to listen to us"; Miner: "Our shiftboss [supervisor] knows when we mean business. If he wants the muck [ore], he's going to do what he can"); (2) they thought there would not be significant negative consequences from management and/or workers (Autoworker: "He's a fair boss so I know I can say something without them coming down on me"); and/or (3) they assumed they could resist or tolerate any negative consequences (Miner: "I know he'll yell and scream but that don't bother me").

   As will be shown, workers again varied in where they drew the lines on acceptable and unacceptable levels of health or safety risk – that is, where they decided that the security of their body (or their mind) was more important than immediate employment considerations. In some cases, they were worried enough to act even when they were certain that it would fail to make a difference (Autoworker: "I doubt they'll do anything but this is my health") but, more often than not, some hope or prospect of change was an important component behind resistance (Miner OHS Representative: "We may not get what we want but I'm pretty sure they'll do something").

   As such, an understanding of resistance to risks revolves substantially around the examination of the circumstances in which workers were more or less likely to perceive resistance or contestation as both possible and nec- essary, and then tracing how and why these different forms of resistance developed or failed to develop into an effective and sustained challenge to hazardous conditions and practices at both the individual and collective lev- els. As with the judgement of health and safety risk, workers often judged the politics of the workplace based on various past experiences, talk among workers and other signs, and then strategized around when to contest and how to contest, with whom and through which channels or relations. On the other hand, there were also times where workers responded with anger and frustration without necessarily thinking through the consequences, remind- ing us that emotions also play a role in shaping risk responses (Lerner and Kelter, 2001; Zinn, 2008a). What will also be made clear in the analysis is that whether workers consent, comply or resist in any given context, is not a simple function of the structural constraints and opportunities operating in that context. Different workers entered into these contexts with somewhat different dispositions (i.e. habitus) and resources (i.e. capital) which often yielded different actions and outcomes.

Consent, compliance and resistance: a constructionist framework

Although products of ongoing social interaction, varying with context and subject to constant change, these worker explanations are conceptualized as collectively constituting a "snapshot" of the total subjective bases of worker consent, compliance and resistance to health and safety hazards, which un- derlie the health and safety politics within a given workplace at a given point in time. Individually, these explanations reflect each person's knowledge, beliefs and dispositions (habitus) towards risk in their workplace. Although the specific rationales or explanations provided by each individual are un- derstood as often being incomplete, ambiguous or not entirely accurate narratives of any given response or practice (see later in this chapter for further explanation), they offer a window into the range of understandings and assumptions that worker construct when acting in and responding to their workplace conditions.

As represented in Figure P1.3, consent by individual workers to a given condition or practice is first contingent on whether the worker defines the condition or practice as a hazard. If they do not, acceptance of that condi- tion or practice, at least with reference to health and safety, is assured from those workers in the given context as long as that is the case. However, if the condition or practice is recognized as a hazard, consent is then contingent on whether the risk is judged to be low or under control. If it is not judged as low or under control, consent is then contingent on whether taking the risk is constructed as being necessary or in their interest. Finally, a known unnecessary risk may still be taken if they see themselves as being coerced

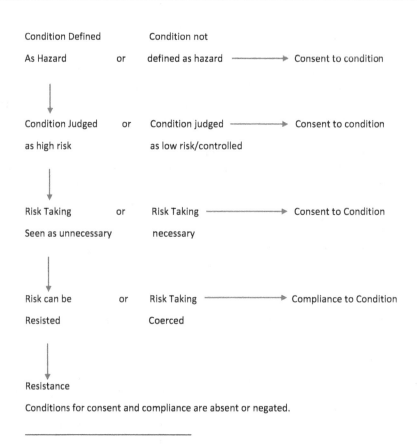

Figure PI.3 A model of worker subjectivity and action: consent, compliance and resistance.

or powerless to do otherwise. Resistance or challenge occurs when all four conditions underlying consent or compliance are negated – workers define the condition as a hazard, judge the risk as unacceptably or unnecessarily high with limited personal or external control, see their interests in contesting the condition as outweighing other conflicting interests, and view themselves (or allied others) as having the power to challenge and/or change the condition.

In this framework, consent, compliance and resistance to workplace hazards are seen as the social and political outcomes of ongoing and continuous interactive social relations and social processes involving human actors (workers/producers, supervisors, managers, employers/contractors) with overlapping but still distinct individual habitus (assumptions, knowledge, beliefs and understandings), which they've constructed while working in politically structured physical, social and cultural fields (mines, factories, farms). These structured environments are conceptualized as existing in the present in the sense that they have certain identifiable characteristics that organize and condition human perceptions, responses and practices, as well as having a history which has involved and continues to involve changes which introduce new demands and conditions as part of the dynamics of capitalist production relations (Bourdieu, 1977; Burawoy, 1985; Thompson, 2010). With further reference to the significance of time and space, workplace actors are recognized as having particular histories which have shaped their individual habitus, which are central to how the actors in any workplace space and time know, perceive, judge and respond to those objective conditions and, if they resist, how and why they do so. These individual histories are especially important in explaining both worker *and* manager variability in risk constructions and responses within the same workplaces and jobs (Bourdieu, 1977).

The role of knowledge is also clearly important in the framework being developed here, drawing from both social field and governmentality theory. Knowledge of hazards and ways of judging and controlling risk are significant bases of worker consent and resistance, while the development and negotiation of this knowledge becomes an integral aspect of workplace politics and power (i.e. embodied cultural and symbolic capital). Knowledge is also the means through which outside influence is realized when these negotiations involve consulting outsiders such as family, friends, doctors, books, newspapers and the web (i.e. social capital). A significant distinction that will be made here is between common-sense knowledge, which is largely experience-based versus *scientific* or technical knowledge, much of which is nested in technological designs and is often management-controlled and valued accordingly Although a governmentality approach implies an adoption of Foucault's argument that power and knowledge are intertwined (Foucault, 1977), knowledge is not equated with power in this book – that is, to know that a risk exists is a precondition to act, avoid, control or resist that risk but, knowing the risk exists is not sufficient to its control, especially in the context of a workplace (Tucker, 2003).

As noted, Bourdieu's notion of habitus (1990) is central here in that common-sense dispositions or orientations to risk are understood as underlying practices, largely based on risk assumptions and immediate perceptions rather than consciously calculated, but which are also subject to change as positions, interests and power resources change. As these latter points imply, while hazard responses are presented in Figure P1.2 as being contingent on certain constructed meanings beginning with the definition of conditions and practices as hazards, this should not be taken to mean that individual workers go through a conscious process of rationally calculating risk and interest every time or even most times they encounter a working condition or practice. As Bourdieu (1990) explained:

> The conditions of rational calculation are practically never given in practice: time is limited, information is restricted, etc. And yet agents *do* do, much more often than if they were behaving randomly, 'the only thing to do'. This is because, following the intuitions of a 'logic of practice' which is the product of a lasting exposure to conditions similar to those in which they are placed, they anticipate the necessity immanent in the way of the world.
>
> (Bourdieu, 1990, p. 11)

Thus, Bourdieu's concept of *habitus* is employed to help explain how and why certain subjectivities and structured practices reproduce management control and risk-taking by workers, while offering at the same time some basis for understanding when and how challenges and change can develop. For Bourdieu, the working class develop cultural knowledge (perceptions, rationales, beliefs, theories) through a lifetime of socialization within and outside the workplace which dispose them towards certain strategic responses to their environment, the rules of the game (e.g. that certain risks and injuries are part of the job, that deference and concessions to management are needed to sustain employment) and the demands of the dominating class (i.e. employers and managers) (Bourdieu, 1977, p. 72). Following Bourdieu, responses to risk are not determined in any mechanical way but neither are workers making conscious choices from all the objectively available alternatives. As Bourdieu (1977) states, the habitus is the source of "[a] series of moves which are objectively organized as strategies without being the product of a genuine strategy..." (p. 73). Given certain objective histories involving work relations, environmental conditions and bodily injuries, workers have constructed what they see as reasonable and unreasonable, necessary and unnecessary, expected and unexpected, and normal and abnormal risk, insecurity and bodily damage, not as absolutes but as informing what is seen as relatively possible and impossible, improbable and probable. Understood in this way, the different kinds of explanations that workers provide for their responses to different conditions and hazards are not the determinants of their behaviour, but rather are understood as reflections or expressions of their conscious *and* unconscious understanding and acceptance of their position in the objective world – in particular, how the workplace works and how much power they have and don't have within that context.

Using aspects of risk theory, I also recognize that workers are often dealing with uncertain situations, grounded in their limited control and power, which means that there will always be a certain amount of slippage in the extent to which workers can perceive and judge risk directly and through other sources, yielding a variety of less-than-rational strategies for coping

(Hall, 2016; Kosla, 2015; Zinn, 2008a,b). While acknowledging that workers often take leaps of faith, this approach brings the analysis back to my focus on control and responsibility as the central meanings underlying worker and management constructions of health and safety practices in as much as even these leaps revolve around the perception and/or implicit acceptance of control and power differences.

Moreover, these dispositions towards risk may only last as long as nothing happens to contradict the underlying assumptions on which consent, trust and worker habitus is based, including of course repeated experiences of "unnecessary" injuries, close calls or health problems. Shifts in power relations among and between workers and managers may also be critical in revealing new objective opportunities or openings for resistance, which can also mean that the standing rules of the game regarding the acceptance of conditions are changed. As Bourdieu (1990) also theorized, a consent-based habitus may still be operative for a time despite these changes, something he refers to as the *hysteresis effect*, a lag in which actors are unable to see the opportunities arising from structural change and continue to fall back on dispositions or understanding which are no longer necessary (p. 78). Again, these understandings may not be explicit or conscious at the time of their application, but that does not make them any less real as meanings which underlie workers' responses.

In this regard, I should acknowledge that in many of my interviews, workers and farmers were often hesitant or unable to explain their acceptance of certain conditions, and it often appeared that they were literally constructing for themselves explanations that made sense to them only *after* they'd acted. As I spent more time with some workers it also became clear that sometimes they were not entirely certain how to explain their actions (Farmer: "I'm not really sure now that I think about it. Maybe I just didn't think it through at the time"), reinforcing Bourdieu's (1977) argument that people often don't fully know why they do what they do, but often try to make sense of it after the fact. From this perspective, it is sometimes difficult for workers to explain why they respond in certain ways to work hazards in as much as these are things that workers often do not think about in any active way or may actively avoid thinking about (Emirbayer and Johnson, 2008) – that is, why they continue to work in conditions that are making them tired, sick or unhappy. Yet, following Bourdieu (1990), risk theorists such as Zinn (2004), and labour process theorists such as Burawoy (1979), it would be wrong to then assume that all or most workers are operating on automatic with little or no conscious attention to the conditions around them. Again, as Bourdieu put it, "people are not fools; they are much less bizarre or deluded than we would spontaneously believe because they have internalized, through a protracted and multi-sided process of conditioning, the objective chances they face" (Bourdieu and Wacquant, 1992, p. 130).

As such, while risk habitus is seen as quite durable and again difficult to change even when the "social field" (i.e. workplace) changes, risk-related practices can change as taken for granted assumptions are challenged. In this research, organizational change and labour process transformations are seen as central to shifts in habitus to the extent that they contradict the underlying assumptions regarding control and responsibility, and lead to changes in power relations (e.g. workers gain or lose employment security). This gives us a means of theorizing the links between subjectivity, individual practice and labour process and management restructuring in as much as we can focus our analysis on the organizational or workplace "fields" in which the workers are operating. However, these understandings of how work and the body connect are grounded in early life experiences brought with workers into the workplaces, meanings which are nested in the social class, gender and race relations in which people grew up, as well as the particular family, community, school and friendship network experiences that they now have. As such, this approach also forces us to look beyond the organizational and labour process contexts to explain worker actions. Such a perspective also gives us a further basis for explaining both similarities and differences in the ways workers respond to similar workplace circumstances and organizational transformations. This is especially important for understanding why certain individual actors within organizations can generate change, while others act in ways to sustain relations of domination by dividing workers against each other. The similarities that will be most important to uncover here will be those common dispositions towards power and risk within the organization which encourage the acceptance of certain hazards and in the process ensure production and profit goals.

Using the above framework, Chapters 2 and 3 show how consent and compliance to hazardous conditions are grounded in the limits, forms and variations of workers' knowledge, perceptions and beliefs about specific hazards, *and* in their understanding of their interests in and control over those hazards within the context of concrete social relations in production. While addressing the issues of relative worker management responsibility and worker insecurity, I attempt to trace the different ways in which these subjectivities and the accompanying politics develop within the context of day-to-day practices, discourses and interactions. In subsequent chapters, I will then seek to link these subjectivities and the associated politics of risk to restructuring in the three industries.

Chapter 2

Identifying hazards and judging risk

This chapter first examines which conditions are known, suspected or unknown as hazards, and then considers how the risk levels of those known and suspected hazards are perceived and judged. In terms of management control, the key research question for this first section of the chapter is: to what extent were workers "unknowingly" accepting actual or potential risks because the conditions or activities were not recognized or understood by them as OHS hazards? In the second section, the critical question is: how many hazards were being accepted because they were judged to be low risk or under control? In addressing both questions, I also want to understand where workers nest responsibility for identifying hazards and judging and controlling risk – in themselves, other workers, management or the state.

Constructing OHS boundaries: the definition and identification of specific hazards

The history of occupational health and safety in any nation or industry is littered with injuries and diseases caused by hidden, unknown or officially unrecognized hazards (Ashford, 1976; Barth and Hunt, 1982; Berman, 1978; Brown, 1985; Derickson, 2018; Leyton, 1975; MacDowell, 2012; Nichols, 1986; Perrow, 1984; Smith, 1987). Given disease latency periods and the challenges associated with "proving" to the authorities the links with specific working conditions, health hazards in particular often did their damage for decades if not longer before finally being officially recognized by scientific and government authorities (MacDowell, 2012; Rennie, 2005; Smith, 1987). As documented by the work of labour and medical historians, these histories usually involved lengthy struggles by unions, activists, sick and injured workers and their relatives to get hazards recognized, as well as the efforts of medical and other researchers to provide the proof needed to make their case (Asher, 1986; Beaumont, 1983; Calhoun, 1988; Doyal and Epstein, 1983; Elling, 1986; MacDowell, 2012; Smith, 1987; Storey, 2006; Tataryn, 1979; Walters, 1985). Politicians and regulators have often played vital roles in affecting the time it took to designate these hazards, although frequently not in a good way (Carson, 1982; Hilgartner, 1985; Sass, 1979; Tucker, 1990).

Mining, farming and auto manufacturing in Canada are no exceptions to this history (Leyton, 1975; Hall, 1998a, 2007; MacDowell, 2012; Storey and Lewchuk, 2000; Wells, 1986). By the time the three industry studies were conducted, all three industries had accumulated lengthy "lists" of officially recognized hazards (see Table 2.1 below for examples) contained in legal regulations, corporate OHS standards and practice manuals, OHS training programmes and government and industry accident prevention association communications (e.g. Inco, 1981; OFSA, 1985–95; OHSA R.R.O., Reg. 851, 1990; OHSA R.R.O., Regulation 833, 1990; OHSA R.R.O., Reg. 854, 1979, 1990). In an era where government narratives on OHS legislation and prevention emphasized the importance of education and the workers' right to know about workplace hazards (OMOL, 1989a,b, 2010), it is also less than surprising to find that workers in the three work settings were able to identify most of these officially recognized hazards. Observations of joint health and safety committees, interviews with committee and union representatives and reviews of committee minutes in the mining and auto studies

Table 2.1 Some officially recognized physical hazards in mining, farming and auto production

Mining	Farming	Auto
Dust – Nickel – Silica	Grain/Hay Dust/Field Dust	Uranium/Metal Dust/ Other Particulates
Diesel Fumes	Diesel Fumes/Manure Gas/Silo Gas	Welding Fumes/Carbon Monoxide
Ammonia Exposure	Pesticide Exposure	Various Chemical Exposures
Explosives		
Oil Mist		Oil Mist
Noise	Noise	Noise
Drill Vibration (White Hand)		RSIs/MSD Hazards
Lifting	Lifting	Lifting
Rock Falls/Collapses	Silo/Grain Storage Hazards	Warehouse/Storage
Floods/Mud & Rock Slide	Manure/Other Ponds	
Falls – Falling to Ground – Falling in Deep Holes	Falls – Ladders, Silos	Falls – Ladders etc.
Contact with Heavy Machinery	Heavy Machinery	Lift Trucks
Vehicle Accidents	Vehicle Accidents	
Fire/Electrical	Fire/Electrical	Fire/Electrical
Temperature Extremes	Sun Exposure/Heat Stress	Heat Stress

indicated that these official "lists" of hazards constituted the focus of joint committee discussions and inspections, and worker complaints, in effect representing the consensual boundaries of health and safety discourse and politics in the workplace.

However, in virtually every workplace, there were areas of disagreement among workers and between workers and management in the hazards they listed. In family farms, there were differences between spouse co-owners or with other working family members such as adult sons and daughters. Within the different mines and auto plants studied, there were often worker representatives and/or workers seeking to get their employer and/or the government to officially recognize some condition or activity as a hazard, or as a possible hazard warranting at least study if not action (Hall et al., 2006). Many of these conditions represented new substances or processes recently introduced in the workplace such as in the auto case study where some workers believed that a new soap they were using to wash down equipment was causing rashes. Sometimes, long-standing concerns continued to be contentious. In the mining context, one of the best examples of this was a struggle over lighting. Going as far back as the 1974 Ham Commission, and probably further, the miner's union (USWA Local 6500) had been arguing that the lack of fixed lighting within mines was a major safety hazard in itself. As the Local stated in one Commission submission,

> in dark unlighted areas, the senses are not sharpened but rather they are dulled…poor lighting or no lighting lessens the sense of danger and so actions are performed in this environment which would never be attempted in broad daylight or in adequately lighted areas.
>
> (As reported in Clement, 1981, p. 228)

However, while the union and some worker OHS representatives were still arguing for safety-related regulations in lighting in the 1980s (USWA, Local 6500, 1981, 1984), very few miners cited the lack of better lighting as a safety hazard in their interviews. Some acknowledged in discussions I had with them that darkness may have played a role in some specific accidents such as falls, but most tended to accept darkness as an inherent fixed feature of the UG environment. For the most part, there were still no regulations in Canada at the time of writing requiring fixed UG lighting (ILO Encyclopedia of Occupational Health and Safety, 2011).

Although farmers often worked at night with limited lighting, darkness and the need for additional lighting in fields was also largely missing from official and farmer lists of farm hazards. However, a better example of a farm condition largely excluded from the health and safety discourse in farming then and still largely now, are the health risks of chemical fertilizer exposure. Although farmers using pesticides were required by the time of the study to take a course on pesticides, the course and the Ministries of

Labour (OMOL) and Agriculture (OMAF), and the farm safety organizations (OFSA, CFSA) were silent on whether there were any risks involved
in the inhalation, ingestion or skin exposure to most chemical fertilizers,
with one notable exception, anhydrous ammonia. Anhydrous ammonia was
understood officially and among farm operators and workers as presenting
both significant safety and health risks which included a risk of explosion
(Ontario Farm Safety Association, 1990). However, other more commonly
used fertilizers such as urea or different forms of phosphate were not addressed in any meaningful ways as potential health risks, and accordingly,
almost all the farmers took no steps to reduce their exposures or the exposures of their family or workers.[1]

Not surprisingly, the extent of the overlap (or disagreement) of official,
management and worker lists also differed with workplaces. For example,
while the auto case study (Study 5) exhibited very little conflict over hazard
identification, the OHS representatives (reps.) from several other auto parts
companies in Studies 4 and 7 reported that their employers provided limited
information to workers on hazards and often contested the reps.' efforts to
raise new concerns about conditions or activities. One OHS rep. in an auto
parts plant described his continuing effort to get his management to recognize workers' chrome exposures in one of their processes:

> It's probably the most toxic thing I've ever seen... but they [manage-
> ment] didn't want to admit it and change the process...I brought pic-
> tures of it to the company and they're still saying it, that the process
> creating chrome sludge doesn't happen.

In the mining case study, there were differences between mines within the
same firm. For example, in one of INCO's mines, which happened to be
the company's most advanced mine in terms of mechanization and mining methods, remote-controlled devices for ore loaders called 'scooptrams'
were identified as unreliable after several close-call incidents. The local
mine management resisted this characterization blaming carelessness on
the part of the operators. Supported and encouraged by their health and
safety representatives, the miners eventually responded with slowdowns and
legal work refusals under the Occupational Health and Safety Act (OHSA,
1979, 1990). Management relented in this particular mine, known as South
Mine, by constructing safety stations in which workers could safely operate the machinery. However, although I found examples in the company records of runaway scoops in one other mine, the issue was never identified
by miners or the worker reps. at other mines in the company and remote
operators continued to work without safety stations. Efforts by the union
to gain company-wide adoption and government orders to that effect failed.
As I observed in one joint health and safety committee in another mine,
the complaints from South Mine were dismissed by both management *and*

worker committee representatives as being caused by operator error and troublemakers in South Mine's joint OHS committee.

As these examples indicate, consensus and conflict over the definition and identification of hazards varies with workplaces, although there may be patterns specific to some industries. For example, work stress was clearly recognized as an occupational hazard by the Ontario Farm Safety Association (OFSA, 1985d), and some farmers noted it was a concern. But stress was not raised as an issue in the mines or by any of the miners or their reps. as a legitimate OHS hazard. For the mines, this may have reflected the time when the study was done, but even though emotional stress has become more acknowledged at the government level since the late 1990s (Hall et al, 2018), it was never mentioned by any of the autoworker reps., workers or managers. By the same token, however, every workplace exhibits a certain level of disagreement among and between individual workers and managers regarding which conditions are recognized *as* health and safety issues. In effect, there were distinct "unofficial" lists of hazards constructed by each individual worker and manager which differed to varying extents from each other and from the official state, corporate or union lists. In those contexts where workers and managers diverged in their definitions, and especially when they were not recognized in official or government policy and egulations, workers and worker representatives were faced with the added challenge of legitimizing their claim that there was potential risk, as well as showing that the risk was substantial enough to warrant management action of some sort (for excellent examples of similar struggles in the Canadian mining and auto industries, see MacDowell, 2012; Rennie, 2005; Storey and Lewchuk, 2000).

However, official acknowledgement of an industry hazard did not necessarily lead to recognition in practice among managers, supervisors and workers within the workplace itself. For example, in the mine study survey, miners were asked specifically about whether they defined certain "known" conditions as hazards, all of which had been identified in company safety materials (All Mines Standards Manual, 1981), the Mining Accident Prevention Association publication and audit materials (MAPAO, 1985), and/ or the OHSA mining regulations as presenting health or safety risks. As Table 2.2 shows, officially acknowledged conditions such as noise, drilling, dust, blasting fumes and diesel fumes were widely defined as health hazards by a clear majority of workers (80%+) but, there was more disagreement about oil mist, ingested dirt and temperatures/humidity, and indeed, the majority did not identify dirt and temperatures/humidity as health hazards.

The greater uncertainty around oil mist likely reflects the increased visibility of the problem caused by the company's shift to larger drilling equipment, again linked to its move to bulk mining (see Chapter 4). However, hand-held jack leg and stope drills had been generating oil mist for decades

Table 2.2 Conditions defined by miners as hazards present in the workplace

Present Conditions

	Yes (%)	No (%)	Uncertain (%)	Known Health Risks
Drilling, Vibration	96	2	2	Carpal Tunnel, Musculoskeletal Injuries (MSDIs)
Noise	100	0	0	Hearing Loss, Tinnitus
Airborne Dust	81	15	4	Silicosis, Asthma, Lung Cancer
Diesel Fumes	84	10	6	Lung Cancer, Chronic Obstructive Pulmonary Disease (COPD)
Blasting Fumes	81	10	9	Asphyxiation, Chronic Obstructive Pulmonary Diseases, Kidney Failure
Oil Mist	51	39	10	Lung Cancer, COPD, Asthma
Ingested Dirt	42	54	4	Stomach and Colon Cancer
Temperature/ Humidity	40	60	0	Arthritis, Asthma, Heat Stress, MSDIs

while research had documented the health problems associated with oil mist since the 1960s and 1970s (e.g. Suthers et al., 1986).

Temperature and hygiene issues had long been acknowledged by the union, the company and the Ministry of Labour as significant OHS issues in mines (Royal Commission on Health and Safety of Workers in Mines, 1976, p. 13), acknowledgements which had pushed the development of lunch-rooms in the 1960s and 1970s with heat, and running hot and cold water. Still, many miners routinely ate their lunches in their stopes or development headings, arguing that they didn't have the time to walk to the lunchroom given production bonus reductions. This meant they were continuing to ingest dirt while sitting on what was often wet ground. Mines were damp and often cold in the upper levels requiring heavy winter coats, yet these conditions were also not seen or treated as occupational health issues by all miners. As one miner put it: "I don't like the damp, it gets pretty cold down there when you're near the [air] intake…But it's not what I would call a health concern. It's just damn uncomfortable." When workers defined these conditions as affecting their comfort rather than their health, they did not raise them as health issues with other workers, their supervisors, their union OHS representatives or management. Those who did were more likely to raise their concerns with management as compared to those who saw them as issues of comfort or convenience. As one miner reported, "I'm concerned about sweating and the cold for arthritis. I've complained five or six times in the last month but they just tell me there's nothing that can be done."

As observed in several workplaces and situations, without a worker consensus on the status of these conditions as legitimate *health* issues, it was much easier for supervisors and managers to dismiss or ridicule the complaints of miners who failed to characterize their concerns in health terms (MacEachen, 2005; Tarasuk and Eakin, 1995). As one supervisor dismissed my question on the health aspects of mine temperatures, "if you want a *comfortable* job, you don't work in a mine." Another miner reported that when he asked his supervisor about oil mist, he was told, "what are you worried about, it'll make your prick hard." This unofficial denial or belittling of the legitimacy of certain OHS issues by front-line managers and supervisors may partially explain why so many workers dismissed these issues, or at least felt uncertain about whether there was anything to worry about, and ultimately, why the company was able to minimize demands from workers even as emerging problems such as oil mist got worse.

Although I don't have similar survey data in the auto or farm case studies, observations and qualitative interviews in both contexts suggest equivalent findings that some officially "known" health and safety hazards were not constructed as such by substantial numbers of autoworkers and self-employed farmers and farm workers. For example, farmer operators reported that despite official warnings on pesticide labels, some chemical company reps., commodity contractors and even some government extension people would tell them that commonly used herbicides (as opposed to insecticides) represented *no* risk to their health regardless of exposure level, which they then communicated to their farm workers and family members. While later discovering otherwise, sometimes through information brought to them by their spouse, neighbours or adult children, many had accepted these definitions without question for several years.

Several autoworkers in the auto case study had similar tales about individual managers and supervisors denying *any* risk associated with uranium dust or metal particles in their workplace, although according to joint OHS committee minutes, these issues were clearly being discussed within the committee. Heat stress had been officially recognized as a critical health issue in industrial plants and farm fields since the early to mid-1990s. At the time of the 2004 auto parts study (Study 4), virtually every plant had heat stress language in their collective agreement and the problem was also officially recognized in Ontario Ministry of Labour guidelines (OMOL, 2003, 2008). Although non-union, the auto parts case study also had a heat stress policy which the company claimed was introduced before the Ministry Guidelines in 2003. Management and the health and safety reps. in the case study acknowledged in their interviews that heat stress was a known health issue but there were still some autoworkers in the plant who defined heat as a comfort issue rather than a health risk.

Regardless of what management, the union or the state defined as hazards, ultimately the presence or absence of worker definitions was critical in shaping whether conditions were detected or addressed in these contexts as health or safety hazards. If most or all workers failed to define a condition such as lighting or temperature as a potential hazard, then there was little to no political pressure for change or corporate/state investment on that issue (MacEachen, 2005). Moreover, even relatively small splits among workers in the recognition of certain hazards, as was the case around mine diesel fumes or dust (see Table 2.1), could have profound effects on the politics of health and safety in a workplace, particularly when those conditions were related to significant changes in technology, jobs or the labour process.

For example, if we look back at Table 2.2, recall that there were some miners, consistently around 15% of the sample, who denied that dust, diesel and blasting fumes presented any risk to their health. These were essentially the same group dismissing not one but several known hazards. The term that miners often used for this group was "haywire" because they were seen as also being more likely to ignore many safety procedures. In interviews, these "haywire" miners insisted that these conditions were simply "nuisances" or "annoyances" which they acknowledged may have some temporary physical effects such as eye or throat irritations but did not present long-term health risks or permanent health damage. Accordingly, they did nothing to protect themselves and largely ignored company safety rules and procedures. While some supervisors and managers cited these workers as illustrating worker carelessness as a major cause of accidents and disease, other supervisors and managers were quite willing to rely on these workers to do the "dirty jobs" or jobs that needed to be done quickly, while also giving them options for dealing with workers who refused those assignments. As one mine shift boss (supervisor) put it, "if someone doesn't want to do it fine, I don't push them as long as I have somebody who is willing to do it."

During the mine study field work, I recall one incident, in particular, where I was with a blasting crew locating and loading holes with Amex explosive (Field Notes, Book 5). I had identified this particular miner as a case study precisely because he had dismissed in his survey interview a number of health conditions, including dust, as non-health issues. There had just been a blasting operation from the previous shift, and as was always the case after a large VRM blast, there was dust and ammonia fumes everywhere, the latter being because the Blasting Agent Amex (Orica, 2012) was made of fertilizer (ammonia nitrate and diesel fuel). The ventilation tubing had also been damaged. The standard company procedure which I had observed in other crews was to water down the dust and re-establish a fresh air supply by repairing and reconnecting ventilation tubing. These tasks usually took about two hours. However, in this case, the miner and his partner immediately began digging for the holes which had been partially buried in the blast. Within ten minutes, the air was so full of dust and fumes, we could barely see five feet in front of us. Neither of them were wearing dust masks (although I was). When I asked one why he wasn't following the rules, he simply grinned and said "it's not worth the trouble." He repeated his claim from the survey interview several times during our conversation during and at the end of the shift that for him he didn't believe the dust was a danger to his health. He also claimed complete ignorance of any harmful chemical residues in the dust: "I'm exposed to dust and diesel fumes all day [but] I don't worry about it. I'm ignorant of all that stuff... I'm not concerned."

As he and others taking similar positions admitted in interviews, they were aware that others including the company and their union had defined these conditions as hazardous, but they either expressed a frustration that there were just too many conditions defined as hazardous to be meaningful, or they acknowledged that it was just "their way" not to worry about it. One miner even expressed the view that health procedures like watering down dust were just a way of interfering with his control over his job and slowing him down. Some also expressed bitterness over interference by union health and safety representatives or company safety inspectors. As one miner stated: "They try to catch us but we look out for them. They're telling us how to do our job. But screw them, I know what I'm doing."

However, as some of the interviews with "haywire miners" suggested, there was often acknowledgement that they refused to see the hazard or believe what they've been told by other workers, managers or experts, because to do so challenged what they understood as their more important interests, including getting the most income out of their job. As one miner put it, "I consciously don't think about health things- that only makes it worse." These dispositions towards ignoring the presence of several known hazards, can be understood as one form of a risk habitus in the sense that they had come to understand over their years as workers that they could not get the full economic value out of their job if they worried too much about their health. Their reluctance to recognize known hazards allowed them to earn economic capital in the form of production bonuses but also yielded bargaining power or capital which they used to gain other advantages and favours from supervisors and managers, which in turn offered them a degree of job and financial security that they would not have had otherwise. Yet, since health is another aspect of security, they also realized at some level that getting sick or seriously injured would undermine their capacity to work and earn money.

To deal with these conflicting interests, workers often coped by "not thinking" about it or by denying that it was a hazard, responses that Dorman (1996, 2006) would likely identify as illustrating "cognitive dissonance". Whether unconscious or a conscious willing not to know, workers supported their practices by explaining away any work connections to bodily signs that could be interpreted as indicating health effects. As one driller acknowledged, "I get a lot of back and shoulder pains now but I figure this happens when you get old. As long as the bonus is there, I'll keep going until I can't do it." Other symptoms were dismissed by these workers citing a variety of individual lifestyle rationales which dismissed work as possible causes, including that they "smoke," "don't get enough sleep" or "don't watch their diet"; but again, there was the sense that these explanations were ways of keeping to the fiction that their working conditions were not responsible and that the conditions were not really hazards. It also helped that individual supervisors and managers were often playing down the risks and rewarding workers for taking them (MacEachen, 2005). While supervisors sometimes protected them to some extent when things went wrong, these workers were vulnerable to blame if they or fellow workers suffered serious injuries. Health risks were easier to dismiss since the effects were "down the road" but the more physically demanding the job, the sooner the accumulated effects loomed as a threat to their job performance.

Parallel patterns were also evident in both the autoworker and farmer groups. Some farmers and autoworkers, again in relatively small numbers, were distinguished from the rest of their occupational group by their general failure to construct certain widely known and officially recognized hazards as hazardous. For example, although most farmers recognized grain and straw dust as both long-term respiratory and immediate suffocation hazards (Denis, 1988; Thu, 1998), there were three farmers who viewed dust as mere irritants or annoyances and, as a consequence, during my time observing them, they or their farm workers never wore dust masks or made any other efforts to reduce their exposure even in very dusty contexts like silos or hay barns (Shortall, McKee, and Sutherland, 2019). Similarly, among the case study autoworkers, there were 4 among the 25 interviewed who seemed blissfully unaware of the dangers of aluminum dust caused by machining in their workplaces, even though, at the time of the study, there was considerable political controversy within the plant around the issue (Peters et al, 2013). Many of the autoworkers in Studies 4, 5 and 7 also seemed unaware of the invisible metal particles that common features of drilling and machining processes in their workplaces, even though again, these hazards were widely recognized in the auto industry and covered in regulations (e.g. OHSA, Control of Exposure to Biological or Chemical Agents, RRO 1990, Reg. 833)

Challenging definitions and exclusions

Although the mine and auto parts studies (see Part 2) suggest that many health and safety representatives operated narrowly by limiting their inspections to the identification and assessment of "known hazards" (e.g. Auto rep. "If it is not in the green {OHSA}book, I don't worry about it"), both workers and their representatives sometimes advocated the addition of new hazards to the list, often in response to reports from specific workers about injuries, physical symptoms or close calls. Some auto plant OHS reps. reported that they got involved in the health and safety committee specifically because they felt a health or safety hazard was being ignored. Mine worker reps. tended to vary as well, with some reps. rarely reporting any attempt to get a new hazard recognized, while others seemed to be constantly trying to push the boundaries of accepted health and safety discourse (see Chapters 4, 8 and 9 for further comparisons). As observed, in some of these cases, managers and/or safety engineers would quickly concur, especially if there were close calls or workers were exhibiting physical symptoms, and/or the advocating reps.(s) were able to provide research and scientific evidence. Often, however, the evidence was less definitive, and the advocates met opposition from management which often took the more benign form of delay tactics backed by the insistence that definitive proof is needed before action can be taken. As one representative described her experience, "it took us over two years with them[management] to finally admit it was a problem, always claiming that they were looking into it [metal particles]." Some worker reps. backed off in these situations, while others worked to achieve something in the long term. However, if certain that the risk was real, there were some reps. who were notable by their refusal to accept management's demand for definitive proof before acting on an issue. As another auto worker rep. stated:

R: People don't know yet and some of the figures were from smaller samples...We don't know why or how. It could be that a single ingredient within the package of metal working fluids people are exposed to. Like I say benzene is included in there, so the studies are done without benzene or without aromatic hydrocarbons. So, you got to wonder. But you can't wait for the research, we recommended caution from the day one. Keeping exposures to the minimum, keep the concentrations to the minimum, keep it all enclosed, or put a barrier between you and the mist, and we have no workplace cancer.

Not infrequently, management responses to worker efforts to introduce a new hazard definition involved threats, intimidation and reprisals. For example, as noted earlier, when workers and worker representatives in one mine tried to push their concern about remote scoop operations, two of the workers involved were removed from their position and the worker committee co-chair was threatened with a transfer.

Locating responsibility for hazard identification

The evidence suggests that workers within and between workplaces varied in the extent to which they expected and trusted management and/or other authorities to identify hazards for them. Veteran miners, autoworker and farm workers (20+ years), for example, tended to place less confidence in management, in part perhaps because their knowledge of hazards was seen as something they had learned on the job, through direct experience and/or through warnings from other workers. Those in the higher-skill jobs, such as electricians or mechanics, tended to insist that managers or even worker reps. could not tell them how to work safely but also had lower tolerance for conditions not specific to their trade (e.g. air quality). For veteran and high-skill workers, most of their official training was well behind them and even though many companies had a steady stream of safety communications, the older and more skilled workers were less likely to pay attention, arguing that they'd heard it all before or knew everything they needed to know.

Many workers expressed trust in supervisors to identify hazards but this also varied with the hazard and the supervisor, in part because supervisors were seen as varying in their reliability and their competence. For example, some miners expressed concerns when their company had moved to promote engineers rather than experienced miners into supervisory positions. In the auto case study (Study 5), some supervisors were seen as "friends" of the employer who had limited competence and, as such, were not trusted as sources of hazard information. Many believed that management concealed knowledge of hazards, often based on previous experiences, while others placed more confidence in management and/or the law to inform workers of hazards. Most fell somewhere in between these two extremes, and most workers accepted, to varying extents, that some hazards were also unknown to management and the authorities, and/or that the health or safety effects would only become evident over time as health problems or accidents emerged.

As will be outlined in later chapters, OHS legislation and regulation can be relevant in shaping where workers look for knowledge and the extent to which they trust management (Luria, 2010). By the time of the mining study, health and safety legislation was broadly requiring companies to identify and control all hazards in the workplace, while workers and supervisors were also required under law and IRS policy to identify and report anything that *they* assessed as unsafe. As miners often reported, their efforts to identify and report previously unrecognized hazards were frequently not accepted by management which also played a role in their understanding of whether reliable information on hazards could be obtained from management sources. However, as I'll show in later chapters, health and safety committees, worker representatives, audits, new technologies and other formal mechanisms and procedures were often set up with the primary function of identifying hazards, and again in varying ways and degrees, became important in shaping how workers understood responsibility and reliability. However, much of their impact depended on how the workers understood their own direct control through skill and knowledge.

Also significant for the analysis that follows in Part 2, new production technologies, inputs and processes were widely viewed by both management and workers as the key sources of hidden hazards.[2] As these understandings also imply, some workers believed that they needed to be on guard for signs of "hidden" health effects or injury risks, especially when new materials, machines or jobs were introduced, while others, again varying in degree, preferred to assume no risk and no need for attentiveness unless officially told otherwise. Most took a 'wait and see' attitude, looking for signs over time that there were problems. While some workers saw the authorities as either willfully negligent in failing to recognize new hazards, others took the position that the authorities, including their union, were overcompensating by classifying everything as a hazard. In either case, workers tended to be more sceptical of official definitions and saw it accordingly as their responsibility or the responsibility of their worker reps. to define what was and what was not a hazard.

Final comments: defining and identifying potential hazards

As outlined thus far in this chapter, each worker had their own "list" of known or suspected hazards, which overlapped substantially with, but was not identical to, other workers' lists or manager, corporate, regulator or expert lists. Together these various individual and institutional "lists" were conceptualized as constituting the boundaries of health and safety discourse and politics at any given point in time within these workspaces, in the sense that workers (and managers) attached certain common or contesting meanings to these conditions and responded to them *as* risks or potential risks to worker health and safety. Although I've noted considerable agreement among workers and management actors on the range of "known" present hazards in any given workplace, at any given point in time some workers or worker representatives were advocating among themselves and/ or with management and the OHS regulator for the recognition of *unrecognized* OHS hazards and their inclusion in risk assessment and control practices. This then became one critical aspect of health and safety politics – the struggle to get a condition or activity recognized as a hazard (MacDowell, 2012). While managers and supervisors often resisted these efforts (as do some workers), some of the definitions of new hazards came from managers, internal or external safety experts or regulators, prior to any recognition or action by workers, especially in the context of labour process changes (Brun, 2009).

Once institutionalized, these definitions fed into policies, rules and procedures for assessing and controlling the risk (Arksey, 1998; Hadler, 1996). Accidents or close-call events were often significant in calling attention to previously unrecognized hazards, although a single close-call incident was rarely enough, and the identification of the causes and therefore the solutions were often contested (e.g. when management and workers dispute the relative role of machine design and operator error). As this rule-making suggests, there were management interests and dynamics which sometimes pushed the recognition and control of hazards just as safety management theory claims (Fingnet and Smith, 2013; Luria, 2010). The problem, however, was that there were also conflicting interests which discouraged management from acknowledging unknown hazards and undermined consistent attention to the known hazards, rules or standards (MacEachen, 2005). The conflict and politics surrounding the identification of "new" hazards were especially important within the context of this book since they tended to emerge during the introduction of new technologies and labour process restructuring.

Finally, the evidence suggests that while many workers were quick to adopt new definitions being advocated by other workers or worker advocates, some workers were slow to accept even the "official" expert or management definition of certain existing conditions or activities as potential hazards. Although often ostracized by their colleagues who saw them as dangerous and by health and safety activists who saw them as undermining their capacity to get changes in conditions, these outliers also allowed managers and supervisors to avoid conflict by assigning them to jobs no one else would accept, thus helping to perpetuate dangerous conditions.

Judging and controlling the level of risk

According to my risk framework (Figure P1.3), once workers identify a hazard, acceptance is then contingent on their perception and understanding of the level of risk within their current work situations and activities. The focus of the rest of this chapter is accordingly directed at understanding how workers *know* and judge *levels of risk* as "safe" or "unsafe." The principal objectives are to show that consent to particular hazards at this level is grounded in two key related conditions: First, there is the importance and confidence that workers place in their own capacity and responsibility to judge and control the risks of specific known hazards in their workplace *relative to* the capacity and responsibility of other workplace actors and their technologies. Second, there is the relative perceived power that workers and their immediate managers or supervisors exercise over the production and control of risk, and any conflicting perceived interests that workers have in controlling the risk. Where workers perceive relatively more power to contest the condition without threatening their immediate interests, they are more likely to judge the risk as unacceptable. If workers have conflicting interests in contesting or exerting more control over the risk, they are more likely to minimize the risks and/or exaggerate their personal capacity to control the risks. As this argument implies, the social construction of risk levels are themselves at least partially grounded in the objective power relations and interests of workers and managers as well the perceptions of those powers and interests. The broader argument is that workers are more likely to consent to risks which they understand as principally their responsibility, when they have confidence in their capacity to judge and control the risks, which in turn is closely related to the control and power they exercise more generally over their work and work relations.

How do workers judge the level of risk?

Along with learning through training, experience and social interaction on the job that certain hazards exist in their job, workers learned how to assess the seriousness of the risk (in terms of both probability and its effects) and, given a certain seriousness, how the risk was controlled (Trenoweth, 2003–05; Xia et al, 2017). What they also learned was that some hazards were assessed and controlled through their own actions, using their own knowledge, senses (seeing, hearing, feeling, smelling), and bodily actions (avoidance, slowing down) and reactions (breathing, pain, mobility), while others were assessed in part or wholly through the actions and claims of others. Workers also reported that they learned, again through direct experience and from others, when and how they could personally control hazards, and when to look to others as key or primary sources of judgement and control. As this implies, as workers learned about how risk was judged and controlled, they were also developing an understanding about who or what had control and who was *responsible* for exerting control – themselves, other workers, supervisors, and/or managers, technologies, or management systems.

Workers in the mines and many of the auto parts firms with established OHS safety programmes reported getting at least some of this understanding of how risks were judged and controlled through formal training programmes or orientations when first employed, or as they moved into certain jobs or occupations within the workplace. Other formal sources were safety bulletins from management or safety talks, or informal warnings given by their supervisor, as well as Industry Safety Association publications, with the latter being particularly important for farmers (e.g. Ontario Farm Safety Association, 1985–92). Fellow workers were frequently cited in all three industries as key sources of initial information about hazards and context-specific warnings. A number of the autoworkers and farm workers, especially where firms or farms had no meaningful training or safety system, cited fellow workers, direct supervisors or farm owners as their primary sources of information. But workers in all three industry contexts pointed to personal experience and the importance of fellow workers in helping them to understand risks in their workplace and in learning how to judge when risk was higher. Stope miners, for example, reported to me that they routinely looked to the previous shift to give them "heads-up" on anything they should look out for, and I observed crews from different shifts often chatting informally about how to control a particular hazard that had emerged in the mining process. Some mine managers instituted a more formalized process at shift change to make certain that miners exchanged such information and shift reports completed by the supervisor were supposed to contain any hazard-related information and passed along to the next shift. Most of the auto plants had similar kinds of formal and informal practices which encouraged an exchange of information on hazards that incoming workers should recognize, especially among machine operators.

Although tending to work alone and in isolation, many self-employed farmers learned from neighbour farmers about hazards and risk levels associated with particular machinery, equipment and chemicals and suggestions for controls, whether during conversations in the local coffee shop, at the grain elevators or through encounters at farm meetings (as observed several times in my field work). If farm owners had workers who were changing shifts with them on the combine or tractor, owners and workers would tend to pass on anything that they thought was a possible hazard (e.g. "One of seeders is sticking, be careful if you have to go under the drill"). Although the grain farms in this research where mainly run and worked by the males, female spouses also seemed to play important roles in passing along safety information to their husbands which they had drawn from official safety newsletters or local farmer networks.

Of course, workers and self-employed farmers also learned about how to judge and control risks through their own hands-on activities and experiences on the job. Regardless of the workplace, and indeed the job, direct worker experiences and observations were always cited by workers and farmers as the most important sources of their knowledge, depending to some extent on how long they had been working in a particular job and workplace (Breslin et al., 2006a; Hall, Gerard and Toldo, 2012).

However, while workers were often informed in the abstract about various workplace hazards and their possible consequences through training, safety materials and oral communications, this understanding was often not made "real" for them until it was experienced on the job (i.e. close calls or injuries/health effects experienced or observed). For example, miners in more conventional forms of mining often talked about their ability to judge the risk of ground falls by sounding the rock with a scaling bar, by listening and looking for cracks and movement and by looking for signs of pressure points and recent ground falls. I remember when I worked underground in a conventional stope,[3] my stope partner showed me briefly how to sound and scale, but it wasn't until I had done it for some time that I began to feel confident that I was able to correctly recognize the visual and audible signs of a "safe" vs. a "dangerous" face (ceiling/walls). I eventually learned through experience that some of this confidence was misplaced in as much as sounding and scaling often didn't reveal or control cracks that were deeper in the rock (an insight I don't believe I was ever told in training).

Nevertheless, my early faith in my own capacity to judge and control the risk was crucial to my day-to-day consent. It was only when I started to work in long-hole drilling (an early form of bulk mining) and was placed in a setting with very unstable ground and witnessed several major rock falls, that I began to doubt the adequacy of those skills in detecting and controlling ground risks.

Along with ground-related risks, miners talked about how they judged whether diesel fumes on a given shift or a series of shifts had reached a "dangerous" level. Although the company routinely assured miners that scrubbers on the machine and ventilation systems were minimizing the risk by keeping exposures below acceptable threshold levels (TLVs), many miners again looked to their bodies to question those assertions, often citing persistent and lasting headaches and dizziness or spitting black at the end the shift (see section below for more on how workers judge the seriousness of bodily effects). At the same time, however, other miners accepted management claims of air quality, either because they were not noticing any health effects or were not interpreting any symptoms as being caused by diesel fumes. As one diesel scooptram operator put it, "I get headaches but I figure I'm just tired."

Farmers and farm workers talked about similar physical symptoms when using pesticides as signalling to them that the health risk was high requiring that they either stopped mixing or spraying until the symptoms subsided, or began using respirators and other protective equipment. Again, however, some farmers did not experience these kinds of bodily signs or interpreted them differently, while accepting assurances they had from chemical manufacturers that their level of exposures were not dangerous, or at least were not as long as they took certain personal protection equipment (PPE) precautions.

Workers often constructed their capacity to judge and control risks as occupational skills. Indeed, some workers responded to my questions about skill level almost exclusively in terms of their ability to judge and directly control risk (Le Berre and Bretesche, 2020). However, some argued that safety was more intuitive and "common sense." As one miner put it, "to me, skill is a good clean blast, breaks the way it should, or a good timbering job, nice and neat no problems. Sure, safety is an important part of skill but some of that is just common sense." Whether understood as common sense or as a job skill, the more confidence workers expressed in their ability to judge and control certain risks, the more likely they assumed responsibility for doing so. Unfortunately, this often meant that they were more likely to blame themselves when an injury or close call happened. This thinking was especially evident when workers claimed that they didn't report injuries because they were "ashamed" or "felt foolish" about a "stupid mistake" *they* had made (Basok, Hall and Rivas, 2014; Hall, 2012, 2016; Hall, Toldo and Gerard, 2012). When workers saw their fellow workers as having control over certain hazards, then they were also more likely to lay most of the blame on them when something happened. As one miner put it when asked about air quality concerns, "eighty per cent of the problem is the working man."

Confidence and trust in technology and others

While the above quote blames workers for creating air quality problems, most miners also laid some of the responsibility on the company, which again varied with miners and with the hazard in question. Along with learning how to control some risks through their own actions (e.g. sounding the face for cracks, checking for loose hay bales, using PPE, etc.), workers also learned that some hazards are assessed and controlled completely or partly through other means or sources, notably through the actions of other workers, experts and management, realized through management safety systems, monitoring and measurement technologies, engineering controls and design. For example, since the case study mining firm had expanded its use of diesel equipment and was doing more large-scale blasting with ammonia-based explosives (Orica, 2013), production would have been impossible without a more powerful ventilation system pumping in fresh air and exchanging the polluted air. Accordingly, as the company expanded its use of bulk mining and diesel mechanized equipment, management was forced to assume more responsibility for providing a technological system that met certain regulative health and engineering standards. Using monitoring and testing technology, the company then claimed that the levels of fumes and dust were normally within the regulated threshold limit values (TLVs).[4] To further support their claim, diesel equipment was equipped with scrubbers which were supposed to reduce exhaust fumes and particulate, representing another engineering solution coming from the management side. Company procedures and rules for blasting and ventilation operation were also established; along with the routine air monitoring, miners were also allowed to request air quality measures although this sometimes led to management reprisals. Many miners accepted some of management's claims on face value. As one put it, "as for the diesel fumes, they must have them under control or the machines wouldn't be there. I've worked elsewhere and got headaches and I don't get them here." Others didn't trust the company measurements or the validity of TLVs as measures of risk, often citing their union's criticisms of TLVs or their own experiences of breathing and other health problems. As another production miner stated, "The [diesel] monitoring is a bad joke. You call them [ventilation department] and it's blue and they stand there and tell you it's okay." (Miner).

While miners often felt they had no option but to accept the situation, some workers persistently challenged management's claims. Citing a diesel operator friend who had died recently, one underground electrician insisted,

"I will not work when there's diesel equipment operating." According to other workers in his area, he was true to his word, implying that he had the power to say no perhaps because of his higher-skill trade position. Others with a similar conviction but less power sought to bid to other jobs or areas where the fumes were less prominent.

In some situations, more collective resistance developed. For example in explaining why they had decided to confront a diesel fume situation together, one group of workers stated that along with noticing some health symptoms such as breathing difficulty, they believed that the key source of the problem was that the equipment and controls were not being properly maintained *by management*. As this suggests, consent and any breakdown of consent with respect to air quality often revolved around the trust placed in management as the responsible actor for maintaining and monitoring engineering controls and the perception that the system was or was not doing what management claimed (Luria, 2010). But trust in technology or in management more generally were not the only sources of consent. Workers also looked to the competence and integrity of specific workers, supervisors, management experts and managers.

> My shift boss told me it [his machine] was okay…he's always been good to me. I got no reason to doubt what he says.
>
> (Miner)

> I didn't get much training but the guy beside me on the line, he helped me a lot. Told me what to look for.
>
> (Autoworker)

In as much as workers also often hear or learn about hazards from their worker OHS representatives and occasionally from Ministry of Labour inspectors, trust and confidence in these actors was also often important in shaping consent, again contingent on workers' direct experiences and "talk" around the workplace.

> I've gone to them and nothing happens. You can't rely on the OSHE. They are all in management's pocket.
>
> (Miner)

> You know Al, I don't have much time for those guys (joint committee worker reps). They come in here and tell me that something needs to be fixed. None of them have ever worked in a stope.
>
> (Miner)

> They (worker reps.) are another set of eyes, so sure I listen to them.
>
> (Autoworker)

For new employees or workers transferred to new jobs, this confidence and trust in others (e.g. supervisors, safety managers, fellow workers) and in management or engineering control systems were sometimes accepted at face value, but this too tended to reflect positive experiences they had had previously in other workplaces or jobs, and were contingent on whether experience over time confirmed their original impressions or assumptions (Hall, 2016; Hall, Gerard and Toldo, 2012; Jeffcott et al., 2006; Luria, 2010).

Some workers were more disposed to being cautious about what they were being told from the start of a new job, tending to reflect negative experiences in previous workplaces. Initial impressions of management, what they said and did when the worker started and their early experiences with other workers would also shape their trust or confidence. For example, one non-union worker had previously worked for a unionized firm which he saw as having a well-organized safety programme with good training and a strong health and safety committee. As he described his new workplace, "as soon as I got in the door, I could see that I had to be careful. There was no training and there were lots of temps." Other autoworkers also expressed the view that the trustworthiness of each individual manager as well as the company had to be considered. "There are some bad managers [but] you have to look at each individual [manager] and the company as a whole... most [managers] are good men who can be trusted to tell you the truth about something" (Autoworker).

Informed by risk theory, we also need to recognize that although the risks associated with some hazards may be easier to recognize and assess, uncertainty tends to be built into the perception, assessment and management of many occupational risks, especially in complex industrial contexts undergoing substantial transformations, and in employment and life contexts where insecurity is heightened, both issues which are of primary concern in this investigation (Brown and Calnan, 2012; Hall, 2016; Luhmann, 2005). Often, workers, and indeed managers and other organizational actors, had limited or conflicting hazard information, with varying capacities and opportunities to reduce their uncertainty (Luria, 2010). This uncertainly could have very different consequences in terms of consent and resistance. Sometimes the lack of clarity yielded immediate responses in the form of worker complaints, protests or efforts to seek clarification, which sometimes led to "shop talk" among workers or worker representatives, which in turn might feed into significant collective confrontations with management and/or concessions from management. These conflicts were partly why management sought to reassure workers and line managers that things are properly understood and controlled through its various OHS management and engineering systems. However, in another one of the many contradictions identified in this book, uncertainty among workers often sustained an acceptance of conditions by playing to a host of worker rationales, interests or motives for saying or doing nothing (see Chapter 3), while again opening opportunities for management to cast doubts on worker and activist complaints. Moreover, to the extent that certain levels of uncertainty were taken as the norm for everyday conditions, something which was perhaps more likely as work contexts and labour processes were being transformed, workers were more disposed to view the lack of definitive evidence on risk as the default for accepting a condition. This was again especially important when workers perceived their interests as being tied to their capacity to do the work without having to worry about the risk to their health and safety.

Variations in risk judgements and understandings of control

Workers in the same workplace often had quite different perceptions of the level of risk associated with the different hazards in their workplace. While some of that variation was likely a function of "real" differences in the level of risk in different specific jobs and work situations, and the different ways in which workers' bodies were affected by those conditions, variations in the judged level of risk were evident even among workers in exactly the same work context. Although there was evidence of these variations in all the studies, the miner study offered the best data on this since I was able to do interviews with and observations of a small number of paired miners (i.e. five pairs of partners) working in *exactly the same* stope or development heading at the same time. Three of the five pairs expressed substantially different judgements of the risk levels of dust and diesel in their work and two of the pairs had different perceptions of the ground risks in their work areas. To quote one pair of production miners talking about ground condition in the same stope at the same time:

> Miner 1: For me it was just *too* dangerous in there – I don't get paid to take those kinds of risks. After the last fall of ground I told my shift boss I won't work in there. It was no problem cause he knew it was bad. He got someone else to do the job with my partner.

> Miner 2: I've worked in some pretty dangerous situations but I know when to get out. I didn't have a problem with the stope that time because I figured we could do it and get out. It was moving and cracking a lot but that isn't always a bad sign.

These were both veteran production miners with similar levels of seniority, yet as their quotes indicate, they perceived and interpreted the same signs of risk differently with distinct consequences in terms of their responses. Although not quite as demonstrative, the mine survey also revealed

Table 2.3 Percentage of miners in different jobs concerned/very concerned
about hazards

	Types of Miner Jobs Non-mechanized			
	Conventional Production (%)	*Mechanized Production (%)*	*VRM/Bulk Production (%)*	*Support/ Services (%)*
Dust	36	44	60	33
Diesel	28	35	55	39
Oil Mist	24	20	40	11
Ground Conditions	48	54	60	44
Overall Accident Risks	56	46	45	28
Overall Health Risks	36	48	40	39

considerable variations among miners within similar job classifications in their risk assessments of certain specific conditions (see Table 2.3), interesting in part because again all these miners had 20+ years of underground experience.

Differences in understanding of what constitutes a serious risk were particularly evident in their judgements of the health risks associated with dust. Some further insight into these health judgements come from simple cross-tabulations of worker concerns about dust and their reported exposure levels. For example, while most miners judged "all day visible" dust as a "serious" risk to their health, 26% of the miners expressed little or no concern about "all day" exposure to visible dust conditions. At the other end of the spectrum, 41% of the miners who reported more limited exposures to dust, as in "every couple of days," judged that these levels were sufficient to cause serious health problems. Interestingly, in light of the earlier example of an electrician refusing to work around diesel equipment, the workers in this latter group were disproportionately trades people in support and non-production jobs such as electricians and mechanics, jobs which tended to involve much less exposure to dust and fumes relative to production miners. This reinforces my argument that power and interests are key in shaping perception as well as in shaping responses. In the mining job hierarchy, the trades were the top tier with higher wage, status and independence, but perhaps just as important, there was no production bonus. However, as will be discussed in greater detail in chapters 4 and 5, the relationship between power, control and interests is complex. For example, most production miners on bonus were much more likely to judge certain risks as tolerable in part because they constructed themselves as having the control they needed to prevent injury and because they had a strong interest in the form of a lucrative bonus.

Judging risk by its consequences

Before moving to Chapter 3, it is important to remember that the perceived need to control a risk is not just based on the judged likelihood of an injury or health problem but also on the consequences which workers associate with those hazards – that is, were the injuries or health effects seen as serious enough to warrant concern and attention. All workers in all three contexts defined a risk of death as serious, as they did the risk of permanent disability, loss of a limb or blindness, but the risk of death was also constructed with reference to time. That is, the "possibility" of premature death in 20 or 30 years did not have the same deterrent effect as the possibility of immediate death (Nelkin and Brown, 1984). Any injury that was seen as preventing them from doing their job tomorrow, next week or next year was defined as especially serious, with the perceived seriousness increasing as the job impact went from short term to longer term to permanent. As one autoworker put it, "I can't afford to be off work for any time so it's always in the back of my mind." While most workers focused their concern on their immediate capacity to work, older workers in particular were aware of the wear and tear, and often worried about how much longer they could do their current job. Yet, many continued to do the job despite this worry: "I never really thought much about it when I was younger but now I wonder how long I can keep doing this. Still I keep going because the money (bonus) is too good to give up (miner)."

As might be expected, most workers define cancer and heart disease as serious diseases and risks, and indeed 56% of the miners in the survey expressed personal concerns about the possibility of their developing cancer and heart disease. Those concerns correlated significantly with their judgement of risk levels around dust and diesel exposures (Pearson $r = .31$, $p < .05$). If workers constructed dust as a cause of cancer or fatal lung diseases such as silicosis, then they were more likely to seek management or expert assurances or protections that exposure levels were below designated safe exposure levels, although again concerns about long-term effects in the "distant future" were not as salient as concerns about immediate injury.

> In June, I was sent home for refusing to work in a heading that the ventilation department had shut down. They installed a new fan and I requested re-testing... They refused to check it and I refused to work... I got my OSHE rep. involved and management sent a ventilation guy down on Monday.

There were also many types and levels of injury and illness which workers widely accepted as not serious, as temporary or as the normal unavoidable consequences of work. Small cuts, bruises, strained muscles, sore backs, headaches and coughs are widely understood and accepted by workers in all three industries as normal events caused by working conditions and work activities (see also Basok, Hall and Rivas, 2014; Hall et al., 2012; Le Berre and Bretesche, 2020; Shortall, McKee and Sutherland, 2019; Tarasuk and Eakin, 1995). For example, while a majority of the miners in interviews reported problems with their hearing, pains and other effects in their hands from vibration disease, aching joints and back problems suggesting musculoskeletal damage, only a minority defined these problems as serious. Fifty-eight per cent reported hearing loss but only 3% defined their loss of hearing as a concern. Sixty-one per cent reported back pains and related symptoms but only 16% defined back problems as serious health concerns. As one miner put it, "you're always banging something or getting a cut – that's part of mining. Sure, I report the really serious stuff but there are some things you just expect." These kinds of responses are in no way unique to miners. Both autoworkers and farmers acknowledged that they had various symptoms or "minor" injuries, especially MSD injuries, without significant concern or without reporting them. Mirroring almost exactly what miners often said, one farmer stated, "[t]here's always going to be something – you bang your knee or pull a muscle or something. You can be careful but some of this is going to happen (farmer)." Several autoworkers in the injury reporting study (Study 7) also talked about self-medicating in order to continue working in pain because they felt there was no alternative, while understanding this may not be something they could do forever.

On the other hand, case study autoworkers (Study 5) were more likely than miners and farmers to view cuts and muscle aches as significant health impacts which needed to be prevented and reported, and were generally more concerned about health symptoms like headaches, coughs, back or muscle pain even though these didn't prevent them from working. Nevertheless, responsibility for preventing these injuries was often still nested by workers in themselves:

> Like when our parts get re-cut, there's always a burr edge, so I've got cuts and scraps from just walking by a rack and not noticing and scratching my arm or something off it. And oh, there's a nice paper cut times ten on your arm. So just pretty much that's one of the main hazards at work is the burrs after the re-cut is off. But that's pretty much your own, like you should be watching, be careful... The parts are sharp and it seems like people just run into them... I mean they're just not careful.
>
> (Autoworker 2)

For miners and farmers, this greater acceptance of some effects and injuries as minor and/or normal was extremely important both from the point of view of acceptance of the conditions as well as helping to reduce the compensation costs experienced by the company (or farmer). As long as a certain amount of damage or wear and tear was constructed in this way, workers made no demands regarding the conditions which they recognized as source of these health problems or injuries, nor did they formally report them as occupational health or injury effects (Basok, Hall and Rivas, 2014; MacEachen, 2005; Novek, Yassi and Spiegel, 1991; Pransky et al., 1999; Tarasuk and Eaton, 1995). As I will try to show in subsequent chapters, these differences between miners, farmers and autoworkers concerning the greater acceptance of certain injuries as acceptable and non-serious were linked in important ways to the greater amount of personal control and responsibility that miners and farmers constructed around preventing harm to themselves or other workers.

Judging risk: summary

As this second section of the chapter indicates, workers accepted a wide range of hazards because they constructed those hazards in most day-to-day work situations as presenting limited and/or acceptable health risks or consequences. While much of that acceptance was grounded in their perception of the control, knowledge and skill that they directly exercised in their work and more specifically over the hazards in question, workers routinely accepted many hazards as low risk or as controlled risks based on what others, that is, other workers, supervisors, managers, safety engineers and worker health and safety representatives, told them. As this implies, trust grounded in a perception of competence and integrity was critical in shaping the extent to which workers relied on these claims (Hall, 2016; Luria, 2010). Although workers may be disposed to trust managers and supervisors, this trust was contingent on both past and current experiences, and as such, was also subject to challenge whenever workers observed or experienced incidents, close calls or actual injuries which brought management claims of safety into question. And again, as in the definition of hazards, workers' judgements were often uncertain and tentative.

Some of the trust reflected and contributed to the development of a work risk habitus disposed towards accepting what they were being told by management, industry or state authorities regarding the hazards in their workplaces and the risks they represented, as well as a greater tendency to assume personal responsibility for some of those risks. Some workers seemed to be less disposed in this respect and more cynical from the outset of their employment, inclined to externalize responsibility on management and the employer. For most, however, in as much as most of these workers and farmers had been working for some time as the time of the study, the importance of work experience and knowledge seemed to be critical in shaping their habitus and what workers accepted or not from management and the employer. Similarly, the workers' perception of personal control over risk was contingent on experience which was especially important in a context of restructuring where workers were working in redesigned workplaces, labour processes and jobs. However, some workers were more disposed to accept management claims of low or controlled risk, and even constructed in their own minds lower levels of risk than the official line, especially in contexts where they saw their interests as being threatened or where they did not see themselves as having the power to contest.

Evidence of variations in risks perceptions and judgement among workers in the same or similar work contexts reminds us that we cannot explain different worker responses by relying solely on an analysis of labour process or job differences. If workers in exactly the same work situations, often with quite similar work histories, came to quite different conclusions about the same conditions, then the workers were clearly seeing and interpreting the same physical and social conditions underlying risk in different ways. To the extent that Bourdieu argued that habitus in any given social field is not reducible to class experience, we should not be surprised to find that workers from similar class backgrounds developed different risk dispositions (Bourdieu, 1990). We also need to recognize that these differences often served management interests more than it did workers, in as much as it undermined the kind of collective consensus that workers often needed to gain substantial systemic changes. Two key theoretical questions this raises for subsequent chapters are how this variability is sustained and what role management plays in this process. However, before addressing these and other theoretical questions of origin in Part 2, the two final areas of risk-related subjectivities still need to be considered – that is, how do workers explain their acceptance of conditions even when they recognize those conditions as presenting significant levels of risk, and how, when and where do they draw lines of resistance?

Notes

1 While there still is no "definitive" evidence linking urea or phosphate exposure to specific diseases, there is documented evidence of temporary health effects while the research on the long-term effects of farmer exposure is limited (e.g. CCOHS, CHEMINFO, Urea and Diammonium monohydrogen phosphate, 2015).

2 While not the case at the time of the mine study, regulations have since placed more legal obligations on employers to identify, test and assess all new technologies for risk in writing; especially the introduction of any new chemical substances need to be reported and assessed in terms of the regulations and existing toxic chemical list, but also the introduction of new machinery and mechanized processes (OHSA R.R.O., Regulation 833, 1990; OMOL, 2016). Although some miner representatives sought to use their joint committees to conduct these kinds of reviews, this was often a struggle since they were not required at the time. Although more common in the auto context given that the newer regulations were in place at that point, this was still conducted unevenly across firms. Farm operations were not covered by these regulations at the time of the study.

3 A stope is the specific section of a mine from which ore is being extracted.

4 Threshold limit values (TLVs) are the maximum concentration of a repeat chemical exposure recommended by the American Conference of Government Industrial Hygienists. Ontario exposure regulations rely heavily on TLVs (Ontario, 1990).

Chapter 3

Taking risks or taking a stand

Interests, power and identity

The previous chapter considered the recognition of hazards and the judgement of acceptable risks associated with those hazards. This chapter examines what happens when workers judge their conditions or activities as involving serious risk or danger.

Taking risks

Although much of day-to-day worker consent to hazards hinges on assumptions and judgements of low levels of acceptable or controlled risk, workers from all three workplaces often accepted certain conditions or situations *even when* they consciously recognized them as presenting high or dangerous levels of risk (Hynes and Prasad, 1999; Le Berre and Bretesche, 2020; Shortall, McKee and Sutherland, 2019; Walters and Haines, 1988). Indeed, the finding that many miners and farmers broadly defined their everyday jobs as dangerous implies an ongoing acceptance of considerable perceived risk within these two occupations. Although less inclined to describe their work as "dangerous," many autoworkers also reported acceptance of hazards which they judged as presenting significant risks to their health or safety.

The survey from the miners is again useful, in this case to gauge the extent to which miners were routinely accepting conditions they constructed as dangerous. Seventy per cent of the miners viewed at least some of their day-to-day exposures to conditions they judged as "dangerous" or "bad for their health." Among those miners who reported they were "very concerned" about the risks of dust and diesel exposure in the formal survey interviews, 30% and 40% respectively reported that they had not taken any action within the previous three months to limit their exposure beyond precautions such as the use of masks or staying out of the area whenever possible; that is, they were routinely accepting daily exposures that worried them without doing anything to try to change the condition itself or contest it.

Among those miners who had taken some action, 60% of those actions were confined to a single complaint or a request for information or air sampling with no further action. Almost all of these respondents indicated that management's response had not alleviated their concern. Furthermore, only 20% of the miners expressing concerns reported any repeated or sustained efforts to achieve better conditions, which in the context of the interview survey meant making multiple complaints to their supervisors; going beyond supervisors to managers, engineers or union representatives; and/or employing various kinds of job-related pressure tactics to resist their exposure to three conditions. Again, the vast majority of the actions defining the 20% who acted in some sustained way to challenge management was confined to using supervisors, which it is worth noting is exactly what workers were supposed to do under the Ontario OHSA and its IRS protocols (OHSA, 1990; OMOL, 2019). However, in as much as miners were still expressing concerns at the time of their interview, their complaints were apparently going unanswered.

Having gained no significant improvements from their supervisor, one legal option that miners had under OHSA was to refuse unsafe work. And, of course, in some cases, this is exactly what some miners did, at least eventually.

> I've been complaining now for months. I'm using 5 to 6 gallons of oil on my jumbo in 21/2 hours. I keep trying to get them to change the lubricator. Nothing happens. I'm coughing black and can taste the oil all shift. The ventilation guy came down and we were 3 x's over the limit (TLV). I went three more shifts, and that was it, I refused to work anymore.

However, both survey and interview data show that most workers never got to the point of refusal even when experiencing such severe conditions (see also Gray, 2002; Hall, 2016; Weber et al., 2018). Official company records on legal work refusals confirm relatively limited use of this right (see Graph 3.1), although it is likely that most work refusals were settled in some way or another without going to a formal report. Still, as is evident in the data, there were periods of higher numbers of work refusals during the 1980s, including 1985–87 when the research was being done. As will be discussed in Chapter 5, most of these took place in particular mines over particular issues.

While other actions taken by miners to contest their conditions, such as slowdowns, collective work stoppages and sabotage, will be examined later in the chapter, these too were relatively rare miner responses to dangerous conditions. Individual case study interviews and observations with other miners and worker representatives further support the argument that miners were frequently, and in some cases routinely, accepting many conditions that they judged as serious risks to their health or safety without contesting them (Glasbeek and Tucker, 1999; Le Berre and Bretesche, 2020.

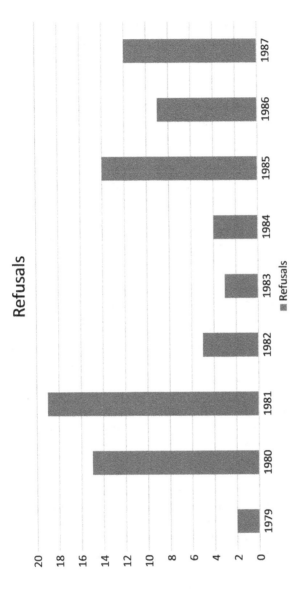

Refusals

Graph 3.1 Official work refusals at INCO.*
Source: Compiled from USWA Local 6500 Records, OHS Chair Office, 1979–87.

As noted, there was also significant evidence of serious risk-taking in the other two industries, especially the farmers (Shortall, McKee and Sutherland, 2019).

> You have to be careful with atrazine (pesticide). People have different sensitivities. My Dad didn't like it. I didn't like the smell and if you get it in your eyes it stings like hell. But it does control a bunch of things, not just one thing. The cost was just too critical not to use it.
>
> (Farmer)

Most case study autoworkers also acknowledged that there were some recognized risks that they accepted including dust, fumes and periodic speed-ups in production. While some did contest their conditions, this was again largely confined to complaints or requests to their supervisor with relatively few attempts to go to their health and safety reps. or no attempts to call in a government inspector. As well, none of the autoworkers interviewed in the case study reported a work refusal, neither one they'd made nor one they'd observed other workers making (see Gray, 2002). The other survey and interview study of workers (Study 7) also revealed a widespread pattern of routine acceptance of hazards recognized as "risky" if not "dangerous" (see also Basok, Hall and Rivas, 2014; Hall, 2016).

Taking known risks: from coercion to consent

Whether routine or situational, the acceptance of recognized risks was explained by workers along a continuum which stretched from freely given consent to forced compliance – that is, at one end of the consent-to-compliance spectrum (see Figure P1.2), workers expressed their acceptance in more consensual terms grounded in personal interests, practical or physical realities, perceived obligations and commitments or valued identities, while, at the other end, they constructed their acceptance as *coerced compliance*, where particular or generalized pressures and power relations were perceived as being brought to bear unfairly by the employer and his/her management representatives.

The key conceptual distinction was not so much whether workers saw themselves as having a choice, because even in some consensual moments, producers saw risk-taking as legitimate and/or as unavoidable tasks which fell to them as workers in a certain occupation. As one farmer put it: "If you are worried about rain, you are out there until the job is done no matter how tired." Coercion, in contrast, is represented by the extent to which workers (or farmers) explicitly presented their risk-taking as being unfairly forced by powerful others and their acceptance as a function of their own relative lack of power within those relations (Miner: "I don't like to complain – look what happened on [level] 1100- those guys really took it [for legally refusing an unsafe situation]. It's not worth it – you can't win." By conceptualizing risk-taking as a spectrum that runs along a continuum of consent to compliance, risk-taking is understood in *relative* terms grounded in different meanings and context. Moreover, as with the definition and judgement of risk, risk-taking may be grounded in a mix of understandings, thoughts and emotions that don't clearly fix blame or responsibility on workers *or* managers. While recognizing the importance of coerced compliance, the evidence presented also suggests that risk-taking was grounded more often in gradients of consent than straight coercion. With these points in mind, let's first consider the different ways in which risk acceptance was explained by workers, starting with direct and explicit use of coercion to gain worker acceptance of risks.

Overt threats, reprisals and harassment

Workers or worker representatives in every study talked about accepting hazards (and injuries) because they had been punished or threatened explicitly by a supervisor or manager (King et al., 2019). Miners and miner OHS reps. in particular had several stories about being "put in a ditch," transferred to another mine or level, taken off production bonus, or sent home for challenging working conditions. Relatively few miners were fired or suspended but management was often successful in intimidating workers into compliance by using the preliminary steps of its disciplinary system (i.e. oral warnings, letters in their file and/or short suspensions). Ironically, underused safety rules were often employed to punish miners who were seen as causing "trouble." Although illegal, some of these actions were reprisals for legal work refusals. However, as workers and worker representatives noted, management would often bide their time so that there was no obvious connection to the refusal (King et al, 2019). Other times managers would act more quickly in an apparent effort to send a message to all workers that refusals invited reprisals. As one explained his unwillingness to employ his right to refuse: "There's not much I can do if my health is being hurt at work. You're alone now and can't fight it – if you do you're labeled a trouble maker and that's mean trouble [for you]."

   Along with specific threats and concrete reprisals such as transfers, reduced hours and "shitty" job assignments, workers reported that some managers and supervisors demeaned and belittled them in various ways. As one miner observed, "their favourite is to call us 'crybabies' when we complain." Workers also claimed that some managers and supervisors will try to use other workers against them if they challenge conditions. As one miner noted,

> if I complain about dust, the shift-boss will just go to someone else and give him shit [for not doing something to control the dust]. So, of course,

that guy comes back to me and gives me shit. After years of hassling back and forth, it's just not worth it anymore.

As this again implies, the acceptance of risk for many veteran workers was often a consequence of several experiences over time in which workers tried to challenge a condition or activity as unsafe with limited long-term gains and painful consequences.

Sure they may do something if you complain enough and raise enough shit. But nothing really changes. I used to get dust when dumping ore cars because the controls were near the flow of air so I complained to them to switch the controls. They did it finally but they only changed that one ore pass. They left the rest of the mine the same.

Intimidation didn't have to happen to them personally to have effects across the workplace, as many expressed fear of reprisals based largely on what they'd seen or heard about happening to others. As another miner replied when asked about why he was not complaining about some worrying ground conditions:

I know of one guy who they fired for not wearing his [safety] glasses. It took the union a year to get his job back but his marriage broke up and he's never been the same. Inco likes to show who's boss every once in a while.

Some autoworkers, and indeed workers in a range of industries in the hazard and injury reporting study (Study 7), also reported similar reprisal experiences or fears as their reasons for not reporting. Although by no means absent in unionized contexts, temporary workers were more likely to report an actual reprisal experience and to express more of their fears. As one temporary autoworker put it, "if I complain, I will not be scheduled at work...so I have to take the pain and look for a way to do the job no matter it hurts or not." Although unionized workers were less likely to report reprisals or threats, some argued that their unions were too ineffective to protect them against reprisals. As one autoworker put it when explaining his reluctance to report OHS concerns, "our union is very weak and you know it is not really one of those unions that you can rely on." Others argued that while they felt they had their union backing, the harassment they faced by complaining was not worth the trouble. As one miner activist put it, "I'm wasn't afraid to address this as much cause I knew had union backing...but no, not being fired but being hassled and you know the stress, not worth it." Others felt less vulnerable but were simply worn down by the harassment. As one of the more politically active miners put it: "There's nothing they can do to me that they haven't done already. I have to watch my ass though because they try to catch me doing something wrong...I still put up with a lot [of bad conditions]...I don't get support from nobody [the union or other workers]. I'm tired of it."

Thus, the political significance of reprisals was not just felt at the level of individual responses to risk. Reprisals were also significant in suppressing worker activists who could be quite central to informing and rallying workers around specific issues. I saw a similar process play out in several cases in the mine study and in the study of auto health and safety representatives (Study 4). In one of the auto workplaces, for example, a worker representative raised some concerns about work pace and, in the next several months as she tried to pursue this and other safety issues, she experienced a steady stream of threats and intimidation, everything from being yelled at and demeaned, to being disciplined for "safety violations" (which she said were contrived). When I returned a year later for a follow-up, she had quit her job indicating she could not stand the stress. A mining steward active in health and safety reported a similar experience, fired for a trumped-up safety violation. It took his union a year to get his job back but he never went back to being a steward or health and safety rep. (King et al., 2019).

Although self-employed farmers don't have to worry about "management" reprisals as such, there were still pressures which they experienced to some extent at least as external coercive powers. One significant source of pressure was contract buyers – that is, when farmers were growing under advance contracts to processors or buyers such as Cargill or Heinz – which was becoming increasingly common at the time of the study – they found that the buyers imposed certain production and scheduling standards which sometimes meant taking extra risks. As one farmer complained about having to use a certain pesticide: "You pretty well have to do what they tell you regardless if you want a contract the next year." Indeed, some farmers reported that they had lost contracts when they had failed to follow fertilizer, pesticide or harvesting instructions. This is reminiscent of the pressures on self-employed contractors in other industries such as construction, cleaning and maintenance as well as temporary workers in factory contexts (Basok, Hall and Rivas, 2014; Lewchuk et al., 2011; Vosko et al, 2020). Although technically independent, subcontracting in effect binds these producers to conditions which were sometimes worse than those imposed on the employed workers in the same workplace, in as much as the self-employed (like farmers and farm workers) were not covered by key labour laws and did not have any form of collective representation (Lewchuk et al, 2011; Quinlan, Mayhew and Bohle, 2001; Vosko et al., 2020). Not being covered by labour or OHS law at the time of the study, farm workers, many of whom were part-time and temporary, were especially vulnerable to loss of employment if they raised concerns about safety or health issues. Although none of the workers in my farm case studies, reported any coercive experiences, there were some contract farm workers in the injury and hazard reporting study (Study 7) who did. As one recalled when she was working on one farm:

> I got tangled and I fell down. I went to a clinic. I had broken my hand... And so I did call him [the contractor] ...and he came into my apartment...and he asked me to make a deal with him, the deal was not to report it [to OWCB [compensation board]...I said to him, 'you take care of your insurance, I take care of my life' right? He wasn't too happy...after my hand got better, he didn't want me to continue working for him.

Another contract farm worker recalled how his reaction to a chemical pesticide led to his dismissal, which he interpreted as meaning that he had to accept future problems without complaint. A temporary autoworker stated similarly, "I have to put up with it right, no other option. I have to pay my bills." Another stated, "If you complain, they (manager) call the [temp] agency, and the Agency calls and tells you "don't go back" and they send someone else. Accepting [risks] comes with the job."

While observations or experiences of explicit reprisals were the strongest indicators of coerced risk-taking, workers often assumed negative management reactions without necessarily having experienced or even observed actual reprisals. As one full-time permanent autoworker stated in explaining why he was accepting a workplace with unguarded machines and no safety training, "one cannot be picky and one does not say no to the job because of safety concerns." When asked if he had been threatened or punished for reporting in the past, he said no. When asked to explain further, he cited his immigrant and racial status as making him a "target" (see also Basok Hall and Rivas, 2014). Insecurity, a perceived lack of power and employment options grounded in race and ethnicity, and one's immediate financial circumstances and obligations all played a significant role underlying his unwillingness, and many others', to take a risk with their employment or job security. Often, they'd heard stories and warnings from other workers and sometimes from friends and family. Sometimes they expressed their fear as simple "common sense" given their status as immigrants or as new or temporary workers. These understandings were in effect a risk habitus grounded in their historical experience of workplaces and other contexts as objectively discriminatory (Brown, 2006; Basok, 2004; Basok, Hall and Rivas, 2014).

Some workers reported an initial reluctance to report any concerns when they first started working but then over time, gained more confidence and security enough to raise at least some issues. As in OHS risk judgements, workers look for various signs or clues in trying to judge whether it is "safe" to report at least some concerns, looking in particular to see what happens to other people when they report, while not necessarily taking at face value management statements that hazards and injuries should be reported. Management treatment of workers more generally may also play an important role in building trust that OHS complaints can be made without reprisals. As one autoworker put it: "I could tell ...the way supervisors order people around – they don't treat people with respect."

Workers also recognized when there were different standards for different workers, and accordingly, were cautious as long as they saw others as being favoured and themselves less so. Again, as one autoworker put it: "I was warned when I first started here to be careful about what I said about safety. There are favourites here. They can say something but I'm not one of them." Again, immigrant status and concerns about discrimination can play a role in shaping the view that they are especially vulnerable to reprisals if they raise OHS issues. As one autoworker put it: "As a black person, complaining about safety only makes things worse." Women autoworkers would sometimes acknowledge that they accepted certain risks because they did not want to look weaker than the men. Although I found little evidence of this in my farm research, a recent study suggests that women who are active producers on family farms take more risks than the men partly to show that they are "authentic farmers" (Shortall, McKee and Sutherland, 2019). Studies in industrial workplaces and construction have found similar tendencies (Breslin et al., 2006a; Paap, 2006).

Reports of reprisals also revolved around the perception of personal, collective and institutional power to achieve the changes and/or protect themselves against reprisal attempts. Although immediate supervisory and worker relations were critical, this perception of power was partly grounded in the assessment of the power and integrity of the union and/or the state.

> [After refusing] The shiftboss told me that if I or anyone else pulled a Bill 70 [again] he'd get that person eventually. That doesn't really scare me but I don't think the ministry will help [anyways], they're on the company side.
>
> (Miner)

> It's not worth it to go to the union – most are in bed with management. and if they aren't they don't know what they're talking about half the time.
>
> (Miner)

As I will also try to show in later chapters, the perception of their power and their vulnerability to reprisals was often based on their understanding of structural changes within their job and employment situation which they saw as having reduced or changed their power. As one miner stated, a new job classification system meant that "If I complain or refuse, they'll take me off the machine and there's nothing the union can do about it...[now]they can move us all over the place, replace me with somebody else. We don't have any protection..."

Thus, several conditions were central to workers' and activists' assessments of whether and to what extent they could challenge high-risk situations: the perception of personal power in day-to-day work relations, the assessment of collective capacity and willingness of workers to resist certain conditions, and the perception of institutionally based power (i.e. the strength and integrity of stewards and OHS representatives, and other union officials, the strength and integrity of Ministry of Labour inspectors, the strength of the collective bargaining agreement if they had one and the OHSA legislation if it applied). However, while coercion was present to some extent in every workplace, it would be a mistake to conclude that coercion was the principal basis for risk-taking. Even in the mines, where 22% of the miners in the survey reported actual experiences of reprisals following OHS-related complaints or actions (see also King et al., 2019), only 7% of the miners in the survey explained their acceptance of risks primarily in terms of reprisal concerns. Many insisted that they were not afraid of reprisals or at least, not enough to keep them from complaining in some situations. It should also be noted that experiences and concerns regarding reprisals varied quite significantly between mines and occupational groups within the mines (see Chapters 4 and 5).

This was not just a miner phenomenon. The data from the hazard and injury reporting survey (Study 7), which also involved construction, education and health workers, as well as factory workers, was also instructive in as much as only 11% of the workers cited threats or reprisals as their first explanation for not reporting an injury or hazard. Again, this shifted with the workers' employment relationship, especially for workers in contract and temporary positions, but being unionized was not a significant predictor of coerced risk-taking or perceived security, and even temporary workers tended to explain their under-reporting other than or equally in tandem with concerns about reprisals (Basok, Hall and Rivas, 2014; Hall, 2016; Hall et al., 2012; Pransky et al., 1999).

Consenting to risks as a condition of employment

Even when job insecurity was the underlying motive for risk-taking, the acceptance was not always grounded in a *fear of reprisal*. Often workers saw the risks to their security, and the need to take risks to protect their security, as being tied to external factors outside not only their own control but also management's (Basok, Hall and Rivas, 2014; Hall, 2016; Pransky et al, 1999). In this context, workers often took certain risks because they saw the risk-taking as a *legitimate* concession necessary to keep their job. As one miner put it: "No, I haven't made any grievances or complaints since the company got into financial trouble. I don't like to bite the hand that feeds you." Several workers in the auto case study stated virtually the same thing when asked to explain why they accepted speed-ups and the resulting increase in injuries. While sometimes resentful that they had tó take the risks, workers were not blaming their management or specific managers for forcing them to take the risks. Indeed, sometimes autoworkers reported that their supervisors or managers would caution them to slow down because they were working too fast or hard but, they continued to do so because they believed it was, as one put it, "it's our common interests to have profits."

Although workers were quite capable of making links between OHS and their employment security on their own, firm management and individual managers worked hard to convince workers that those links existed and re-flected economic realities outside the firm's control (see Chapters 5 and 9). Moreover, supervisors and managers in both the mining and auto contexts often encouraged and rewarded these extra efforts and the added risks im-plicit in those efforts, with praise and recognition, encouraging the workers to think of risk-taking as means of enhancing their security. The risks as-sociated with the introduction of new technologies and processes, if seen as important to worker security, may also be widely accepted on this basis, and again without necessarily placing blame or responsibility on manage-ment. For example, in the mining context, there was considerable contro-versy about whether the widespread use of a new form of bulk mining called Vertical Retreat Mining (VRM) was creating increased risks of major cave-ins and rock bursts. Many workers and union health and safety activists believed that it was, as did some government inspectors who expressed this point in their interviews with me (see also Ames, 1987; Stevenson, 1986), but actions to address these problems were delayed and muted in part because a good number of miners as well as the local union hierarchy believed man-agement when they insisted that VRM was essential to the company's sur-vival. As one union health and safety representative noted: "We know VRM is to blame but I know that there's no way we're going to have any mining around here without VRM [so we say nothing]."

A similar situation in the auto case study involved the introduction of a new packing system for parts which introduced risks of repeat injury and MSD risks. While there was acknowledgement of some increase in RSI injuries and risks, and some effort to moderate the risks by management, both the firm and most workers defined the new process as essential to their ability to meet just-in-time orders which was further understood by them as necessitating the acceptance of those risks. As one put it, "It didn't take long, I'm getting some soreness in my shoulder, but the system works to get the stuff out the door."

Central to this willingness to accept the necessity of increased risks were perceptions of change in the firm's status. If there were layoffs or rumours of layoffs or shutdowns, especially in non-union plants, workers expressed con-cerns that it was important not only to stay out of trouble by avoiding com-plaints but also to demonstrate a "willingness to go the extra mile," in the hopes that this would help them survive the layoffs. Workers in temporary positions often talked about the need to demonstrate a willingness to do whatever they were asked in the hope they would be made permanent. Others argued that they liked their current employer, or certain aspects of their employment for various reasons, whether it was good pay or good health benefits, or even the respect generally given to the workers by the owner or management (see Chapter 9). This commitment meant that they were willing to do more in certain circumstances to preserve their employment.

Self-employed farmers were also extremely conscious of financial constraints, reflecting what many had constructed as the crisis conditions of most Ontario farmers during the 1990s (see Chapter 7; Lind, 1995). As such, they routinely explained their risk-taking with reference to recent drops in prices and the effects on their business finances and survival. Working at night, for example, was widely accepted as a risk that had to be taken during planting or harvest in order to avoid costly weather delays. This was especially dangerous because it involved driving their large tractors or combines in the dark on the road. Every farmer interviewed said that they had either done this at least a couple of times in the past year, or did it virtually every planting and harvesting season. For some, there were certain specific uncontrollable circumstances such as weather that made this and other risk-taking necessary, and for others it was something they felt they had to accept routinely – as one of those things that farmers had to do every year if they wanted to get the crop in or out as early as possible.

Longer-term farm workers, as opposed to the temporary workers, expressed a similar willingness to take risks in these kinds of contexts accepting that these were fixed constraints of the job. Most knew the farm injury statistics and fatality reports as they were widely circulated by the Ontario Farm Safety Association (OFSA, 1993b), and as noted, some had even had a few close calls or accidents themselves, and yet, they still did it. One of the farm case studies had a family member who had been killed on the road. Yet as he stated, "I don't like to do it. My dad was killed on the road when someone hit him while he was working. But with all the different fields and crops, I can't afford to take my time." Compromising health and safety on this basis was usually more evident among farmers with more limited financial resources, especially if the farm's future was in real jeopardy which was not at all unusual in the industry (Hall, 1998a,b, 2007); but, as will be shown in Chapter 7, the wealthier, more secure farmers also made similar kinds of arguments about costs or production priorities.

As will be shown in subsequent chapters, there were other factors underlying the differences between industries and firms than a simple concern about layoffs or business closure, but the dynamics of consent and compliance within and between the industries and firms had much to do with the recent history of concrete job losses within the specific firm, as well as what was happening in the local labour market more generally at the time. Job insecurity is, of course, an often noted consequence of neoliberalism and globalization (Harvey, 2005; Vosko, 2010). However, what the data also suggests thus far is that among relatively secure unionized miners, even in the 1980s, a broader ongoing commitment to profitability and an emphasis on risk calculation within an implicit cost-benefit framework was viewed as a rational response to insecurity – a way of balancing risk to jobs and to health and safety in light of their class position. At the same time, it is worth nothing that a logic of cost and benefit was often present in worker discourse, limiting or muting the potential politicization of these health and safety concessions, precisely by presenting risk-taking as questions of necessity and self-interest. For example, it is interesting to note how often worker OHS representatives acknowledged the constraints and pushback they experienced from workers when the acceptance of some risks was tied to job security and other interests. Consider the comments of two autoworker union OHS reps. who were interviewed together:

> AWR2: You can only do so much for the worker. If it's not important to them, you can get yourself into a bind. The last thing you want to do is, like you say, you shut something down, management might send three hundred people [home] for two days and your name is going to be mud real quick if you're crying wolf or depending on how management puts that back in your lap.
> AWR1: That's exactly right cause we had a heat policy in place two years ago and the company drew up the policy without planning with the committee. We ended up shutting the plant down when it got warm, forty-five minutes an hour and we're just enforcing their company policy and the workers were mad at us.

The (capital) rewards of risk-taking

While job security concerns tended to stand out as a rationale for accepting risks, workers also talked about other interests which they saw as being linked to how they responded to high-risk situations. Of particular importance to miners, for example, was the common view that the acceptance of high risk was a means of gaining financial or other job rewards. The high pay rates in mining coupled with the production bonus/incentive pay which could almost double their wages, often figured prominently in the decisions of production miners to accept certain common conditions which they recognized as involving considerable risk. As one miner stated: "…I should get off the drill. My back is bad and getting worse… my hands too. But it's hard to get off [bonus] when you're making good money."

While their capacity to consume was part of what drove these workers to take risks for bonus, workers were also trying to build financial security in part because some of them were worried about their health and how much longer they could do the job. Concerns about university costs for children or family health problems often figured in their accounts about the need to make as much as they could while they were able. Some of the auto plants in Studies 4 and 7 also had bonus but they were generally not as lucrative. Still, some of these workers acknowledged that they sometimes cut corners or worked faster in order to exceed their quota. Financial incentives for risk-taking were not as evident in the explanations of the case study of auto-workers (Study 5), which figures since wages were low and there was no production bonus (see Chapter 9). Some workers were very explicit in saying that "they were not paid enough to take risks." However, like the miners, there were a host of other advantages or interests that were cited as explanations for accepting particular conditions or circumstances – preferred jobs or job postings, access to training opportunities, or simply just "getting ahead" and "getting favours" such as time off, longer lunch breaks, etc. As noted, autoworkers often argued that they took additional risks such as removing machine guards in order to finish their work "more quickly," as in the case of this autoworker, who explained:

> If I can fix the machine in an hour, I can go sit in the cafeteria and smoke, drink, have a coffee or whatever and not be working... It's human nature to be lazy and do as little as possible. Then you can also get the supervisor who says, you know what? Why don't you go home early today, you were working hard today.

The importance of sustaining a "good relationship" with one's shift boss and partners in the mining context was also a frequent rationale underlying miner accounts of some risky activities, as well as concerns about gaining simple respect from supervisors and other workers.

> Everybody makes deals. So sometimes we get the work done and no on asks any questions. Sometimes it means doing something for him [that is dangerous] and he'll do something for me when I ask [I. Like what?] He lets me off early when I got to go to the hospital to visit the wife.

These findings suggest workers had learned that risk-taking was, in Bourdieu's (1990) terms, a strategy for gaining different kinds of capital (financial, cultural, social and political), capital which could in turn be used to secure greater relative financial or employment security within overall conditions of insecurity (see also Burawoy, 1979, pp. 51–73 for a labour process version of this argument). For workers, this included being able to use the capital gained through risk-taking to negotiate reduced risks in other contexts. Workers sometimes expressed conflicting feelings and resentment about the unfairness of having to compromise safety to make other gains, and the need to make these kinds of compromises were often understood by workers as the way things are for workers or "people like them." Still, while sometimes seeing the objective structures that constrained their options and opportunities, they often internalized blame for placing themselves in these situations. As one autoworker put it, "if I'd gotten an education it would be different so this is what I have to do to make a living."

Risk-taking as an occupational responsibility and identity

Continuing to the far end of the risk-taking continuum (see Figure Pt1.2), consent is at its strongest when workers construct risk-taking as a necessary feature of the job or occupation. These rationales were particularly signifi-cant in internalizing responsibility for risk-taking and blame for injuries. In the mine survey, for example, the most common explanation for the lack of complaints or jobs actions was the expressed belief that it was their respon-sibility as miners to deal with or cope with certain high-risk situations; that is, 75% of those who reported no actions in high-risk situations explained that they accepted the risk because they believed it was "their job" to work with the situation as presented to them. The mine survey data also supports the argument that this understanding of occupational or job responsibil-ities is quite significant in explaining miners' responses to key conditions. For example, the correlation between the level of risk concern and action around diesel fumes went from an insignificant $r = .12$ to a significant $r = .31$ ($p < .05$) when I controlled for workers' general beliefs about the individual responsibilities of workers regarding the risks of mining. In other words, those miners who expressed stronger views regarding individual responsi-bilities were *less* likely to complain or take other forms of action to protest risky conditions.

Some autoworkers were also very quick to internalize and blame worker carelessness for many accidents citing their training and a belief that man-agement had fulfilled its responsibilities. As one autoworker put it, "every-body knows how to lift. They [management] really emphasized back lifting... You have to be pretty goofy to hurt yourself." On the other hand, being less likely to see their work as dangerous as compared to miners and farmers, autoworkers were also less likely to define their job as requiring serious or dangerous risk-taking. While miners often showed little in the way of resent-ment when they took recognized risks, autoworkers were much more likely to see the imposition of higher-risk situations in negative terms. For exam-ple, while case study workers were willing to accept the risks of speed-ups as necessities brought on by exceptional circumstances such as a delivery rush or an unscheduled delay, they were seen more as unfair impositions and co-ercive as they became more frequent. When asked if the work pace was too fast, one autoworker responded as follows:

R. I wouldn't find that there is. Only if say maybe we're behind on parts and we've been a little slow that day they'll just come over and say can you guys think you can pick up the pace? I don't really mind if they say it [nicely]. But if they come over and say it kind of rudely, I would be screw you, but they're usually pretty good with that...

As this quote suggests, the perception of personal responsibility and control over work was related to the worker's understanding of what was reasonable, possible and necessary but there was also an emotional component in the sense she wanted to be treated in the right way.

Oftentimes, workers construct a workplace, job or technology as having certain fixed or natural characteristics. For example, as one miner stated: "Mining is a mechanized job and it's in a closed area. Any way you look at it, it's going to be noisy. That's the way it is." Farmers and farm workers also tended to think in similar terms that many risks in their work were fixed by the natural environment and certain other requirements: "Farms involve a lot of machines, a lot of time pressures, lots of things can go wrong. That's just the way it is." As this suggests, workers constructed certain understandings about what their workplace was like, what their work required, what its basic and necessary characteristics were, and what distinguished their workplaces and work from others.

In a related way, miners, farmers and autoworkers often constructed an image of what it meant to be a miner, farmer or autoworker – that is, they constructed occupational identities which included the understanding they *were paid to be risk-takers* (Bourdieu, 1977; Fitzpatrick, 1980; Kosla, 2015; Wicks, 2002; Wilde, 1999). Miners in particular saw their jobs in this way. The notion that worker health and safety is a commodity for sale is by no means a new concept especially within Marxist political economy (Barnetson, 2010; Navarro, 1982), but this argument also brings us back to the central importance of the way that the workspace and the labour process are constructed by the workers themselves – that is, the idea that mining and farming, in particular are naturally or necessarily more dangerous than other work situations (Fitzpatrick, 1980; Haas, 1977; Kosla, 2015; Wicks, 2002; Wilde, 1999).

To be a miner or farmer then, was to accept at least some significant levels of danger. But this was not a simple fatalism as some authors have suggested (Legendre, 1987). Indeed, miners and farmers constructed identities as "risk-takers" around which a considerable sense of pride was invested. Taking certain risks on the job, being the kind of person who will "tough it out" (Kosla, 2015) was part of how they valued themselves as workers (Kosla, 2015; Lawson, 2009; Paap, 2006). For example, when I asked miners in the formal interview if they were proud of their work, 75% reported in the affirmative. Among those who elaborated on why they were proud, many cited their ability to meet the challenge and danger of the work environment. As one put it, "to tough it out and get the job done" was what really made him feel good about himself and the job (see Fitzpatrick, 1980). Occupational skill in this respect was often understood as the ability to take risks in a controlled way.

For some farmers, reputation or image maintenance was also expressed openly as an important motive behind taking some risks (Shortall, McKee and Sutherland, 2019). Farmers often worried about what other farmers they knew would think of them if they failed to get a job done in a certain period of time, which they acknowledged sometimes drove them to take risks such as working in the dark or when they were too tired. Paradoxically, however, this construction of one's occupational self was also tied in important ways to the control, knowledge and skill that they believed they had within their work and work environment, in as much those images informed a perceived responsibility to take the risks which fell under their control. These dispositions to assume certain risk responsibilities, can again be understood as a form of "risk habitus," orientations which can extend to the very people who are supposed to be contesting OHS conditions – worker OHS representatives. As one miner OHS rep. put it: "It's not the union's position, but for me I figure if you put the [miners' hard] hat on, you accept the risks that go with it – somebody has to do the job." This was a case study miner with whom I spent several days in his mine, and it was quite clear that this was not an idle boast. I recall one incident where he and his partner were drilling a raise (vertical tunnel) in an area with a lot of loose ground. Another crew had refused to work in the area because, as they put it to me later, "we were scared shitless." As I observed when the shift boss first *asked* them to take the job, my case study miner and his partner were not being forced to do the job, nor were there any immediate concessions (see below) granted for doing the job. Indeed, they accepted even when the shift boss announced on one shift that the scissor (lift) truck was not available and that they would have to use a scooptram (adding risk because it was much harder to get out of the way of falling rock in the confined space of a scooptram bucket). Over the two days I was with them on this assignment, we had no less than three close calls where rock falls sufficient to kill us just missed.

As several analysts have suggested, gender may play an important role in shaping these "tough it out" dispositions and responses (Breslin et al., 2006a; Emslie, Hunt and Macintyre, 1999; Fitzpatrick, 1980; Kosla, 2015; Paap, 2006; Shortall, McKee and Sutherland, 2019). As noted previously, there is also some evidence that women in male-dominated workplaces often take on the same risk-taking tendencies of their male colleagues (Breslin et al., 2006a; Shortall, Mckee and Sutherland, 2019). Unfortunately, I do not have the data to explore this particular argument in any detail. What I did find in the mine and the auto studies, however, is that there were times, as noted earlier, when workers expressed their acceptance of conditions that they didn't want to be seen as "women," "whiners" or as "kids." In the mine study, in particular, I observed interactions in which a worker's manhood was questioned by supervisors in an apparent effort to undermine their complaints. And some miners thought that proving one's manhood was part of an occupational culture dynamic underlying risk-taking. As one retired union steward observed: "Mining is a hairy job yet some abuse the idea of risk. The more risk the more macho you are. It's hard to stay away from that."

While the auto and farm data did not reveal the same level of pride associated with toughing it out, a number of the farm women remarked that they often needed to caution their male partners about pushing themselves too hard and too long, and it was clear that they saw this tendency as partly a "male" thing which they did not share. Perhaps, if more had been involved in production at the same level of the males, they may have responded differently (Shortall, McKee and Sutherland, 2019). A couple of female autoworkers in the auto case study also remarked in their interviews that they saw some questionable dangerous behaviours such as reckless lift truck driving or the removal of machine guards as reflecting male egos trying to prove something to others or themselves. Still, none of the men or women in the case described their acceptance of risks explicitly in gendered terms. However, some women in the injury reporting study (Study 7) expressed a need to show some "strength" or at least avoid showing weakness by accepting certain recognized risks, fearing negative effects on their security or standing among workers or with the supervisor.

However, as Haas (1977) suggested in his study of high steel construction workers, there may be some important social functions associated with proving to others that you can be relied upon not to lose your nerve. Consistent with Hass' findings, many miners in my research emphasized the importance of having confidence that other miners, particularly their crew partners, were able to deal with crisis situations and problems without panicking. Miners often talked about moving from partner to partner until they felt comfortable with someone who could be relied on when things got "rough." Although favoured in some ways by supervisors, "haywire" miners were often ostracized and therefore it was often difficult to find partners who would work with them. On the other hand, I often observed miners taking a razing from other miners for making a complaint or taking some position on a condition that they thought was part of the job, in effect insisting that they tough it out.

This suggests an element of cultural disciplining among the miners themselves around the question of danger and mutual protection. However, again I stress that this "culture of danger" as others have characterized high-risk workplaces (Fitzpatrick, 1980; Legendre, 1987), was not about taking every and all risks that were presented. As shown, miners who persistently took above normal risks were labeled as "haywire" and sanctioned or avoided by most miners on that basis. What seems more important from this evidence is that the acceptance of recognized high-risk situations, as in the case of the judgement of risk levels, is tied in contradictory ways to the workers' perception of control over their work and their power to influence management, in as much as these perceptions also shaped a sense of responsibility and identity for taking specific risks and accepting certain conditions. This means, somewhat paradoxically, that an increase in worker power and control over their work, whether real or imagined by workers and management, may sometimes translate into *more* rather than less risk-taking in as much as they are more likely to internalize responsibility (see Chapter 4).

Taking a stand: resistance and its limitations

The emphasis to this point in this chapter has been on worker acceptance of risk. I want to end by reminding the reader that this does not mean that workers never contested or obtained changes in their working conditions. Indeed, as a general trend, the greater the perceived danger, the greater the likelihood that workers challenged their conditions. For example, in the interview survey of miners, the results show that miners who expressed the greatest concern about the levels of risk to their health or safety (i.e. "very concerned") *were* more likely to complain about those conditions (see Table 3.1), and the follow-up miner case studies and the farm and auto studies seemed to confirm the finding that they were more likely to raise or contest a perceived risk the greater the level of perceived risk.

Table 3.1 Pearson correlations between perceived risk and action

Exposures	Level of Concern/Number of Specific Complaints	Direct Job Action	Work Refusals
Dust	.39**	.10	−.13
Diesel	.28**	.08	.03
Oil Mist	.21*	.08	−.02

* $p<.05$; ** $p<.01$

However, as noted earlier, these individual responses, which often were limited to informal negotiations with supervision, rarely developed into more confrontational collective actions such as work slowdowns, or sabotage or more formal legal action such as work refusals. On the other hand, while most miners were relatively limited in their resistance, there were a small number of miners who were almost continuously contesting conditions they felt were unsafe or unhealthy.

> I argue with my shift boss everyday about loading and other safety procedures. Loading is really a problem. We work out the number of holes with the engineers and he's always pushing to take more holes. But we

[he and his partner] know what we can do safely…I'm fighting with him all the time.

I never work in dust cause I water it [the muck pile] down every time I work in an area…There's always pressure from my shift boss – he keeps on saying I don't need to do that [although its official company policy] but I just tell him I won't work in dust. He knows I'll refuse cause I have a number of times…

Some autoworker representatives in the first auto study (Study 4) and some workers in the case study (Study 5) were also clearly more dogged and persistent in pursuing changes and refusing to accept management assurances. As one autoworker rep. stated:

Yeah, I file complaints and it comes back, and they say it's diluted. Well what things are it diluted with what? Does it mean it's diluted with water? The supervisor… he asks me to go do that job drilling over that drainhole and that stuff is leaking out, I would refuse it cause only the fact is that I don't know what's in there. I know it's got chromatic acid for one in varying amounts depending on how much leaked into the bar.

Both the mine and auto studies also revealed several examples of collective action including mass or coordinated work refusals, slowdowns and sit-downs. One particular mine (South) had multiple work refusals in my one year of fieldwork along with several slowdowns and one sit-down strike on multiple levels (see Chapter 5). Although there were several underlying issues, high levels of diesel fumes were frequent points of conflict. Managers and supervisors often complained about unreasonable workers who refused or complained to the ministry without giving them a chance to act on the issue. However, individual refusals or collective kinds of actions often only occurred after several previous attempts where workers attempted to follow protocols by filing written forms or raising safety concerns with supervisors and then sometimes with the worker representatives and the health and safety committee. Consider the following account from an autoworker OHS representative of a sit-down strike over a heat stress issue that happened in one of the auto plants.

It unfolded due to some [OHS] complaints in department [which had not been addressed] and so I told the supervisor I want to deal with some complaints for the guys and rather than going around individually it would be quicker to all sit together. So we pulled the lunch table over and we got the whole department sitting down chatting…So after sitting there for about half an hour chatting, the foreman came by and said are you guys going to be chatting much longer? You know my whole department is shut down. We said, why don't you get the head of personnel and the plant manager here, maybe we can quickly resolve this...They got the message...So they agreed, they weren't happy, they agreed and we got the fans.

Although often missed as an important aspect of how workers contest conditions, my interviews, case studies and observational data also revealed a consistent pattern of ongoing one-on-one interactions and negotiations among miners and between miners and their supervisors regarding the acceptability of conditions and the modifications required to meet their particular concerns. These were not generally expressed nor understood by miners or supervisors as formal complaints nor as signs of conflict or confrontation but rather just part of the give and take of day to day work relations. Autoworkers also reported on these kinds of interactions and negotiations on the shop floor as seen in the following interview:

I. When you see a problem with the way the parts are arranged for you can you change it?
R: Move it on the side, yeah. Usually I'll ask the lift truck driver, I'm like, just switch that around so it's a lot easier for us. They'll do it.
I: They'll do it for you?
R: Yeah. As long as you'll help them out, they'll help you out. But if you get a red tag and stuff, sometimes if people don't write the tags, you won't help them out cause they're not helping you out. So that always makes it easier for me, help them out, he'll help me out.

Regardless of how these interactions are understood by workers, these processes can be viewed as political efforts to directly influence conditions and, as many supervisors acknowledged, they placed important constraints on the kinds of conditions which were imposed on workers by supervision and management. At the same time, however, these negotiations, compromises and exchanges were also critical to the overall reproduction and continuity of consent to working conditions since workers and management were ultimately "working out" the terms for continued cooperation which often included some level of risk-taking as well as some level of control. In Bourdieu language, these were the "rules of the game" which gave some sense of control while also providing opportunities to accumulate and use capital to exercise some effect on their working conditions through negotiation and exchange. At the same time, however, the rules of the safety game and the capital they had to play with placed significant limits on what they understood as negotiable, thus shaping and reinforcing over time a habitus which was disposed to see many risks as normal, necessary or reasonable.

As noted earlier, one of the main shortcomings of worker challenges in all these workplaces was that they were often isolated individual complaints which management ignored, deferred or addressed in very narrow and limited ways. Part of what we need to examine in subsequent chapters is how management and state policies and procedures surrounding the internal responsibility system helped to individualize, channel and choke-off sustained resistance. At the same time, we want to show that more substantial changes in practices or conditions, when they came, were almost always a function of emerging collective concerns or actions which carried weight both because the numbers lent the claims legitimacy, or because they represented a more significant threat to production outcomes. However, the big question for subsequent chapters was whether restructuring altered these relations in ways that changed the workers' capacity to negotiate or leverage changes in working conditions and what this meanst for worker consent.

Part I conclusions

The main objective of Chapters 2 and 3 was to understand the subjective bases of worker consent and compliance and, in that regard, there are a number of main conclusions. First of all, despite the substantial knowledge that workers (and managers) shared regarding the main hazards of the job, there were important exclusions and variations in workers' knowledge and definitions of risk and risk conditions. Second, although workers were broadly aware of the range of hazards in their work, they generally consented to most of these conditions on the understanding that the risks were acceptably low, either "naturally" or after being controlled in some way. Whether they were or were not controlled is another question, but the third key point here was that even when workers perceived significant risk, conditions were frequently accepted just the same.

While this latter acceptance was sometimes understood by workers as coerced compliance, risk-taking was more often constructed as a matter of legitimate necessity, choice and responsibility, built around a complex of assumptions, perceived interests, cultural beliefs and identities grounded in both workers' perceived security and control. When workers assumed they could contest risks without reprisals and/or assume risk-taking as a necessity, they were again acting within the framework of a work risk habitus, in the sense that many farmers and workers understood their power and the necessity to take certain significant risks as built-in elements of their occupation and what they understood as the "rules of the game" (Bourdieu and Wacquant, 1992, pp. 98–99). For workers, these risk understandings can also be recognized as being grounded in a broader class-grounded habitus in the sense that workers were disposed to take certain risks because they had come to understand over their careers and in their current situation, the objective limits of their power within a job and an employment relationship.

However, while workers sometimes saw themselves as powerless, especially in certain employment contexts, I am suggesting that risk-taking was more broadly build into the way workers understood their job and employment positions and the rules of the game within that context. As such, it was not just about passively accepting risks. Workers also saw risk-taking as the means of protecting or achieving employment and financial security, and other interests. While often appreciating that the risk of injury or disease was also putting their security at risk, they were in effect gambling that this would not happen at least in the short term, a conflict which they again dealt with in part by constructing a sense of control and/ or simply not thinking about what could happen. The construction of occupational and family identities and responsibilities were also involved in the sense that the capacity and need to take OHS risks reflected the workers' understanding of their abilities, their obligations, opportunities and their life chances, which helped to explain some of the variability in responses we saw within any given workplace. So, while workers and farm owners may have been disposed to accept certain risks, this should not be taken to mean that their acceptance was automatic or predetermined. Certainly, this is not what Bourdieu (1977) meant when he talked about the concept of habitus (p. 79). Indeed, as I've tried to show and will demonstrate more fully in later chapters, while many risks were accepted routinely, the acceptance of risk was also often a negotiated outcome in which workers discussed the terms of whether and how they would take risks with supervisors and other workers, in ways which often extracted concessions.

What these chapters also showed was that there were important differences in the ways that workers within a given workplace or job site understood, perceived and responded to their working conditions which cut within and between the different workplaces. In demonstrating that workers' perceptions and interpretations varied in substantial ways, we are reminded that each worker's particular work and life history, including potentially their current relations outside of work, was important in shaping variations in worker perceptions, knowledge and habitus in regards to hazardous conditions. At the same time, the similar range of responses between workplaces suggested the significance of broader societal and structural influences including social class, media exposures and state involvement in shaping risk- and health-related beliefs and perceptions.

However, some of the other comparisons, in particular the perceptions of miners in different mines and job categories, indicated that many of these variations were linked to differences in both objective working conditions and work relations within different levels or contexts of restructuring. As I seek to show in subsequent chapters, these patterns and variations provide an important means of demonstrating certain arguments about the impact of labour process, job security, management and state restructuring, but as I have suggested as well, these differences also reflected the varied impacts that similar conditions were having on worker responses – that workers were not some uniform mass who perceived, thought or acted in the same way in response to a given set of conditions or structural constraints. At the same time,

I have suggested that some of these divisions placed some crucial limits on the political power of workers and union activists, and to that extent at least, sustaining and exploiting those differences constituted a key aspect of management control.

These analyses leave us with three key unanswered questions. First, how can we account for the dominant habitus which defines and limits the boundaries of risk construction and action within a range consonant with the interests of production and capital accumulation? Second, why are there differences in the knowledge, habitus and ideology which different workers use in perceiving and interpreting similar work conditions and relations? And third, why do some workers evaluate their relations in different ways even when there are no substantial differences in general work or managerial ideologies which might explain those perceptions? In sum, we now have to account for the dominant forms of discourse and actions identified among workers within the three work sites, and, at the same time, explain the substantial variations in subjectivity and action within those groups which serve both to offer and limit the potential for collective action. To address these questions, the focus now shifts to a consideration of the specific industry case studies.

Part 2

Case studies of health and safety in hard rock mining, family farming and auto parts manufacturing

The critical OHS literature traditionally placed considerable weight on the idea that the lack of worker control over the labour process is a major source of occupational accidents and disease, including of course, stress-related injuries (Grunberg, 1985; Grunberg, Everard and O'Toole, 1984; Lewchuk and Robertson, 1997; Lewchuk, Stewart and Yates, 2001; Navarro, 1982; Novek, Yassi and Spiegel, 1991; Sass, 1986). Much of this literature has also adopted an LPT framework which suggests that various aspects of power relations organized within the labour process play critical roles in shaping workers' responses to hazardous conditions (Barnetson, 2010; Glasbeek and Tucker, 1999; Gunningham, 2008; Hall, 1993; Leslie and Butz, 1998; Walters, 1985; Walters and Haines, 1988). Consistent with this literature, the analysis in Chapters 2 and 3 suggested that the perception of control, responsibility and power were important elements underlying many aspects of workers' constructions of risk levels, their perception of the power to resist hazardous conditions and their sense of personal responsibility and identity in dealing with high-risk situations. At the same time, however, the chapters demonstrated significant variations and complexities in these perceptions and ways of thinking about control and risk both among miners, farmers and autoworkers, and between them. A key aspect of this complexity was the tendency of workers in contexts of conflicting or contradictory interests and limited power, to construct hazards as non-hazards, and high risks as necessary or controlled risks. The next six chapters are aimed at linking and explaining these variations and complexities in risk subjectivities and practices to industrial and labour process restructuring within the three industries in question, while also examining the effects of broader political economic changes.

Chapters 4 and 5: health and safety in mining

In Chapter 4, I show that the transformation of the nickel market and the case study company's financial situation pushed management to transform its mining operations. This transformation in turn heightened demands on management to exercise more control over production disruptions and output and, at the same time, to seek more flexibility and discipline from its workforce. While these changes introduced significant new hazards and accentuated others, the principal focus of the analysis is on the effects on miners' direct control and negotiating power with respect to several key OHS risks. I argue that these shifts in control and power undermined the subjective foundations of habitus and consent operating in mining operations, most notably the miners' internalization of personal responsibility for working in dangerous stope and development conditions (Fitzpatrick, 1980; Gray, 2009; Wilde, 1999).

To identify these effects on risk subjectivities and politics, I examine and compare miners working in different mining processes and jobs to show that there are substantive differences both in the structure of their work, control and power and in their experience, perception and understanding of those conditions and relations, which are reflected in different dispositions with respect to risk perception, judgements and responses. Having linked restructuring to shifting demands on management and to emerging problems of labour consent to risk, I then seek to demonstrate in Chapter 5 that management both anticipates and reacts to the economic, political and social implications of restructuring through management and institutional changes in its supervisory, human resources, labour relations, and health and safety discourses and practices. I suggest that while some of these reactions intensify labour conflict, their overall effect is to re-establish consent within the restructured workplace, with particular reference to the judgement of risk, both by providing enhanced assurances of management competence and integrity and by giving workers new ways of constructing a sense of personal control over work and power in relations with management. As I show, risk-taking compliance is achieved by exploiting increased employment insecurity, while consent to risk-taking and risk management are presented as means of protecting their security. At the same time, I demonstrate that the restructuring of management and OHS management systems serve to extend management control over workers and in areas of production and risk that management did not previously exercise in the conventional mining process.

In addition, I examine how variations in management and worker orientation to these changes, some driven by internal splits within the corporation and workplace, and others by broader external political economics, constantly tear at worker compromises and rationales. As I reveal, these tensions limit management control, leading to various forms of worker contestation. An historical analysis also presents evidence that health and safety activism among miners is largely absent until mechanization and bulk mining methods are introduced in the late 1980s and early 1980s (Clement, 1981).

Chapter 5 also considers more broadly the roles of the state and the union in contributing to **both** consent and resistance by reflecting, moderating and mediating the changes in worker and management control and power. These institutional actors offer part of the explanation for persistent differences in responses from miners even among workers in similar job situations, while also being recognized as framing the new forms that labour consent took. Finally, these various arguments are then elaborated and demonstrated through a comparison of OHS politics in three mines.

Chapters 6 and 7: health and safety in agriculture

A central argument in Chapters 4 and 5 is that INCO's restructuring efforts are increasingly tied discursively to health and safety through the claim that economic and health and safety goals can be integrated as common and coordinated objectives of change, as represented by the company's frequent use of the phrase – "productivity and safety go hand in hand." I argue that this was a shift away from the long-standing corporate phrase "safety first," reflecting a cultural turn within neoliberal health and safety discourse to the idea that risks can be automatically controlled through market forces without the need for state oversight. I see this as being particularly consonant with and tied to the state's implementation of its so-called internal responsibility system (Tucker, 2003). However, underlying this change is a broader managerialist discourse that risk, whether in the marketplace or in safety, can be effectively calculated, integrated and managed with the right tools and technologies, the right knowledge, and the right controls. This bringing together of safety and productivity discourses represents the core theory underlying INCO's approach that *management* can effectively control economic, technical and social risks as one coordinated set of knowledge-intensive *management* tasks. In other words, the dual tasks of controlling risk and reproducing labour consent to risk are achieved through management and information technologies which claim to simultaneously address the two key aspects of worker insecurity in a neoliberal context – health and employment. Yet, as shown, whether this discourse actually worked effectively to reproduce miner consent is often in question and often contingent on the particular dynamics and orientations of the various actors.

This brings us then to two central questions which I want to pursue in the two chapters on health and safety in farming. First, to what extent were these new "managerial" orientations to risk also evident within the restructuring of agriculture and farm production processes? Second, to the extent that they are, do they help to further explain the links between labour process restructuring, restructuring discourses and risk subjectivities? In selecting agriculture for this analysis and by focusing mainly on self-employed farm owners who are directly involved in production, the foundations of the analysis thus far are being shifted. Perhaps most obviously, when farmers are making owner/manager decisions about how to produce, harvest and sell their crops, they are exposing themselves and oftentimes their family members to the risk-related consequences. At the same time, we are no longer dealing with the same power, knowledge and interest differentials between owners, managers and hourly wage workers, differentials which have been central to most critical theories within both the LPT and OHS literatures (Barnetson, 2010; Navarro, 1982; Tombs and Whyte, 2007). Even when the self-employed have been studied, especially within the precarious employment literature (Basok, Hall and Rivas, 2014; Vosko, 2010), they are usually conceptualized as being "like workers" in the sense that they are working under restrictive contracts in physical places often owned by the contracting firms, which accordingly means limited power and control.

While the significance of contracted production work will be recognized in these chapters, most of these farmers were selling all or most of their produce in the global marketplace, often using quite sophisticated sales and marketing tactics such as crop diversity strategies, commodity futures contracts, long-term storage etc. And unlike most own-account contract construction workers, cleaners, or for that matter the mass of contract workers in the so-called GIG economy, the farmers owned much more than a computer, cleaning equipment or carpentry tools. They owned and sometimes rented substantial amounts of land (up to 4,000 acres) which they worked using very expensive mobile machines (tractors, combines, flatbed trucks) and other equipment (tillers, planters, balers etc.). While not employing a larger number of workers on a full-time basis, these farmers were employing and directing workforces, often temporary or contract workers, which also raises interesting questions about their OHS management and employment relation practices relative to other small business owners (Eakin, 1992; Eakin and MacEachen, 1998; Frick and Walters, 1998). In short, if I can show that these farmers made risk-related decisions in many of the same ways as both mine managers and miners, based on similar restructuring and cultural rationales, and if these were related in comparable ways to the restructuring of the agricultural labour process and the industry, then I have some evidentiary basis for arguing that labour process transformations and shifts in broader neoliberal cultural discourses on risk and production were cutting across industries, business types *and* class positions. At the same time, I try to show that family farmers, similar to the increasing numbers of self-employed more generally (Vosko, 2010), also had limited degrees of freedom in terms of production decisions once they adopted particular technologies and models of production (Eakin, Champoux, MacEachen, 2010).

Chapter 6 uses the archival data and interviews from industry and farm organization representatives from both studies to provide an analysis of the main restructuring pressures in agriculture and the development of different production and management responses to those pressures. Along with economic restructuring, I emphasize here the influence of environmental and social movement pressures as shaping the different forms of farm production and management, and their approaches to production and environmental risks, while also noting the increased public attention being drawn to farm safety. Agriculture in the 1990s was ideal for this analysis because the industry yielded two competing streams of thought and practice as far as restructuring is concerned, one that was neoliberal, globalist and managerialist and in that sense consistent with the restructuring logic governing the mining company examined in Chapters 4 and 5; and another that challenged the neoliberal model on environmental and social grounds. I identify both threads of thinking as operating to different degrees within conventional *and* organic farming and suggest that once again we see within the neoliberal model an effort to merge economic growth, environmental and safety issues as common coordinated objectives which can be jointly achieved by simultaneously controlling all three types of risk within a new intensive integrated management framework. As in the mine analysis, I also show that there are underlying structural tensions which continued to tear at this effort to get workers, or in this case, self-employed producers, to buy into this way of thinking.

The subsequent Chapter 7 then presents a comparative analysis of the 20 farm case studies, again by combining the data from both studies – 10 conventional farms (counting the mixed conventional/organic) and 10 organic farms. At the outset of the chapter, I suggest the 20 farms can be divided into 5 groups, distinguished principally by their commitment to a neoliberal versus alternative model of production, management and marketing. By looking at and comparing family farmers in this way, that is, owner producers who were organizing, managing and carrying out the work in somewhat distinctive ways, I examine how the risk subjectivities underlying their responses to health and safety hazards varied according to their adoption of the different restructuring models. Two main arguments come out of this comparison. First, although farmers following the neoliberal management model addressed safety and health hazards in a much more systematic calculating way, and in that respect, a more conscious preventive way, risk-taking as cost-based compromises were built into this thinking with a consequent negative effect on OHS practices Second, alternative-oriented farmers did not generalize their environmental concerns or social critiques of conventional or productivist farming in addressing health and safety issues – that is, they tended to underestimate the personal risks and, as a consequence, took relatively few precautions. I argue these findings point to the significance of both structural and cultural pressures which justify and necessitate the exchange of safety for production, further illustrating a point made in the INCO study. Even when workers have significant control and power over their work, as conventional miners did, and certainly as farmers did relative to most wage workers, there were persistent cultural and economic influences that reproduced substantial personal risk-taking. The farm analysis in Chapters 6 and 7 also confirms the mining findings that the neoliberal and post-Fordist emphasis on leaner forms of production and management were often accompanied and enforced through an audit culture which carried with it significant influences on how health and safety risks were understood and managed at the production level.

Chapters 8 and 9: auto parts manufacturing

To understand more fully how these orientations, and the considerable variations within them, play out in terms of labour consent and workplace politics, the analysis moves in Chapters 8 and 9 back to an industrial wage labour context where the class and labour/management distinctions between the owners, managers and workers are less ambiguous and, in particular, where differences in power, control and interests are more distinguishable. Turning our attention to the auto parts manufacturing sector, these chapters focus on the role of worker participation in mediating the politics of health and safety within the contexts of lean production and performance-based management.

As the mine study will show, the company's efforts to achieve leaner and intensified forms of flexible production had a range of complicated and oftentimes contradictory effects on health and safety, while the auditing and performance technologies influenced workplace politics and consent in similarly complex and conflicting ways. Part of the problem at INCO was that senior management were unable to fully sell both line managers and workers on the idea that this new approach to management really worked to control risk without threatening their respective interests in production and production bonus, which explains why many miners did not accept corporate management's claims that its various mechanisms, technologies and techniques were legitimate or effective. I will argue that this lack of consensus over both worker participation and performance audits among INCO managers and workers had something to do with the historical context. When the mining study was done in the mid to late 1980s, neoliberal shifts in management thinking were in their infancy, and certainly for a company that had a long history of quite adversarial labour relations and approaches (Clement, 1981), it is perhaps less than surprising that INCO's transformation to a softer, gentler labour relations approach was not entirely smooth or consistent. The more participative IRS model in Ontario was also in its early days, with enough political controversy across the province in the 1980s to suggest that INCO was not the only firm experiencing difficulties in transforming its management of labour relations and OHS (NDP, 1983, 1986; Storey, 2004; Storey and Tucker, 2006; Walters, 1985). By the turn of the century, however, the IRS model of OHS governance had been in place for some 20 years while much of the Ontario auto industry had been restructured around lean production models emphasizing quality audits, continuous improvement, just-in-time and participative team management (Lewchuk, Stewart and Yates, 2001; Lewchuk and Wells, 2006). At the same time, we know that the auto parts sector is a complex one with considerable variability in production and management systems, offering a further basis for assessing how these governance ideas and mechanisms affect labour consent and resistance (Lewchuck and Wells, 2006). The shift in time frame allows me to recognize changes in the public and workplace politics of health and safety compared to the 1980s and 1990s, with both negative and positive consequences in terms of worker capacity to challenge their conditions.

In combination, these findings bring the discussion back to the original debates first introduced in Chapter 1. Does the declining robustness of a public politics in health and safety in Ontario reflect a more consensual workplace politics, and if so, is this consensus grounded in the greater effectiveness of the IRS system within the context of a more consistent neo-liberal post-Fordist approach to management across industries; or, does the consensus reflect the depoliticizing and disempowering effects of the IRS system within the broader context of union busting neoliberal discourses and post-Fordist restructuring? At the same time, to the extent that there are persistent conflicts over health and safety within the industry, can we explain this persistence with reference to the same arguments developed in Chapters 4–7, that is, the failure of firms to reconcile the claims of worker involvement and empowerment with the efforts to intensify management control through labour process changes and increased performance and information technologies?

With these questions in mind, the next two chapters on auto parts production continue my examination of the links between the politics and subjectivities of health and safety and the three main elements of neoliberal post-Fordist management – lean production, audit-based management and participative management. Chapter 8 relies on a comparative study of 36 auto parts plants in south-western Ontario (Study 4) supplemented by the second study of worker reps. (Study 6) and the study of injury and hazard reporting (Study 7). The focus is on comparing the politics of health or safety changes as seen mainly through the eyes of OHS representatives working in the plants in the context of different management approaches to production and labour relations. One central question is whether there are links between the adoption of lean and flexible production principles and the management's reported approach to health and safety – with particular reference to whether management is perceived as exhibiting more or less commitment, responsiveness and cooperation with workers and the union in dealing with health and safety issues depending on the level of lean production being pursued.

This line of analysis returns to the argument that came out of the mine study that as firms transform their labour processes towards lean and flexible production models, there are increased production pressures placed on workers coupled with an increased corporate concern with managing safety incidents and injuries as costs and disruptions to production. This double effect then yields management efforts to achieve worker commitment through more extensive worker involvement mechanisms while at the same time seeking to exercise more control over worker practices.

This chapter also identifies three different forms of worker representation operative in the different workplaces, one of which represents an internalization of the IRS discourse and limited possibilities for OHS improvements. The other two are presented as oppositional responses to the limitations of the IRS partnership approach, explained partly as the contradictory outcome of neoliberal forms of OHS governance.

Chapter 9 elaborates these same arguments but in the context of a case study of a *non-union* auto parts company which underwent significant changes in its health and safety approach from 1998 to 2005 (Study 5). I demonstrate how the company's adoption of continuous improvement principles, audit management, worker involvement and modern human resource practices combine to both shift the management approach to health and safety *and* reproduce worker consent to hazards in the workplace. However, as in previous analyses, the chapter also seeks to identify and explain some of the contradictions of this process with reference to consequences which limit worker consent while asking whether the lack of a union makes a difference in the conduct of health and safety politics.

Chapter 4

Transforming the mining labour process

Transforming risk and its social construction

Restructuring mining methods, processes, jobs, relations and internal labour markets

When INCO entered the 1980s, it was the third largest multinational producer of nickel and copper in the world with mines and surface operations in Canada, Indonesia, Caledonia and Guatemala. While a still a major nickel market player, INCO had lost its historical monopoly by the late 1960s, in large part due to the Soviet Union's entry into the market (Clement, 1981; Swift, 1977). In an effort to meet global competitive changes in demand and supply as well as increasing labour costs, INCO adopted a globalization strategy in the 1970s, investing millions of dollars in mining projects outside Canada such as Indonesia and Guatemala. As global competition intensified, again especially from the Soviet Union, its diversification and expansion strategies not only failed to restore INCO's capacity to control nickel prices but also added a considerable burden of debt.

In this context, INCO began in the 1970s to make significant changes to surface plant operations and mining labour processes with an initial emphasis on transforming its surface operations (Clement, 1981; USWA, Local 6500, 1987). Similar to the case of farming (see Chapter 6), some of the pressure for these surface plant changes were environmental in nature, in as much as INCO's smelter and refinery were identified as major polluters – in particular, as **the** central source of acid rain in Ontario and north-east US in the 1970s (Clement, 1981; Ontario/Canada Task Force on INCO Limited and Falconbridge, 1982; Swift, 1977). However, INCO's failure to post profits in 1982, something that had not happened since the 1930s depression, signaled a definitive break from the company's historical control over markets and prices.

It was in this latter context that the company greatly accelerated its transformation of underground mining methods and technologies, and ultimately, its entire approach to nickel mining production and management. While the labour process at INCO had been moving towards increased forms of mechanization since the late 1960s and the company had experimented through the 1970s with a more capital intensive and productive form of mining called bulk mining (Clement, 1981), the large scale move to mechanized bulk mining did not occur until 1982–85, when the proportion of ore produced using bulk mining techniques shifted from 6% to 80% (USWA Local 6500, 1987).

Given the speed of the shift and the lack of any hiring, most of the miners working in bulk mining in the 1980s had been transferred from conventional or mechanized stope miners and, in that sense, had directly experienced the labour process transformation. However, for a variety of technical and economic reasons, INCO also continued to use conventional mining methods in some of its Sudbury operations through most of the 1980s. As such, when this study was conducted in 1985–87, miners within and between different mines were working in substantially different labour processes which varied in terms of mechanization, specialization, labour intensification and integration. Among other things, this offered an opportunity to compare worker responses to their working conditions within these different labour processes, while holding constant the corporate health and safety programmes and discourses which were present to some extent across all its mines at the same time. Accordingly, in this chapter I look at four critical areas of labour process changes and their OHS and political effects,, and then in chapter 5, consider the management efforts to manage the risks and worker responses.

Mining methods

In 1985, the ore being produced in Sudbury was derived from three main techniques: undercut and fill (UCF), cut and fill techniques (CF), and a form of bulk mining called vertical retreat mining (VRM). During the 1960s and most of the 1970s, the majority of INCO's mines relied on the "cut and fill" techniques (UCF and CF) (Interviews with INCO Senior Management; see also Clement, 1981; INCO, 1980–82; USWA, Local 6500, 1987). When INCO made its major shift to VRM in 1982, it closed several mines and mining areas that were not suitable to VRM or could not be mined profitably using cut and fill techniques. However, some of INCO's largest older and deeper mines (i.e. Levack, Creighton, Stobie), still had a good number of cut and fill operations (INCO, 1986; INCO Triangle, 1983, p. 4; USWA, 1987). Most of these operations were involved in mining out pillars[1] and remnants of nickel ore and precious metals, the latter of which had become more valuable during this period. These two procedures are usually referred to as "pillar recovery" and "farming." Pillar recovery had become more important in the 1980s because sluggish nickel prices had pushed INCO to rely more on the mining of rare metals such as platinum. Prices for these metals were high enough to warrant the continued use of small-scale conventional operations since they were better suited for working in the areas where such ores were found.

In some mines, mechanized cut and fill techniques were being used for "virgin" ore because of ground stability problems, some of which had been caused by the earlier version of bulk mining called "long hole" mining (Ames, 1987; Stevenson, 1986). As the company went deeper underground, it encountered increased problems with ground pressures which mitigated against using bulk mining techniques (Ames, 1987; INCO, 1985). While this meant that the older mines had mixed mining methods, there were other "newer" mines such as McCreedy, North and South Mines using almost entirely VRM, offering a further basis for comparative analysis within different mining contexts (see next chapter).

Levels of mechanization

Until the mid-1960s, conventional mining methods were limited in terms of mechanization, in that workers still used handheld drills (jacklegs and stopers) to drill holes, and slushers to remove or "muck" out the ore (air-pressure, cable-drawn shovels). Electric or diesel trains would then be used to "tram" the ore to the ore chutes which would deliver the ore to ore cages or elevators. As noted in Chapter 2, in the late 1960s and early 1970s, INCO began to introduce large mobile diesel trackless equipment which could drill (jumbos) and remove ore (scooptrams) at a much faster rate than had been possible using handheld drills and slushers. INCO introduced its first "scooptram" (a front-end loader on rubber wheels) in 1966. By 1969, INCO had 77 scooptrams operating in its Sudbury mines. By 1975, there were 225 of these machines being used and by the 1980s this number had tripled. A similar pattern of growth in technology was evident in the use of jumbo drills, the use of which doubled in number between 1969 and 1975 (INCO, 1980–82; Kitchener, 1979). Trucks were also increasingly used to transport ore instead of trains, while jeeps were also introduced to make it easier and quicker for shift bosses (supervisors), managers and engineers to move through the mine. A system of ramps was also introduced in most of the newer mines to aid freer movement of vehicles and workers to different levels.

During the 1980s, the company also began to introduce a new technology called "continuous mucking machines" which could draw ore continuously from the blasted face onto a system of conveyor belts. These kinds of machines had been operating successfully for some time in coal mines but, at the time of the study, the company was still having significant problems with this system. However, the company had moved to introduce conveyor systems on some levels replacing trains which were meant to enable the company to move the more significant volumes of ore coming from bulk mining production.

Since the size of the mining area and ground support system in UCF did not lend itself to large machinery, the level of mechanization involved in the

undercut and fill method had not changed appreciably since it was introduced in the 1950s. The regular cut and fill method (CF) was more amenable
to significant levels of mechanization and, accordingly, with the introduction of mechanized equipment and the development of new ground control technology (rock bolts and screens), the cut and fill method became
increasingly mechanized and much larger in scale over the course of the
1970s (USWA Local 6500, 1987). Where conditions permitted, stopes were
increased in size (from 10 to 40 feet wide and from 30 to 250 feet long), as was
the size of the crews working them (moving from two to five workers and,
in some cases, as many as eight). These large cut and fill stopes, euphemis-
tically referred to as "GO GO" stopes, represented the first major intensification and reorganization of the mining process since the introduction of
cut and fill methods (Clement, 1981; USWA, Local 6500, 1987). "Captured"
scooptrams, jumbo drills and eventually scissor trucks (lift platforms), that
is, mobile equipment taken apart and reassembled in an enclosed area, were
introduced in large stopes. While not all cut and fill stopes used jumbo drills
and scooptrams, most cut and fill stopes were highly mechanized by the
1980s (Interviews with Mine Superintendents and General foremen at Levack and Creighton Mines; See also Clement, 1981).

The use of trackless equipment and rock bolt technology also changed
development work – that is the drilling of drifts (horizontal tunnels), raises
(vertical tunnels for workers/ventilation), chutes (vertical tunnels for ore)
and ramps which are used to access ore bodies and support a mining production operation (USWA Local 6500, 1987). The use of jumbo drills and
scooptrams significantly reduced the amount of time required to develop
new drifts and made it possible to expand the amount of development in
any particular area, without the need for extra crews; that is, by moving
equipment around from heading to heading (drifts in development) in the
same area, crews could develop multiple headings with the same equipment.
Prior to trackless equipment, development work relied entirely on handheld
drills, slushers, rail-based mucking machines and ore trains, requiring a significant amount of preparation work and physical effort (e.g. laying railway
tracks). When I worked for INCO at Thompson Manitoba in the mid-1970s,
I worked on this kind of pre-trackless development crew for several months.

The re-organization, specialization and flexibilization of mining jobs

Until the early 1980s, production miners were posted to specific stopes or
development headings under relatively broad job categories such as UCF
or Development Driller. In practice, especially in the less-mechanized
UCF and CF stopes and development headings, this meant that the conventional miners were often involved in a *full cycle* of production involving
drilling, blasting, mucking and ground control (cribbing, rock bolting etc.).

As outlined in the collective agreement with the union, all stope and development heading jobs were posted and certified and qualified workers bid on them with the most senior qualified applicants getting the job (USWA, Local 6500, Union Grievance Log Book, 1975–1978; Interviews with Union and Management Officials). This meant that management could not easily replace the miners in these stopes, temporarily or permanently, unless the miners bid out of these jobs (or they were successfully charged with a major safety infraction). If there was some delay in the stope or development heading, management was also unable under the collective agreement to temporarily reassign these miners to other jobs. While mechanized cut and fill and development miners had already become more specialized by the 1980s, miners were still attached to particular stopes or development headings via the seniority bidding process, and there was often a greater level of involvement in a variety of work activities, although this tended to be structured by the miners themselves.

Vertical retreat mining (VRM) is substantially different from the cut and fill methods, and as organized by INCO, resulted in a major restructuring of production mining jobs (Ames, 1987; USWA Local 6500, 1987). In particular, there was no longer a "stope crew" which was responsible for mining out a section of the ore body (e.g. a cut and fill stope). Rather, the process was completed by a number of functionally specialized separate one to two person crews (development crew, ITH driller, blasters and scooptram operators), who followed each other through the mine, working at different stages or points in the production process. Although the miners were tied to each other as elements within a more integrated process, the specialized miners or crews work independently of each other. With the exception of the McCreedy Mine, where a full mine team bonus approach was used, their respective production bonuses were not linked in any way.

While the actual jobs in the production process became more specialized, the company also moved to gain more flexibility by radically reducing the number of distinct job classifications and postings. As one senior manager explained the "need" for change:

> We realized we had to change our job structure in 1980...the number of grievances we were getting, trying to use our work force [efficiently and flexibly]. We had more than 120 job categories and a maturing CBA [collective bargaining agreement] and people were using it to protect their rights to jobs according to seniority.

Although conventional mining jobs and trade classifications changed very little, virtually all jobs were reviewed by a joint committee and expanded under the auspices of a "cooperative wage study" (CWS) programme, which the union and the company had negotiated in the 1979 collective agreement. This agreement was proceeded by a bitter eight-month strike which was fought largely over the issue of job classifications (Clement, 1981; Radecki, 1981). By 1982, 120 job classifications had been consolidated into 40. The creation of "mechanized miner" and "development trackless" job classifications basically merged the separate classifications of conventional drillers, utility drillers, drift drillers, jumbo drillers and scoop operators, allowing the company to utilize workers with these classifications in all these jobs without having to post the positions (USWA, Local 6500 Union Records, Job Classifications Binder). In three of its newer mines, South, North and Mc-Creedy, the company also experimented with a two-category job structure, production miner and support miner, which allowed even more flexibility.

Although the local union had fought a long strike on the classification issue, a new more conservative union executive, elected after the strike, cast the CWS as a progressive means of increasing wages and balancing inequities in the existing wage structure (Clement, 1981; Interviews with Union officials). According to many union stewards, worker OHS representatives and miners, the key outcomes were that miners no longer had control over their work assignments via posting requirements, and accordingly, were much more reluctant and less able to question or challenge OHS conditions.

One other feature of the work reorganization is important to recognize here. Until the 1980s, all the production miners (UCF, CF and VRM miners; scooptram and rail tram operators) were on a crew bonus incentive system, while the support miners (e.g. cage operators, shaft crew) and trades workers (mechanics, electricians) were not. VRM and mechanized production crews had lower incentive rates relative to conventional miners which also meant that VRM and mechanized miners had to work more hours in the day to earn anything close to the conventional miners. As one ITH driller put it, "you have to have that drill going eight hours a day to make any bonus." This can be contrasted with UCF non-mechanized miners who, according to my interviews, reported an average of five hours per shift in actual working time, while making as much 60% to 70% bonus.

Lean production

The company also sought to achieve efficiencies in other ways. Although company management rarely used terms like "lean" or "flexible" production, they had implemented just-in-time (JIT) delivery and supply systems along with flexible production planning and scheduling regimes. Senior and mine managers explained that the company was seeking the capacity to respond to shifts in demand and market prices. Miners and shift bosses often commented on the impact that this latter approach had on mining practices. As one shift boss reported: "We used to think in terms of very long-term development plans but suddenly things could change over night and we'd be moved somewhere else or we'd suddenly change direction because the price or demand for something had changed." What this meant then was that as demand and prices shifted, stopes and development headings were often being opened and closed depending on the kind of ore that was available and the methods that could be used to access the ore. In that sense, there was also a constant management process of calculation going on around the returns and costs associated with production and market demand. These shifts often also translated into considerable pressure to speed up production in these sites in order to capitalize on market shifts, although the capacity to shift gears in a mine was somewhat more restrained than in a factory context (see Chapter 9). While mechanized and VRM miners were subject to more speed-ups, the older conventional and mixed operations also often suffered from disinvestment as the company sought to minimize the costs of its more labour-intensive processes by cutting maintenance, as one miner described it, "to the bone."

In many respects, the company's efforts were reflected as much by what Russell (1999) called producing "more with less" than a fully integrated lean production model. The company streamlined its maintenance and housekeeping programmes in large part by reducing staffing in these areas. Basic housekeeping on levels used to be the responsibility of various support mining jobs such as a "powderman," a job with which I'm personally familiar, but most of these positions were eliminated by the 1980s, and production crews were given nominal responsibility for maintaining drifts, lunchrooms, powder storage areas and cage stations. In reality, as both shift bosses and workers acknowledged, production miners rarely had the time to do routine maintenance and cleaning so both tended to deteriorate. The only times I observed some attention being placed on cleaning and maintenance was just before a Ministry of Labour inspection or MAPAO safety audit, or a visit by some other dignitary. Reduced warehousing of parts and supplies was another feature of the shift but, in many respects, this was not so much an effective JIT supply system as an effort to reduce costs by delaying parts and supply orders as long as possible, which meant that it took much more time to fix and maintain things. In practice, this meant that miners were often operating malfunctioning diesel machinery and drilling equipment for months at a time (Miners, Mechanics and Supervisor interviews). Although the company began to implement various worker involvement schemes (see Chapter 5), it did not have a formal continuous improvement programme and there was no concept of "kaizen" as often used in the auto management discourse (Rinehart, Huxley and Robertson, 1997).

In sum, the labour process analysis points to three distinct groups of production miners operating at the time of the study reflecting stages in the firm's labour process development, conventional non-mechanized miners, (conventional) mechanized miners and VRM miners. At the same time, all three groups were subject to the effects of reduced maintenance and housekeeping. This co-occurrence provides a means of comparing the effects of different labour processes and, in so doing, allows me to draw inferences about the impact of the transformation that was taking place at the time. What I'd like to do next then is first consider what these differences meant in terms of occupational hazards and risks, and second, compare the risk subjectivities and politics of these groups. Although there is a fourth group of workers who I've categorized as trades (electricians, mechanics) and other support workers (e.g. cage operators, crusher operators), for the sake of brevity I say very little about this group. However, for the trades in particular, I argue that while their actual skills, independence and tasks did not change much, the transformation, given the increased use of machinery, electrical equipment etc., did increase their numbers and the demand for their skills. At the same time, the emphasis on cost control within lean production translated into major increases in maintenance problems and pressures from management to keep machines running without proper parts.

The impact of technological change and restructuring on mine hazards

Before addressing how miners responded to hazards in these different labour processes, let me briefly describe my understanding of the *objective* differences or changes in hazards and risks, relying in part on my own experiences in these three different mining processes and my research observations; statistical and expert evidence; and interviews with miners, OHS representatives, managers, engineers and government inspectors. There are two central points I wish to make. First, the shift to VRM reduced miner exposures to several serious safety hazards and altered others. The work of the conventional miner was much more physically demanding than VRM since the former involved working with heavy, hard-to-operate, hand-held "jack leg" drills which vibrated causing "white finger", carpal tunnel syndrome and a range of MSD problems. In contrast, the VRM as well as the jumbo drill were mounted on tracks or wheels, which meant no direct vibration, while the operator's lifting – and indeed their only routine physical task – was largely confined to attaching and detaching drill sections. Conventional CF and UCF miners were also working in much more confined spaces with more unpredictable machinery, most notably the "slusher" they used to move the blasted ore into the chute. In mechanized conventional stopes that used scooptrams, there was the added risk of working around large moving machinery in a very confined dark space. Conventional miners were also usually in closer proximity to deep holes and relied heavily on ladders to climb significant distances into and out of their stopes. The increased use of mechanized trackless equipment in VRM and mechanized conventional mining areas, as well as the use of conveyor systems, also eliminated many of the hazards associated with underground trains, which routinely went off the rails with serious risks associated with getting them back on; again, risks which I experienced first-hand when I worked with a development crew tramming ore. However, as noted, an added complication was that when INCO began to move away from trains and conventional production methods, it spent less on maintaining the tracks, the levels and the equipment in the older mines.. As such, for VRM and mechanized workers still working in the older areas or mines, there were the additional risks of having to work with poorly maintained machinery and physical environments.

While VRM significantly reduced the lifting and physical effort of most mining production jobs, and eliminated the problem of hand-drill vibration, the larger explosions used in bulk mining methods and the more rapid displacement of rock generated more ground movement and, as a consequence, increased the potential for larger-scale and, in some cases, catastrophic cave-ins and collapses (Ames, 1987; MAPAO, 1985a; Stevenson, 1986). However, since most mines were either full VRM or mixed VRM/conventional, all the mines and therefore all the miners were being exposed to more ground movement and risk of falls. CF and UCF methods also had some significant ground control risks without these added pressures. The use of remote-controlled devices allowed VRM drillers and scoop drivers to stay out of "uncontrolled areas" where rock falls presented a significant danger, and INCO claimed that further engineering and geological innovations reduced these risks further (Ames, 1987; INCO, 1985d). However, blasting crews were still often being exposed to open ground. Specialized ground control crews were at the strongest risk as they were charged with installing rock bolts and nets. At the time of the study, there were growing concerns which led to several government studies and a special Ontario government commission of inquiry after a number of significant ground fall events where entire levels were lost (Ames, 1987; Inco, 1985d; Stevenson, 1986). An added additional risk was certainly that workers were being pushed to work faster and harder, at least within the VRM and mechanized contexts.

However, although accident statistics are clearly problematic, it is worth noting that the more serious injuries, usually referred to as lost time accidents (LTAs), declined quite substantially as VRM was introduced; that is, from 1981 to 1985, LTAs at Inco declined from 8.6 to 4.8 per 100 employees (Inco Direct Line, March 7, 1986; Inco Annual Report, 1986). A printout I obtained through a special request from the Mining Accident Prevention Association also identified conventional miners as having a disproportionate number of lost time (LTAs)) injuries compared to VRM miners (MAPAO, 1986).

While this data is suggestive, it should be noted that after fairly steady declines in the 1970s, the number of fatalities actually increased in the industry from 1982 to 1987 from 9 to 19 fatalities with INCO contributing substantially to this increase (Ontario, 1988, pp. 10–11). This may mean that the company was concealing its LTA injuries through questionable injury management practices, including of course persuasion and overt intimidation. Relating to this point, an Ontario legislative Standing Resources

Committee concluded in its 1988 investigation of safety in the mining industry that the apparent improvement in lost time counts from the 1970s to 1980s likely reflected an intentional shift in industry reporting practices and a concerted effort to conceal lost time injuries by making liberal use of modified work or light duty (Ontario, 1988, p. 13). Certainly, many workers and union activists I interviewed would have concurred with this conclusion with respect to Inco's practices. However, the fact that 12 of the 13 fatalities at INCO during this time period were in the 4 older, less-mechanized mines also suggests some support for the argument that the older mines and mining methods presented more serious acute injury dangers than the newer operations.

The second main point about the OHS effects of the transformation was that VRM and mechanization produced an increase in several health hazards. The introduction and increase in diesel fumes were perhaps the most notable added risk. VRM scooptram and truck drivers in particular were exposed to high concentrations of diesel fumes, somewhat more than operators in mechanized conventional stopes, since a single VRM blast could keep several machines operating continuously at the same time. On the other hand, ventilation was often better in the mucking drifts than in the production stopes and VRM specialized mines generally had superior ventilation systems than the older mines. VRM drillers and blasters had somewhat less exposure since the scooptrams were usually operating on levels below them but this would vary. However, as noted already, the company's new emphasis on "producing with less" translated much more broadly into reduced maintenance across all the mines, which accentuated certain risks such as diesel and oil mist exposure, which again was more prevalent in mechanized mines and VRM production areas.

Dust levels were also intensified in the mine as a whole given the larger size of VRM explosions, although management again claimed that this problem was addressed over time with better ventilation systems and blast scheduling. At the same time, dust levels in conventional stopes were often just as severe as in mechanized stopes during mucking and, of course, after blasts. Reduced maintenance and speed-ups also had generalized effects across all the job categories but again perhaps more substantially for VRM and mechanized miners since they were often working with large mobile diesel equipment which were not being properly maintained.

In sum, I would argue that the shift to VRM reduced some hazards and risks and added others. On balance, I'm argue that conventional miners at the time of the study were generally at greater risk of serious injury or death, while VRM and mechanized miners were perhaps at somewhat greater risk of serious long-term health problems given increased exposures to diesel fumes and oil mist. The one major exception perhaps in terms of VRM safety risks was the effects on ground fall hazards, although again there were mixed effects and exposures since conventional mining also presented some very dangerous ground conditions (Ames, 1987). Certainly, INCO argued publicly that its new approaches and mechanization were safer (e.g. Inco, 1986, p. 8; Kitchener, 1979). The question then is how did the miners in these jobs view these risks?

The differences in risk subjectivities and politics

Differences between mining methods

Chapters 2 and 3 demonstrated substantial differences among miners in how they judged and explained their responses to risk. Here I want to show that some of those differences can be explained through a comparison of conventional, mechanized and VRM miners. As will be recalled from Table 2.3, VRM miners in the survey were more likely to express concerns about specific hazards such as ground conditions, dust, oil mist and diesel fumes than the conventional miners, with the mechanized miners tending to fall somewhere in between. The only area where conventional miners were more likely to express concern was in terms of overall accident risks – a measure of their overall assessment of the dangers they faced in the job. While these distributions fit in some ways with the above descriptions of differences in exposures to risks, it is hard to explain the more frequent specific concerns among VRM miners since their diesel or oil mist exposures were not likely to be consistently greater than the mechanized miners. I've also argued that significant dust exposure was a common problem across the spectrum, so again, it is not clear why VRM miners would be more likely to express concerns in this area based on exposure levels alone. While the smaller differences between the groups on ground conditions is somewhat consistent with the argument that ground fall risks are significant across all groups, on balance, the use of remote-controlled equipment by VRM miners reduced many of the ground risks to which miners in the other two groups were routinely exposed. On the other hand, the greater concerns about overall accident risks among conventional miners was in line with my argument that they were exposed to a much wider range of significant safety hazards. However, the more frequent concerns about health risks among mechanized miners are a little harder to figure out. These general concerns may derive substantially from greater specific concerns about diesel fumes as mechanized miners often emphasized this issue in interviews, but it is interesting that in response to the direct questions, only 35% of them expressed significant specific concerns about diesel, proportionally fewer than among the VRM miners (55%) and the trades/support workers (39%). This suggests that at least some of these mechanized workers were judging their health risks on other grounds. Although somewhat speculative, one of the more common complaints that came from mechanized stope miners, especially those working in so-called "go go" stopes, was that the pace and pressure to produce was placing them under considerable stress. Conventional miners in contrast said very little about these kinds of pressures or stress more generally, and at no time, did I observe stress as being a topic of discussion in health and safety committees or in management or union OHS communications.

The differences in concern between the three groups of miners also extended to their responses, in as much as VRM and then mechanized miners were more likely than conventional miners to report concerns to their foremen, engineers, managers or health and safety representatives (see Table 4.1). They were also more likely to report work refusals which certainly signals a level of resistance and conflict with management.

For their part, conventional miners reported a greater tendency to accept high-risk conditions and reported far fewer political actions aimed at changing those conditions. For example, among conventional miners who expressed major concerns about dust, less than 10% reported taking any political action to address that concern. In contrast, over 50% of the mechanized miners reported taking some actions to change the conditions which usually meant approaching a supervisor or filing a formal written complaint via company procedures. It could be argued that this reflects the different levels of dust within the two situations but, as I've suggested, both groups were regularly exposed to significant levels of dust.

The argument that these differences reflect distinct job-related risk subjectivities is strengthened by considering the perception and response to safety hazards such as ground conditions. While conventional miners were more likely to express concerns that their jobs were often dangerous, they were also more likely to accept those risks without challenging management or supervision. And perhaps most importantly, they were much more likely to explain this acceptance in terms of personal responsibility, occupational necessity or bonus trade-offs; and, were much less likely to explain their acceptance in terms of any fear of reprisals or job loss.

Table 4.1 Forms of actions taken by miners over hazards[a]

	Conventional	Mechanized	VRM
Complaints to Shift Boss (SB)	36%	50%	61%
Multiple Complaints to SB	12%	31%	50%
Dust Complaints	12%	29%	39%
Multiple Complaints	5%	19%	30%
Diesel Complaints	8%	17%	9%
Multiple Complaints	5%	8%	4%
Complaints to OSHE	36%	50%	65%
Multiple Complaints to OSHE	8%	19%	30%
Unsafe Work Refusals	27%	58%	65%
Multiple Refusals	12%	24%	43%
N	66	48	23

a Trades and support miners not included.

Table 4.2 Miner perceptions of work control[a]

	Conventional	Mechanized	VRM
Control over Daily Decisions	80%	31%	39%
Control over Health Conditions	52%	25%	25%
Accidents Caused by Workers	35%	19%	26%
Disease Caused by Workers	48%	25%	13%
N	66	48	23

a Trades and support miners excluded.

The varying tendency to externalize or internalize responsibility is further demonstrated in Table 4.2, in that conventional miners were more likely to agree or strongly agree with the statement that accidents and health problems were caused by the careless behaviour of individual workers. Conventional miners were also much more likely to agree that they had personal control and responsibility for their own health.

The miner case studies which sampled miners in the different job classifications also provide additional support for this argument. Conventional miners reported more in-depth knowledge of the range of safety conditions in their workplace and more confidence in that knowledge. They were also more likely to explain and interpret specific hazard-related events in individual terms and explicitly report personal control and responsibilities as a more important rationale underlying their acceptance of high-risk situations. As one conventional miner put it when asked if he was exposed to dust or fumes in his workplace:

> Health and safety is a state of mind – it's up to the individual. You're always going to have some little accidents but the worker makes it safe. [Q. What should the company or union be doing to protect health?] According to the news [TV], smoking is the major cause of people's health problems... It all depends on the individual and how they look at their lives.

In contrast, VRM and mechanized miners were more likely to point to technology, machines, management or engineers as the sources of unsafe or unhealthy conditions and as the means through which control can be exercised. Consider this exchange with a mechanized scooptram operator in response to the same question about dust and fumes:

> I. Are you exposed to dust and fumes?
> R. Well I work on the scoop so there's dust from the road and when I'm dumping, and there are fumes since it is a diesel scoop.
> I. What can be done about this? What can you do?

R. Well as long as *they* [management] keep the fans going, the ventilation system, it will be better. If *they* did the roads, watered or put some dust retardant that would help.

What this suggests then is that direct worker control and decision-making played a more critical role underlying consent within the conventional context, while VRM miners and, to some extent mechanized miners, were more likely to judge safety and define responsibility with reference to management systems, practices, and control technologies.

Another critical difference between the groups of miners is that the conventional miners reported relatively little pressure or direct orders from their supervisors or engineers. They presented themselves as being largely in charge of what goes on in their stope. If they encountered difficult or dangerous conditions, they might consult with their shift boss or an engineer who would give what they interpreted as "advice," but they thought of themselves as being in control and making most of the day-to-day decisions, which included decisions about safety and health conditions. As one conventional miner put it: "I'm good at what I do. I make good money and I'm my own boss. That's what I'm paid for. It's like running your own business. You got to work hard."

Reflecting their sense of greater vulnerability to management power, mechanized and VRM miners were more likely to express the view that management was telling them what to do giving them limited choices and more limited control over the OHS practices and decisions. They also saw management as increasingly pressuring them to ignore health or safety in the favour of increased production. As one VRM scoop operator put it:

They're more muck crazy, they push you all the time – they want you to work even if the equipment is down, and they always blame you, they want you to break the rules and take responsibility for it. I complain to my shift boss but he just tells me to keep mucking and that's there's no problem.

As this suggests, the workplace politics of risk within the transformed jobs and mines (see Chapter 5 for a mine comparison) were grounded more in the experience of coercion and mistrust, which was then reflected in a greater degree of overt conflict and resistance. This also speaks to the different capacity of managers in the new processes to exert this kind of production pressure. As several shift bosses and managers acknowledged in their interviews, given conventional miner isolation from supervision and the limited levels of technological control, they had little choice but to leave the mining process to the conventional miners, while relying on lucrative bonus arrangements, worker commitment and worker skills to get the levels of desired production without disruptions. Indeed, managers and supervisors understood in part from past experience that if there were more substantial management efforts to direct or pressure conventional miners, the miners would simply push back and production would decline. As one shift boss stated: "You have to work with your people. They get you the muck. If they think there is a problem, I listen to them."

Management also recognized that supervisors and managers had less knowledge and understanding of the risks within conventional stopes, and therefore less capacity to control outcomes. Management concerns in these contexts were less about preventing injuries than ensuring that the stope continued to produce ore at a profitable level and cost, but the key point is that management recognized a higher degree of dependence on worker knowledge and commitment to get the job done. As one mine manager acknowledged: "[W]e've tried various things in the conventional stopes including closer supervision but it just doesn't work. We tried reducing their bonus because of the cost but that didn't work either."

However, as we move away from conventional mining to large mechanized stopes and development headings, and then to VRM, the increased capacity for direction supervision, the increased technical controls achieved through machinery, standardization and engineering plans, and the limitations placed on worker knowledge and control through specialization, reduced miner knowledge and control over their conditions. With respect to OHS, VRM miners had less direct involvement in the range of activities relevant to the perception, judgement and control of their immediate working conditions. As three VRM miners put it:

> I have lights from the scoop so I can see ahead of me. But I'm running up and down the ramps all day – a half a mile easy one way. I have to keep my eyes on the road. I have no time to check the back [roof].
>
> (VRM scoop operator).

> I used to be proud to be a [conventional] stope miner cause I did a job there. You had your own stope, you controlled the place, like running your own business. Now all I do is install rock bolts [for VRM drillers] all day long.
>
> (VRM Development Driller)

> I move around a lot. it's not like being in the same stope for months or even years. You don't *know* the areas as well, you don't have time. It's also difficult to see with the mist from the [ITH] drill. It's bad in there sometimes –I'm redrilling a lot. There's no screen over my head because the blasts have tore it all to shit.
>
> (VRM driller)

As this also suggests, the VRM miners had less direct knowledge of many of the risks and causes of risks within their workplace and had to rely more on sources of information other than their direct perception of hazards in detecting and judging risks. It also means that since they were less directly involved in various daily practices which addressed the full range of major safety hazards in the workplace, VRM and to some extent mechanized miners were relying more on other sources of control and information such as specialized ground control or development crews, engineers, technical devices such as ground monitors and so on.

As I outline in detail in chapter 6, it was in the context of these labour process changes that managers, supervisors, formal corporate safety programmes, and specific control and monitoring technologies played a much greater role in controlling and shaping what workers "knew" about risk in the transformed context. As managers and engineers also noted in their interviews, they were conscious that they were less dependent on workers in judging certain safety matters and what generally was happening in a given work area. These shifts in knowledge and judgement capacity clearly offered opportunities for concealing or distorting what miners knew, but it also suggests that the worker perception of management and its approach and commitment to health and safety were much more important elements to the judgement and responses of VRM miners than conventional miners, since the former group was forced to rely more on technology and management claims of safe conditions. When management and management OHS controls were perceived as trustworthy, this perception translated into consent. As one VRM driller stated: "VRM's not the problem. It's much safer for the miner because he's not exposed to the open ground like in the cut and fill stopes. The remote control keeps you out of trouble." However, as some of the previous VRM quotes also indicated, not all VRM drillers agreed that their safety was assured by institutional and technological assurances, and it was in those kinds of contexts where consent began to break down.

Along with the reduction of direct hazard monitoring and control practices, the movement to VRM involved an overall reduction in the range and amount of miners' decision-making regarding a whole range of production and organizational issues. As Table 4.2 shows, there was some further support for this argument from the survey interviews in that 80% of the conventional stope miners reported significant control over their day-to-day work decisions, as opposed to only 40% of VRM and 31% of mechanized miners. Thus, to the extent that decision-making and direct control practices played an important role in shaping the social construction of working conditions and accident as products of their *own* decisions or actions, the implication is again that VRM miners were less likely to define the working conditions and accidents as personal responsibilities, and more likely to look to management and engineering assurances.

The impact of these shifting conceptions of risk origins, responsibility and control played out in important ways. For example, if we look at Table 4.3, we see that for VRM miners, the level of concern for health and safety conditions was significantly correlated with the perceived competence and integrity of company management, the health and safety committee and the union health and safety representatives, whereas for conventional miners there were much weaker relationships, none of which were statistically significant.

While all miners tended to follow company policy and channel their initial complaints and actions through their relations with their supervisors, miners in VRM and mechanized contexts were much more likely to report a need to go beyond supervisors to management, the health and safety committee or the union. And as noted earlier, reflecting the lack of success in using those channels, VRM miners were also more likely to report using OHS law in the form of legally prescribed work refusals.

Another distinct area of difference relative to risk perception and compliance relates to the issue of security and the perceived capacity of miners to negotiate, resist and/or pressure for changes in their working conditions. Although many studies have demonstrated that workers can exercise power after deskilling through various collective bargaining mechanisms (Thompson, 1983, p. 107), one of the implications of Burawoy's (1979; 1985) research was that the post-Fordist shift to more flexible job structures and classification amounted to a significant loss of collective as well as individual worker power (see also Thompson, 2010). The changes to the CWS system at INCO amounted to much the same process of flexibilization and, in that sense, all miners were to some extent disempowered both through deskilling and flexibilization. The loss of the long strike in 1978–79 and the shorter 1982 strike over these very issues was also an indication of a downward trend in union strength, while the strikes themselves did much to weaken member resolve and militancy. However, within this weaker union environment, VRM and

Table 4.3 Pearson correlations between concern over health and safety and the perceived integrity of management and the union

Level of Concern for OHS Conditions	Conventional Miners	VRM Miners
Trust Supervision	.09	.34
Competent Supervision	.15	.27
Fair Treatment by Management	.24	.58**
Company Respect for OHS	.14	.37*
Company seeking to improve OHS	.04	.25
Competent Health and Safety Committee	.09	.41*
Competent Union Representation on JOHC	.16	.51**
N	25	68

* p <. 05; ** p <. 01

mechanized miners were also much more vulnerable to intimidation and threats around disputes over working conditions than conventional miners, and less capable of using their own positions of power in the production context to negotiate or pressure for changes through supervisory relations and direct individual job actions. It is not surprising then to find that VRM miners were more likely than conventional miners to report coerced compliance or explain their consent in terms of employment security when taking known risks.

Conventional miners on the other hand, were more likely to cite the necessities of the job and their occupational identities as their basis for accepting significant risk. For example, conventional and, to a lesser extent, mechanized miners were more likely than VRM miners to proudly recount events in which they had used their knowledge and skill to manage a high-risk situation "on their own." They also more frequently used the imagery of the "tough" and "skilled" as their basic construction of themselves and as explanations for why some risks "just had to be accepted as part of being a miner" (Fitzpatrick, 1980; Haas, 1977; Paap, 2006). Moreover, their definitions of skill and the sources of enjoyment in their work were tied more to the demands of risk judgement and control. VRM miners tended to report that their challenge was to cope with the fact that they had little personal knowledge or control over conditions which concerned them.

What this suggests is that in the absence of informal arrangements and an ability to negotiate or pressure supervision directly, miners in mechanized and VRM contexts were encouraged to seek other more formal means of addressing concerns, which means in turn that their perception of their collective and institutional power as workers with formal rights became much more important to their sense of control over risk, which speaks in turn to the greater significance of the union, the law and the state in the reproduction of consent and whether or not workers comply on coercive or insecurity terms. The survey data again provides support for these arguments regarding differences in power relations and interests. Within the context of the survey interviews, conventional miners were more likely to express the view that they had the personal power to influence management and supervisory decisions relevant to their health and safety (58% vs. 26% mechanized and 39% VRM). When miners were asked whether they should have the individual authority to stop unsafe work, the vast majority of miners agreed that they should have that right. However, conventional miners consistently replied that they already had that authority within their existing relations. Moreover, VRM miners were more likely to explain their acceptance of high risk as concessions to their limited power, while conventional miners were more likely to understand their acceptance as "fair exchanges" or negotiated compromises. Conventional miners were also more likely to refer to bonus or reciprocal arrangements with their shift bosses as the critical interests underlying their acceptance of high-risk conditions, rarely citing concerns about management reprisals or threats. In contrast, many VRM miners expressed concerns about reprisals and their vulnerability to reprisals as a basis of their acceptance of conditions. Indeed, VRM and mechanized miners were more likely to express the view that they were fundamentally powerless to do much about the major risks in their present working conditions.

The different power relations between miners, supervisors and management are also relevant to the extent to which miners in the different jobs reported various arrangements and understandings with their supervisors, in particular those which involved exchanged acceptance of risk and unreported injuries for favours and good treatment on issues such as bonus hours and assignments, equipment requests etc. The supervisory relationship was very important to the conventional miners because of bonus arrangements and the acceptance of risk was more likely to be explained in these terms among conventional miners. As such, conventional miners were also more likely to cite an obligation to a partner or to a shift boss as a basis of their actions in responding to risks. It is not surprising to find then that it was conventional miners who were more likely to make the argument. When asked about his supervisor, a UCF miner put it this way: "We get along good. It's generally understood…Everybody makes deals. Sometimes we get the work done and no one asks any questions."

However, as might be expected given the relative isolation of miners from each other in the labour process, all miners tended to share a perceived lack of collective power in terms of their capacity to act together to resist or force changes. Interestingly, miners from all groups tended to view direct actions such as work slowdowns as legitimate forms of protest and political action, but VRM miners were more likely to report a belief that these would not be effective and that formal official channels were generally more appropriate (and less likely to yield management reprisals). This often translated into greater VRM use of the worker representatives and the joint committee as vehicles for making complaints.

> I've used OSHE [what Inco called the mine joint OHS committee] a number of times. A few months ago there was bad loose in our area from blasting and I talked to OSHE and it was closed down. I always tell him [worker representative] to check certain things that I know about when he's doing an inspection. He puts them in his inspection reports.
>
> (VRM miner)

This difference in the perceived importance of the health and safety committee was also reflected in their reported use of the joint committees when concerned about risks. In the survey interviews, for example, only 38% of conventional miners reported any contact with a union worker committee representative, while 64% of the mechanized miners and 53% of the VRM miners reported using them to address a concern. While most conventional miners insisted they'd "never gone to them" and were often very critical of their OHS representatives, VRM or mechanized miners would be more likely to say something like the following:

> They are good to have around…My shiftboss tells us to do three things and we don't like it, we do another…He [the shiftboss] may get mad but he doesn't do anything because of the union…I just tell him I'm going to my steward.. [Q. What about the OSHE committee and reps.?] The OSHE, they come in and it's the same thing, maybe they'll see something I didn't see…We're working a number of slots in here [large mechanized stope], no one's perfect.

In summary, the analysis thus far suggests that the subjective bases of management control and labour consent, and the conflict and politics surrounding OHS, were substantially different within the different labour processes, and that those differences reflected substantive variations in the level and type of work practices, skills, decision-making, autonomy, job security and power relations as structured within the different labour processes and jobs.

In social field theory terms (Bourdieu, 1977, 1990), the findings suggest that by nesting considerable control and power in workers, the conventional mining labour process had shaped a work risk habitus which internalized stronger miner identities, commitments and responsibilities for working with dangerous working conditions in exchange for independence and lucrative bonus (i.e. the rules of the "game"). The miners in the transformed jobs were *less likely* to accept their working conditions, not because they were demonstrably worse in terms of overall risk but because they had lost much of that control and power over their work and working conditions, and ultimately their capacity to judge and control hazards. Using Bourdieu's language, mechanization and VRM shifted miner control and power, changing the "rules of the game" and disrupting a conventional *habitus* or disposition to risk which was based at its root on a mutual understanding that workers were in control of the work process and environment and thus responsible for identifying, judging and controlling risks with limited management and expert input.

While grounded in the substantive control that workers exercised, the conventional habitus included assumptions and understandings about individual responsibility and identity as a miner, which meant the routine acceptance of considerable risks understood partly as an exchange for the production bonus as well as their relative independence as producers. Management periodically sought to challenge this conventional habitus but, this was largely unsuccessful until the labour process restructuring substantially reduced the individual control and power that conventional miners exercised. In a context of reduced control and power, mechanized and VRM miners were pushed to learn the new rules of the game including more supervisory/engineering oversight and demands; limited independence and decision-making; more specialized jobs; more machine controls; more direct production pressures; less knowledge and capacity to directly judge risks; more dependency on management and engineering definitions and judgements of risk; less protection against job changes, threats or reprisals; and a bonus that required more work per shift. With limited individual power resources or capital at their disposal, some mechanized and VRM miners looked to other institutional sources of power or influence, notably the union, the stewards and the worker health and safety representatives, and the state OHS inspectors. Often finding those sources wanting, reflecting weaknesses in the law and the union, those with hazard concerns often looked to supervision, management and expert assurances that their concerns were unwarranted or were being addressed. Others sought to gain power through other means, ironically in some situations, by seeking to exchange a willingness to take risks in the new environment, either generally (i.e. haywire miners), or more selectively, for other favours and capital. And as noted, miners in these transformed jobs and processes were more likely to be accepting risks on compliance rather than consensual grounds, which also meant more experiences of coercion and direct conflicts with management.

However, while the increased conflict and resistance among mechanized and VRM miners, and evidence of coercion by management, suggests a disruption of the habitus underlying consent in the conventional mining context, we need to remember that many of the mechanized and VRM miners expressed relatively limited concerns about their health and safety conditions and even fewer were actively engaged in confronting management on these issues (see Tables 2.3 and 4.1). While there were some significant collective challenges in some mines, these were also relatively rare. In other words, the company was largely successful in getting most VRM and mechanized miners to accept the new work realities with relatively little visible resistance or conflict. Some of the consent among miners in the new processes seems to be grounded in their perception of management and expert reliability, knowledge and trustworthiness, but some was also grounded in renewed claims of personal control and responsibility, similar to although not the same as the control and responsibility expressed by conventional miners. As this suggests, along with the conflict and coercion, mechanized and VRM miners, supervisors and managers were beginning to establish a new habitus or consensual frameworks around which risk was constructed, approached and accepted. The questions we need to address then are how was the conflict and resistance within the new labour process managed, and how was compliance and/

or a consensual habitus reestablished? To address these questions, let's move to the next chapter where I consider a number of other significant changes that were introduced by INCO during this period – in particular, changes in its management structures, its labour relations and human resource practices, and the introduction of a new management health and safety administration and program.

Note

1 Pillars are support sections of the stope that are left in cut and fill mining.

Chapter 5

Reconstructing miner consent

Management objectives and strategies

Chapter 4 showed that as compared to miners in conventional stopes, miners in mechanized and bulk mining jobs were more likely to express concerns about their health and safety conditions and to contest those conditions increasingly through reference to their legislated OHSA rights. However, the evidence also demonstrated that INCO was still successful in achieving its restructuring with relatively limited labour resistance. In this chapter, I consider how management reproduced its control within the context of labour process restructuring, while also showing the limits of those efforts.

Management control strategies: pre and post 1980

Following Littler and Salaman (1982; 1984), I use the concept of "management control strategy" to capture the ways in which corporate and local mine management seek to exercise management control within the restructured labour process. Management control strategies refer to an *intentional* organization of management structures, practices and discourses aimed principally at the reproduction of worker consent and compliance within contexts of productivity and cost containment objectives. Four core areas of management strategy are examined, concluding with a comparison of worker consent and resistance in three mines. While examining each area, I also show that front-line management practices vary in important ways often limiting or contradicting stated corporate management policy.

I. Constructing a Security crisis

In 1971, there were over 18,000 hourly workers in the Sudbury operation. By 1986, this number had been reduced through layoffs, early retirement plans, buyouts and attrition by almost two-thirds to 6,500 workers. Along with layoffs and internal transfers, an eight-month production shutdown in 1982 (USWA, Local 6500, 1987) and several permanent mine closures laid the foundation for heightened worker insecurity. In what was essentially a example of what Buroway (1985) called "hegemonic despotism", management communications encouraged workers to accept this insecurity as a new economic reality which could only be alleviated by a common management and labour commitment to technological change and other productivity and efficiency changes. As Chuck Baird, the chairman of INCO, put it in a 1983 INCO employee publication:

> With technological change, some jobs have to be redesigned to make sure our employees are working more productively....We have been through difficult times. I believe that Inco is on the road to recovery but no one can guarantee it. Only a competitive and profitable company can provide assurance of rewards – monetary and nonmonetary. Therefore, we must be efficient to ensure the company's survival and return to profitability. That is the biggest incentive for us all. Otherwise [all] our jobs are at stake....
>
> (INCO, 1983, pp. 18–19)

INCO routinely held "information meetings" and circulated written materials for workers aimed at explaining the causes and implications of market changes. These included a monthly "President's Newsletter" focused on the "state of the company," frequent company magazine articles on technological changes and the nickel market, and the introduction of annual information meetings called the "face to face" programme, where senior management met with the entire mine workforce (Inco, 1983, 1985a).

By 1985, front-line and senior management were expressing confidence that the messages had sunk in. As one mine manager put it: "[P]eople were voicing the perception that we were hiding things in the past. I don't hear that anymore... there's this new understanding of the business of nickel mining and its problems." Mike Sopko, the company president at the time, communicated a similar message in 1985 in a regional newspaper: "The reductions in employment have not been the wild slashes of an executioners' act...there is an understanding among employees that what has been done *had to be done*..." (Sopko, 1985).

In my miner survey, 86% agreed that at least some of the changes in working conditions and new technologies were necessary to "compete in the present nickel market" and virtually all miners acknowledged that VRM was vital to the company's profitability and their jobs. Shift bosses uniformly argued that their workers understood the current risks to their employment and the need to accept certain conditions they would not have previously:

> We all know that there are some bad conditions down here but my men understand I don't have the men and the company doesn't have the money to do the things the way we used to do – it's as simple as that.
>
> (Shift-boss)

Yet, as the nickel market improved into the mid-1980s, and the company's financial situation improved, many workers also believed that the company was intentionally overstating the crisis. In the 1985 mine survey, for example, 83% believed that the company was consciously using the "crisis" as a threat to get workers to work harder and to accept working conditions without protest, and 66% believed that the company's financial situation was not "as bad" as claimed. While these latter findings suggest that this strategy was wearing thin over time, concerns about job security were still very prominent in miners' explanations of risk acceptance.

II. Cooperation, participation and responsibility: INCO's HRM approach

Beginning largely in 1980, INCO made several major changes to its human resources, supervision and employee relations approaches. This shift, which can be broadly characterized as a move to a softer human resource management (HRM) approach, had four central components: one, a more intensive and proactive employee communication strategy, which was partly described in the previous crisis section; two, a move away from command-and-control supervision; three, the creation of various employee involvement and management participation mechanisms aimed at improving productivity and the quality of work life; and four, a more "cooperative" approach specifically aimed union relations. As framed by Mike Sopko:

> We were paternalistic, we [thought we] knew what was best for our employees. Our labour relations were not good. We are [now] much more aware of the importance and value of our employees at all levels. We must develop them properly and treat them well. We will do the utmost to remove the adversarial attitude from our union relationships.
>
> (Sopko, 1985)

Along with the "face to face" programme and weekly mine newsletters and monthly magazines described earlier, local mine managers and superintendents held regular information meetings and worksite tours, and instituted open-door policies, suggestion-box programmes and reporting procedures. As senior management understood this effort, the communication strategy was aimed at building worker trust in management. As one senior manager put it:

> This came about as a result of criticism that the company was not telling workers enough about what was happening and why. So, we started around 1980 after the strike...It was clear then that there was a lot of suspicion and mistrust.

In backing up its claim of a new approach to management, INCO also sought to revise the way in which workers were directed and supervised. Prior to its reforms, INCO had encouraged its supervisors, who normally had moved up from hourly paid status, to "keep their distance" from workers in a clear command-and-control system. However, as Chapter 4 suggests, supervision within a conventional mining context was in practice often much more hands-off. As one senior manager expressed in his interviews, the tendency of miners and shift bosses to form informal agreements and alliances was a major concern behind these changes. As he recalled:

> One of the reasons we wanted to change things [in the 1980s] was that when we started looking at who was doing what and who was rated to do what, we found that people were all over the place – we [central management] had no control. They [shift bosses and miners] had all kinds of agreements, people were getting driller rates and doing nothing, or working a train...It was a mess.

INCO fixed this "mess" to a large extent when it introduced bulk mining in as much as VRM miners no longer had the same level of control over the production process (see Chapter 4). However, in most of its mines INCO also shifted many production decisions to mine management, management-controlled planning committees, joint labour management committees, mine engineers and geologists. Supervisors retained some enhanced powers to move workers around within these new job processes and internal labour market arrangements but, as described by the shift bosses themselves, their work time was increasingly taken up by the growing mountains of reports required by the audit- and information-intensive managerialist approach which INCO had adopted (see below), which included daily reports on footage (drilling) and tonnage (ore), injuries, equipment and maintenance problems, and safety incidents.

As shift bosses struggled to get their job done with limited resources and personnel, and felt the increasing pressures to "produce or perish," there tended to be two very different responses (USWA, Local 6500, 1984, 1987). Some shift bosses came to realize that their only job security lay in their relationships with their crews. If they wanted to survive, they had to establish new alliances with their workers and develop new tactics within the constraints and pressures of the transformed labour process and their revised responsibilities and powers. What this often meant was that miners and shift bosses would conspire to control and keep information from middle and senior management, for example, by distorting production and injury reports as well as safety audits. Sometimes these actions limited worker exposures to poor conditions but, at other times, they would work in the reverse where shift bosses and workers conspired to hide injuries and unsafe conditions so workers could continue to achieve production goals and earn bonus. In these contexts, then, mutual relations of interdependent interests and trust between supervisors and miners were often established, frequently translating into compromises which shift bosses brokered into labour acceptance of risks. Telling in terms of the implications for consent was a miner who described his shift boss as being "one of them now," or as another who stated more fully, "they [shift-bosses] are like me...fighting for their jobs like everybody else."

However, there were other shift bosses, and this often corresponded to differences in the managers of mines (see mine comparison later), who responded to the pressure with directive and coercive methods of supervision, essentially ignoring the corporate call for a softer form of supervision. As one shift boss observed:

> We no longer have the authority or the tools to make deals with the workers, and in a mine that's always the way things were done. Some of us have done our best but more and more supervisors refuse to meet workers halfway and end up resorting to strong arm tactics to meet production and cost quotas.

As some supervisors moved in a more coercive and directive direction, miners tended to report more substantial confrontations and conflicts and were more likely to report work refusals or collective actions such as production slowdowns or informal sit-down strikes (e.g. see discussion of South Mine in this chapter), which in turn often led to reprisals:

> I get threats from my shift-boss all the time cause I'm always complaining about the dust on the ramps, the water and the ground conditions...I finally pulled a Bill 70 [work refusal] cause of an unsafe haulage truck. They weren't doing nothing about it so I wasn't going to drive it anymore. After the Bill 70, they stuck me in a ditch
>
> (VRM scoop operator)

Some front-line managers also expressed a shift in their thinking towards giving workers more room to make their own decisions even as the safety management system seemed to be tightening the rules. As one manager described it:

> [Y]ou can't make decisions for real people in real situations just by going with the law...You have to talk to people in a way that gives them input, not telling them what to do...give them the opportunities to make their own decisions. If you have confidence in them, they'll have confidence in you.

This participative orientation worked to draw some workers into taking responsibility for their work conditions in as much as some mimicked this idea that management gave them room to make their own decisions, but other workers were more suspicious. As one miner put it, "the company preaches safety but says it's ok for you to do what you want, you decide, but then they'll put you in an area with no ventilation and see what you do." Others complained when rules were imposed and enforced, saying that they were being treated "like children."

In the context of these different supervisory and management tactics, the third main thrust of the HRM approach was the creation of new worker participation programmes. Although INCO's union had negotiated joint health and safety committees well before they were legislated in 1978 (see OHS management section later), INCO had not embraced any other forms of participative management within the production sphere until this time. The impetus for this particular step may well have come from a government inquiry into mine safety in 1980–81 (Burkett, Riggin and Rothney, 1981). In their report, the Burkett Commission as it was known at the time, admonished the Ontario mining industry for retaining an old management control system which failed to address the issue of worker involvement:

> The hierarchical control systems which typify industrial organizations have remained the norm in the Ontario mining industry. In the underground operations in particular...it is difficult to understand why the industry has been reluctant to consider new approaches...In a situation where the work group is quite autonomous already and has a significant measure of control over output, the logical working arrangement in our view is to shift responsibility for day to day production to the individual crew, or at the very least, to involve the crew in all aspects of planning.
> (Burkett, Riggin and Rothney, 1981, p. 53)

Within two months of the Commission's report, INCO management introduced a programme called "stope planning groups." Workers from all the cross shifts from each stope or development heading met as a group at the beginning of a job or section of the job with their shift boss, geologist and engineer to plan the work and stope from beginning to end (Strutt, 1986). In many respects this was simply a formalization of what was often happening unofficially in the conventional system, and certainly many miners, especially conventional miners, saw this as a minimal change. For some mechanized and VRM miners, however, it was often seen as a meaningful opportunity to discuss the mining plans, conditions and progress.

Several other kinds of worker involvement programmes were introduced after this by the company which were entirely new and not specific to health and safety, some of which were task-oriented management/worker committees working on the introduction of new technologies, efficiency plans etc.

while others were referred to as "problem-solving committees." When I requested a list of the different committees from Levack mine management, 15 separate "problem" committees were reported for a 2-year period (1985–86). Some mines also implemented a type of quality circle programme which they called "employee involvement groups" organized around the production activities of each level (Mine Managers and Superintendents, Levack, Frood and Stobie Mines).

However, as interviews with mine managers and superintendents revealed, there were very different perspectives within management on the value of involvement programmes and the manner in which they should be conducted. Some clearly believed that workers could be motivated by these programmes, while also giving management access to information which could be used to make the production process operate more smoothly, efficiently, and in some cases, safely. Others assumed that workers were ill-prepared and unwilling to take on decision-making roles. As one mine manager put it when explaining his limited support for group programmes at his mine: "I think most Canadian companies have had the same experience... Our workers are not like Japanese workers. They are more individuals and don't take to sitting around talking in groups." Another mine manager reported that he had stopped the programmes in his mine complex entirely because there were "too many troublemakers." A manager in another mine acknowledged that his move away from stope planning groups reflected his view that mechanized and VRM jobs were so standardized and fixed with limited decision-making at the point of production that there was simply no need for them. As he put it: "We used the group programmes a lot for cut and fill mining but we just don't need to do that with VRM and continuous process mining. What's the sense given the nature of those jobs?"

However, corporate management was also experimenting with different team-oriented models in some of its mines. For example, in two of its more mechanized mines, McCreedy and North, the normal positions of general foremen for each shift were eliminated and the number of shift bosses were reduced to one per shift for the entire mine. Considerable decision-making and responsibility were shifted to self-managing mine teams, which were made up of two distinct production groups each with its own hourly paid unionized elected team leader. Each group or team was provided with its own equipment, budget and supplies and an area of the mine as its responsibility. Long-term production objectives and schedules and engineering plans were largely set by management but it was left to the workers and their leaders to organize, assign and coordinate the work on a day-to-day basis. The job classification system was also completely restructured reducing the CWS 40 category job structure to 2 basic categories – production and support miner.

Not surprisingly, the impact of teams and other worker involvement schemes on risk-related concerns, control, responsibility and consent depended quite substantially on the extent to which worker involvement was consistently practiced in its various mines, which I would argue further explains some of the variations in labour consent demonstrated in previous chapters (see also mine comparisons later). A shift boss in one of the least participative mines (South Mine) had perhaps the greatest insight with reference to this: "They [team approaches] only work if you are willing to let them make some major decisions, and that wasn't there in most cases." Or, as a union steward put it: "They [management] could kill us with these [participative/team] programmes if they did it properly but as it is, they always end up screwing it up by trying to push us around."

Given these frequent limitations, many miners quickly learned to say as little as possible when asked for input while at the same time trusting very little of what they were told. This in turn reinforced the beliefs of managers who were disposed to assume the Canadian workers were not suited for participative management approaches. For many miners, and this was again evident in some mines more than others (see mine comparison later), the worker involvement strategy simply accentuated their concerns and their sense of limited information or control regarding their working conditions (see Rinehart, Huxley and Robertson, 1997 for similar findings in the auto industry). Still, as the survey indicated, there were a substantial number of mechanized (32%) and VRM (46%) miners who believed that these programmes demonstrated that the company was moving to give miners more say in planning and decisionmaking, and perhaps for those miners, this offered an important sense of security with respect to risk information and control. However, it is noteworthy that fewer conventional miners shared this view (12%). Conventional miners tended to argue that they didn't need worker participation programmes or worker reps. as they exercised significant individual control at the points of production. Interestingly, electricians and other trades people, also relatively powerful occupational groups, expressed similar sentiments.

The fourth main HR initiative was framed as a general attempt to improve management relations with the union, relations which were quite strained after the long 1979 strike (Clement, 1981). The main thrust of this approach was to encourage more open communication lines between union officials and managers and a greater willingness to seek to settle grievances informally or at least without arbitration. As one senior union executive put it:

> Things have been improving between the union and management since 1982. Because the company was in a tough spot, they realized they had to get honest with us. Old guys retired and we got a new breed of managers, more educated. We haven't had one firing discharge since 1982. Most of those problems are now dealt with outside the grievance procedure. I just call up the plant manager or whoever and get things done, cleaned up without too much trouble.

As noted, the union executive elected in 1982 was much less militant than the one that led the 1979 strike (Clement, 1981), and certainly seemed to adopt a more cooperative stance in its relations with the company. However, some miners interpreted this new relationship as the union "being in bed with management," which meant concretely that many miners and worker reps. no longer trusted the union to protect them.

III. The OHS management system

Building competency, control and responsibility

The third main INCO initiative revolved around major changes in its OHS management system. However, before outlining the features of the new OHS management system, it is important to understand some of the history behind Inco's approach to OHS. To help in this regard, I interviewed long-term and retired health and safety activists and other union representatives (N = 18) with a focus on the history of health and safety in the Sudbury mines. I also spent about two months in the Union Hall going through old union OHS and grievance committee files. According to these sources and several management interviews, up until the 1960s INCO had viewed health and safety as an exclusive management right, which meant that their safety management programme was confined to a façade of command-and-control safety rules coupled with a classic "safety first" discourse which emphasized worker awareness and carelessness (Interviews with Retired Miners and Union Officials). In terms of safety technologies, workers relied mainly on limited personal protective equipment while supervisors and managers selectively enforced rules through a coercive disciplinary step system, often used as "weapons" to punish recalcitrant workers (see Clement, 1981, p. 230). Health issues with perhaps the exception of hearing protection were largely absent from the management discourse. And as noted in Chapter 4, management relied heavily on conventional miners' skills and knowledge, pushed by a lucrative bonus system, to work around safety rules and procedures to meet production goals, and then blamed workers for breaking the rules when injuries and fatalities occurred.

Although this weak command-and-control approach to safety was partly retained right through to 1980s, a crack opened in 1966 when the company and union negotiated a collective agreement containing provisions for a joint management/union health and safety committee structure, with union-appointed worker representatives. Anticipating what became Ontario law some ten years later, it acknowledged the right of the union to contribute to health and safety decisions through committee participation, if only in an advisory capacity. Between 1966 and 1975, the collective bargaining provisions around health and safety and the joint health and safety committee structure were expanded considerably from a single committee with limited rights to a four-tier structure of mine committees, area committees and a central senior management/union committee. In 1975, the company and union also negotiated a joint occupational disease research programme called the Joint Occupational Health Committee (JOHC), in which a group of medical researchers from McMaster University were contracted to conduct ongoing epidemiological research and respond to specific demands from the JOHC for specific shortterm studies.

According to my interview (both management and workers) and archival sources, the joint health and safety committees (1966–79) operated in practice much like grievance committees. Complaints were made by workers to union representatives who presented the workers' case. There would often be yelling and threats and, if no agreement was reached, a formal grievance was often filed. Most of the health and safety representatives at the time were past or simultaneously union stewards, who had been trained by their union to approach health and safety in an adversarial manner, as they did other labour relations issues. They had little or no formal health and safety training until the 1980s. While grievances offered a means of challenging management, they were not an effective way of gaining changes in health and safety conditions, in part because they simply took too long. This encouraged stewards and workers to rely more on direct political pressure from workers or public pressure outside the workplace if the issue was seen as significant enough; and as such, some health and safety committees as reported by both workers and managers, became sites of significant conflict and animosity as the company began to ramp up its mechanization (Clement, 1981; Various Interviews).

While some OHS committees became increasing points of labour/management conflict in the 1980s, most of the early pressures for and from joint committees at INCO, did not come from its miners and, in particular, its conventional miners. Although much has been made in the Canadian literature about the "struggles of miners" in the 1960s and 1970s to improve their working conditions and the effect this had on emerging state reforms (MacDowell, 2012; Storey, 2005, 2006; Walters, 1983), the early activism at INCO came principally from the surface refinery and smelting plants, *not* the mines. According to my interviews, conventional stope miners in particular were actually quite resistant to their efforts to bring health and safety issues to public or state attention. Most of these miners were also not supportive of the push for joint health and safety committees – this again came from the surface plants and, to a lesser extent, a small but growing group of mine activists concerned about mechanization. As one long-time union OHS activist put it: "We never got much support from the miners...if anything, the problem we had with the early [OSH] committees was their resistance to them as threats to their control or their bonus." Consistent with my argument about the OHS impact of labour process changes, the surface plant concerns and activism emerged around the same time that INCO was automating its surface operations, which preceded most of the major VRM mine restructuring as noted in Chapter 4 (Clement, 1981).

While mine health issues emerged as important political issues in Ontario in the early to mid-1970s, these were fueled largely by developments in Elliot Lake, notably the 1974 uranium miner walkout over radiation exposure and cancer (MacDowell, 2012). The Ham Commission was formed as a result which in turn shaped key elements of the 1979 Ontario Occupational Health and Safety Act (Ontario, 1976; OHSA, 1979; Walters, 1983). Given their common membership in the USWA, union officials and OHS activists from Elliot Lake and Sudbury began working together to promote proposals for a new OHS legislation, more research on health hazards and a more activist committee system, and then later in the 1970s a small number of mine activists at INCO coalesced around the health effects of mechanization, most notably the increase in diesel fumes (OMOL, 1983; USWA, Local 6500, 1975; Interviews).

Changes in OHS management: the 1980s

While INCO's injury record showed some modest improvement in the late 1970s, the Ontario mining industry and INCO had an extremely bad year in 1980 with 18 fatalities in the first 6 months of the year (Burkett, Riggin and Rothney, 1981). This was a major embarrassment to the government, the industry and INCO given the recent introduction of the new OHS legislation, leading to yet another government inquiry into mine safety (Burkett, Riggin, and Rothney, K. (Burkett Report), 1981). As noted, within INCO, health and safety was becoming, by this time, a greater source of internal labour problems and public controversy, as evidenced in part by a major increase in work refusals in 1980 and 1981 (see Graph 3.1). Within this context, and just *before* the Burkett Inquiry was completed, suggesting that at least some of the motivation was overtly political and aimed at the government, INCO announced a major effort to improve its health and safety management programme.

As in its previous OHS reforms, several of the programmes implemented during and after 1980 were quite consistent with past developments or practices. For example, it announced a further strengthening of its internal safety inspection system, adding an additional safety foreman to most mines (Inco1980a, b, c). When the company negotiated worker inspectors in 1985, it began to shift more of this responsibility to the worker inspectors, while pushing as much as possible to limit the connections between the inspectors and the union (e.g. prohibiting inspectors from holding any union office). Qualification standards for specific mining jobs were also "strengthened" and reviewed on a more regular basis. A "job observation" programme was also implemented which involved supervisors engaging in periodic specific observations of workers performing a specific job task, which, interestingly in terms of the "safety first" roots of this programme, were broken down into its basic components in terms of safe practices. Many of these changes seemed to indicate the company's traditional concern with exercising control over workplace accidents through standardization, rules and technical and procedural controls. Indeed, as noted in Chapter 4, management was conscious of the effects of its restructuring on miner skill and control, believing that the move to mechanization, automation and bulk mining "had removed the element of worker judgment." As one manager explained, the training programmes and standards were an effort to exert more direct control over the remaining aspects of worker judgement through better training, specified procedures and direct monitoring. While management assured workers that they could gain more control through the strict adoption of these procedures, miners also recognized that they were often being pushed through new production standards, bonus requirements and cost containment measures to break those rules and procedures, much as they had in the conventional system. As one VRM driller put it: "[T]here's no way to meet production targets or get any bonus without breaking the rules, they [management] know that."

However, other OHS governance changes were more clearly concerned with building a picture of management and worker cooperation and joint or shared responsibility. To begin with, the company signalled its greater recognition of health and safety as a "people" issue by merging the departments of safety, occupational health, environmental control and employee relations within a single human resources department (Management Interview). Unlike past safety appointments which drew from the company's crop of engineers and experienced mine managers, a "human resources" professional rather than an engineer was placed in charge of this new department. Asked to define his new position and the reasons for the combination of the four departments within a "human resources" department, the new director stated:

> We have excellent employees and the four areas with which I am concerned affect them very significantly. They need to know that safety and productivity are linked inseparably. They need to have responsible employee relations policies. They need to be assured that every effort is being made to protect and monitor their health....
>
> (INCO Triangle, 1981)

A number of new safety "communication" and "worker participation" programmes were subsequently introduced including regular "crew safety meetings," "daily foreman safety talks," superintendent's monthly safety meetings, and planned personal "safety contacts" with shift bosses (Ashcroft and Taylor, 1982; INCO, 1980, 1985b; Ross, 1984). Worker participation and safety communications were also encouraged within the context of

safety suggestion programmes and improvements in existing safety compe-
titions (prizes for no accidents and low mine accident rates), safety poster
programmes and safety newsletters (Ross, 1984; Strutt, 1986).

Coupled with the new HRM emphasis on worker involvement and participation, these OHS-specific components may help to explain why many
miners in bulk mining (48%) and mechanized jobs (42%) expressed the belief that they had access to information and control regarding health and
safety conditions. However, these mechanisms were not just about making workers feel more secure about their health and safety, they were also
about changing workers' attitudes towards management and their work.
Many managers were also clear that they viewed participative programmes
including audits as means of gaining more control over production and
safety problems in tandem. As one put it: "[W]e are not generating paper safety here, we are collecting information and acting on it to prevent
problems." And indeed, safety participation and reporting mechanisms
were also being used by management to gather information both about
health and safety *and* production more generally, helping them to monitor
and assess the success of the new production systems and technologies.
At the same time, however, these communication and worker involvement
mechanisms increased the direct contact between management, supervision and workers on health and safety issues, which as observed numerous
times, allowed a presentation of management-constructed information
and ideas, including technical and numerical frames of reference with respect to risk assessments.

As the scope of the OHS programmes increased, the use of safety audits
was also greatly expanded (Ashcroft and Taylor, 1982; Guse and Gunn,
1980; Inco, 1985b). Many of these audits were delivered by a so-called independent non-governmental agency – in particular, the Mining Accident
Prevention Association of Ontario (MAPAO) which, although entirely industry controlled, was assigned the role of prevention under the auspices
of the Ontario Workers Compensation Board (OWCB) – now called the
Workplace Safety Insurance Board (WSIB). Along with the long-standing
MAPAO "five-star audit" system which the company had been using since
the 1970s, the company announced a series of internal and joint union/
management safety audits aimed at reviewing safety procedures and standards (Guse and Gunn, 1980; Inco Triangle, 1982). One key example was
the adoption of a packaged safety audit management programme from the
South African mining industry (of all places) called Total Loss Control
(Bird, 1974; see also Russell, 1999). A senior corporate manager defined it
in the following way:

> It is a management technique to minimize incidents. The philosophy
> is by minimizing the little incidents you control the big incidents. It's a
> philosophy, not really any particular objective set of activities. It's the

molding of a person's mind, to structure their thinking to the idea that we are all part of a team that cares about everything that happens in a mine.

As this manager indicated, the approach was intended to change the way that workers think about safety, in part by redefining safety and injury incidents as costs or "losses," so they are seen as being no different from any other kinds of cost, including damage to equipment or material, or the environment or lost production. Changing workers' attitudes towards their *responsibilities* in production and in health and safety is accordingly a key objective underlying loss control. As one senior manager put it:

> If everybody had a good attitude we wouldn't need the OSHE committee and Loss control but we need to develop the attitude to eliminate problems when they occur by the people who create them. Workers and some foremen only do safety because they have to.

While the Loss Control manual emphasizes the need for worker involvement and participation in decision-making as important sources of motivation and commitment to safety, the sections on supervision are largely oriented towards methods of identifying problem workers and getting them to behave in a responsible, standardized and coordinated way (Bird, 1974). Moreover, while workers were being encouraged to take on more responsibility, many of the concrete changes introduced as part of loss control, in particular the new reporting requirements and monitoring activities, reflect an elaborate extension of management surveillance.

There were also other initiatives where the company was clearly seeking ways to fix more responsibility on workers. An excellent example of this responsibilizing discourse was evidenced in the company's reintroduction of a safety check procedure called the Neil George Safety System. This was an old "safety first" programme which had fallen into disuse since the 1960s, but in 1981, it was "reborn," interestingly as a *joint* recommendation of a union/management safety audit (Inco, 1985b). The basic procedure was that workers were to follow a five-point safety check-off using forms called "Neil George safety slips" which miners were supposed to sign confirming that they had "checked" and "certified" their workplaces as safe. As the Neil George system manual points out, the basic safety principle underlying the system is that safety is in the "minds of workers" rather than a function of people–environment interactions (Inco, ND, p. 27). Much of the procedure relates to activities such as housekeeping and ground control actions such as checking and scaling the face and back of one's work area.

One of the ironies here is that this programme was developed for conventional mining operations when miners did indeed exercise much more control. However, as one shift boss observed, "the Neil George is a farce in the new mining system – if they [miners] really followed it, they'd never get any work done." As he explained, the VRM pace is much too fast and areas in which they work are too large to make most of these checks meaningful. Many VRM and mechanized miners were also acutely aware of these contradictions and saw the Neil George system as a blatant attempt to blame workers for accidents, since workers had to certify that their workplaces were safe with their signatures even though they were being told on the other hand that they needed to get the footage or the muck as their first priority. As one long-time union activist put it, "it's just a way of playing the blame game."

Coupled with its loss control discourse, the company also made increasing use of the phrase "safety and productivity go hand in hand." As the president of the Ontario Division of INCO, Mike Sopko wrote in a letter to the Ontario Minister of Labour:

> It has always been our belief that productivity and safety go hand in hand. In essence, it has been our longstanding experience that a safe workplace is also a productive workplace, and at all times, we place safety and productivity as equal partners.
>
> (Sopko, 1986)

Through this discourse, miners were encouraged to embrace the idea that health and safety were built *into* the restructuring process almost as a given or natural consequence of technological change and restructuring. Yet, individual managers acknowledged in their interviews that they understood that there were budget and production priorities which conflicted with safety. As one mine manager put it:

> My job is to try and get the best profit from this mine, but I also have to maintain safety. My biggest problem given the company's situation is to get the dollars. And if I want to argue for some new equipment or improvements, I have to make an argument on the payback [likely improvement in production *and* safety costs within a two year period]. It all has to be costed out to make sure you're getting something for the investment. Otherwise I just have to do with what I've got.

Many elements of the corporate health and safety system can be understood as giving assurances to miners, particularly within the new bulk mining labour process, that engineers, managers, other workers, and control technologies and procedures were competent, sufficient and reliable as sources of information and control over hazardous conditions. This was done not just through management claims of technical control but also by expanding the number of specific technically oriented prevention and monitoring programmes including respiratory- and hearing-testing programmes, the epidemiological research programme, toxic substance control and monitoring programmes, ground control monitoring and research programmes, a respiratory protection programme, and so on (Inco, 1985b). These programmes involved the expansion of medical personnel in some areas, and special engineering functions related to OHS issues – ground control mechanics, air quality and ventilation engineering, in others. With respect to labour consent, the training programmes, the increased emphasis on "qualifications" and stricter procedures for qualifying for jobs, the use of audits, the increased safety inspections programmes, the creation of worker inspectors, the revisions and expansions of standards, the development of health and air quality monitoring programmes and the introduction of management and independent safety audits (e.g. five-star audit programme) were all important symbols of *management* commitment, competency and control over risk.

Just as clearly, however, when workers questioned the validity of these programmes such as management's ground control or air quality monitoring program, as many did, these programmes also became major sources of concern since managers and engineers were perceived as "hiding" something, especially when they acted defensively or coercively against workers seeking answers. As two mechanized miners stated:

> I'm worried about my breathing but I had an awful experience with the company lung test monitoring. I had my x-ray and then they wouldn't tell me what it was. They don't treat you like people. You see people at work and no one seems to know they're sick and then they are dead. It seems we should have more control.

> I complained about black smoke coming out of my scoop and the ventilation guy came down with a draeger meter and he says to me 'we don't test black smoke cause the scientists say it's no bad for you, so I don't check for it. Then I got in trouble for calling him down on my own.

Yet, it is important to remember that these new OHS programmes and technologies were not just about achieving labour consent. They were also substantive efforts to exert more management control over the frequency of costly and disruptive safety incidents. As in any shift to higher-speed continuous production systems, INCO management had heightened interests in preventing accidents or close-call incidents since the resulting delays were much more costly in equipment and lost production than when they occurred in isolated conventional stopes.

Perhaps the best indicator of the increased corporate concern with reducing the frequency of accidents and close calls was the adoption of specific health and safety indicators as part of performance assessment for line management and supervisors. After 1982, the performance of senior and mine management was assessed in reference to accident and compensation rates in addition to other cost indicators. Although never as important as production targets, the evaluations and careers of managers and supervisors were more closely tied to their capacity to control accident rates and/or reported injuries as well as their success in meeting safety audit requirements (Inco, 1985c; Various management interviews). Corporate management also indicated the increased weight it placed on health and safety by including health and safety indicators or measurables as part of its monthly and annual IV reporting procedures (Senior and Mine Management Interviews).

A key problem however was that although corporate management was insisting that productivity, cost controls and health and safety could be readily achieved and balanced, mine managers and shift bosses often saw little alternative in a context of leaner production systems, but to subvert the safety management processes in some way, either through coercion, deception, corruption or collaboration; or, in the end, to play the numbers game fairly loosely, if they were to have any chance of meeting production *and* safety targets. These various accommodations, broken promises and compromises often undermined worker and ultimately supervisor confidence in the integrity of management and the value of these technologies and controls. As one shift boss stated angrily: "Five Star [audit] is bullshit. Garbage can lids are more important than loose [rock falls]. It's just a pat on the back for managers."

As this latter evidence further indicates, there were a number of significant tensions and contradictions in the company's approach to health and safety management which both reproduced and undermined consent. As will be shown next, the mines with the most conflict were often the newer most mechanized VRM mines with the higher production and efficiency quotas (e.g. see mine comparison later).

IV. Controlling the politics of complaint and protest

INCO's OHS history as presented thus far suggests that its health and safety committees and worker representatives had become more significant points of conflict and resistance to mechanization and VRM by the late 1970s and early 1980s. While the events of 1982 (failed strike, shutdown and layoffs) did much to weaken resistance, management also moved in other explicit ways to undermine the growing political challenges coming from worker OHS representatives. This began in 1982 when INCO announced that it was placing a priority on developing better management/union relations *within* the OHS joint committee structure. The company openly acknowledged that despite its earlier efforts to move OHS committees away from confrontation to cooperation in 1975, many of the committees were still major sources of conflict and resistance (INCO, 1980b; 1981). As reported by union representatives, this meant increased management efforts to moderate labour/union conflict by exchanging more information, consulting more informally and regularly, developing some joint training, and establishing closer personal relations with union OSHE representatives.

Consistent with their general labour relations strategy, there were pointed efforts to clearly depoliticize the health and safety committees and the network of worker representatives. The company did this in various ways but what was particularly blatant was its attempt to get the union to agree that the health and safety representatives could not hold any other positions in the union, especially a steward position. Although never entirely successful in gaining union approval for the exclusion of stewards from OHS rep. positions, much of the effect in terms of depoliticizing the health and safety committees was still gained by excluding OHS committee participation in any grievance disputes and by creating an extensive set of technical responsibilities and practices for OHS representatives, making it almost impossible for stewards to simultaneously function as OHS representatives (Hall et al., 2006, 2016). By 1985–86, a separation of OHS representatives from the union and the steward network was occurring, and, as the more militant stewards retired or moved away from OSH committees, a new generation of OHS representatives took positions in committees without experience as stewards.

Through their certification training for representatives, the government and like minded corporations such as INCO promoted what I and colleagues have called elsewhere a "technical-legal" (TL) orientation to representation (Hall et al., 2006; 2016). Emphasizing technical skills and basic legal knowledge, the training reinforced the legislative and policy framing of committees and representatives as auditing structures and advisers with highly bureaucratized inspection and reporting practices. As observed at INCO, managers often used the committees, the increased direct personal contacts, and joint health and safety training programmes to immerse representatives in cooperative and technical discourses of risk assessment, cost-effectiveness, loss control and efficiency. By the time I had completed observations of committees and interviews with OSH co-chairs and representatives in 1986, it was clear that most of the committees and the worker reps. were functioning in a limited passive capacity as technical observers and advisers, largely mimicking pro-management and collaborative ideas well beyond the norm for miners more generally. As one of these depoliticized TL representatives stated:

> Management knows what they are doing. They put something in here they don't want it to hurt anybody. My job is to look for the little things like glasses or equipment lying around.

When in the workplace, worker representatives were accompanied by management or shift bosses limiting their capacity to independently interact with workers. Inspection requirements meant a preoccupation with technical duties defined essentially as processes of data collection. As another worker co-chair stated, "we find safety violations and write them up, that's really what we do." As acknowledged by several worker reps., they saw their primary reporting responsibility as being to management rather than to the union or workers.

To the extent that some OHS representatives retained a political view of themselves as *worker* representatives, management also employed various tactics and strategies to limit the ability of those OHS representatives to get things done. One common management tactic was to insist that worker representatives dealt with each issue or complaint as a distinct and isolated problem unrelated to the rest of the mine or the rest of the company. Thus, OHS representatives were faced with constant efforts in which they had to slog through each specific case of a problem without ever really getting at the source of the problems and without dealing with them in a systematic way (Elefterie, 2012; Facey et al., 2017; Rae et al., 2018). As one politically oriented activist (PA) put it:

> You need organized and planned action but all we do is fight fires. There are plenty of big things we should be doing underground – overhead protection for scoops, moving to electric scoops [from diesel], slowing down the [pace of] of vertical retreat mining...that would make a difference... but we spend our time checking fire extinguishers.

Workers too would notice this and express frustration when trying to gain changes through the committee. As two miners reported:

> I complained to my rep about a valve in an ore pass and they fixed it eventually. But the problem was in every ore pass in the mine and they didn't touch anything else.

> You can't rely on them [OSH reps] – they're useless. The company is smart and picks out people who aren't too bright, who believe what they are told.

The accounts of some workers and activists also revealed that management tactics sometime backfired by politicizing workers, OHS representatives or entire OHS committees, which in turn led to a significant groundswell of support from workers in the form of refusals, complaints, reports to media and sabotage. During the two years of my research, two of the six committees had major disagreements between management and worker representatives, prompting a number of actions by the representatives and the workers, including the organization of slowdown campaigns, coordinated work refusals and slowdowns, and repeated calls to the Ministry of Labour and media involvement (e.g. Pender, 1986a,b,c; see discussion of South Mine later in this chapter for further illustrations). Not surprisingly, in these contexts, management often employed threats and reprisals to shut down these protests including the tactic of transferring an entire committee to different mines.

Again, this recognizes the contradictory features and effects of management control strategies. To the extent that management was successful in co-opting or coercing the union and OHS representatives to "cooperate," there was also enhanced potential for other workers to construct the union and its representatives as corrupt, powerless or incompetent. While these assessments discouraged some resistance, they also weakened the potential hegemonic impact of company claims of union involvement in health and safety decisions, and ultimately undermined consent as workers realized that neither management nor union assurances could be trusted. As two miners stated:

> There is some fighting back but they're [local union] letting too much slide. Some of the stewards are burning the candle at both ends. People are starting to talk about the union, saying the company is laying down extra money for them [union officials] to back off and leave things alone.
>
> (Miner)

> I advise a lot of people to skip the grievance because I know the union hall will drop it. There's no use getting a man's hopes up. If it's a health and safety thing, I call the ministry of Labour directly.
>
> (Steward/OHS rep.)

While the integrity of these TL reps. was important, their competence was also often questioned, especially by the trades workers such as electricians and many conventional miners. From their perspective, these reps. did not have the background or knowledge to tell them what was safe and what was not.

The political impact of law

As one of the above quotes suggests, the union is not the only means through which workers and worker representatives pressured management on health and safety issues – there were also the rights provided in the Ontario Occupational Health and Safety Act (OHSA, 1979, 1990). As the following illustrates, workers and workers representatives often outlined how they had been able to use their right to refuse unsafe work, or a threat of refusal, to challenge OHS conditions.

> Once I complained about a leaking and fuming scoop. First I complained to my shift boss and nothing happened so I went to the superintendent and nothing happened. So I threatened to pull a Bill 70 [work refusal] and then finally they did something.
>
> (Miner)

When we get a major decision or win an appeal on some work refusal, the company knows we are around. I have to believe that it makes a difference because the way things are [with jobs], we need the ministry badly. We're not going to change the world but bit by bit something comes out of it.

(OSH Rep.)

On the other hand, the adoption of the "internal responsibility system" (IRS) philosophy meant that Ministry inspectors rarely acted as the enforcement arm of the state, thinking of themselves more as "partners in prevention" (Bittle and Snider, 2015; Snider, 2009; Storey and Tucker, 2006; Tucker, 2003). Two Ministry mine inspectors confirmed in their interviews what they saw as the limitations of their enforcement powers work:

If the company wants to hide something from me, I won't know about it unless some OSHE guy [worker rep.] pushes it. I rely on them and if the company tells me it's going to fix something, I don't write an order. I leave it up to the OSHE to tell me if it hasn't been done... I don't follow up orders, I get a note from the company saying it's done or will be done. I leave it to the internal responsibility system.

The union says that the legislation is not being enforced and in many ways that's true. Our tradition is not to come down with a hard fist but to give time [to companies] to correct things...I have a major workload problem and with designated substances, I can't do it all.

Accordingly, worker representatives and workers were often disappointed and thus discouraged from raising issues with the Ministry. Moreover, since INCO was relatively successful in bureaucratizing and pacifying most worker representatives and joint committees by 1985, they were unlikely to raise issues with the Ministry even when there were substantial problems.

Yet, as I've noted, there were some important points of resistance that emerged in the different mines (also see Mine Comparison below) and, at least occasionally, some of that resistance was recognized by Ministry inspectors in orders against the company. These local points of resistance also reflected a larger province-wide politics and a burgeoning OHS activist movement against the adequacy of the new OHSA and its weak enforcement of workers' rights (NDP, 1983, 1986; Storey and Tucker, 2006). These larger political developments ultimately led to Bill 208, the first major reform of OHSA, as well as several government announcements of increased enforcement (Hall, 1991; OMOL, 1990; Storey and Tucker, 2006). The controversies and public pressure surrounding the law were also provoked by persistently high injury and fatality rates in mining, which, as noted, resulted in several public inquiries which were very critical of the industry (Burkett, Riggin, and Rothney, [Burkett Report], (1981), 1981; Standing Committee on Resources and Development, 1988; Stevenson, 1986). It may well have been this political controversy as much as anything that pushed INCO to introduce major changes to its corporate OHS programme through the 1980s. Certainly, the company's submissions to these inquiries and commissions, and the government during the Bill 208 review period show that the company was seeking to ensure that the government would continue with its IRS approach without further empowering workers and without imposing more restrictive regulations or enforcement (Inco, 1980, 1984). For the most part, the company and corporations more generally largely succeeded in limiting the reforms and any shift to meaningful deterrence measures (Storey and Tucker, 2006). Nevertheless, these outside political developments provided health and safety activists with leverage they would not have had otherwise. It is likely as well that this political pressure pushed senior management to try tomake its cooperative OHS approach appear more substantive – that is, it became important politically to offer meaningful compromises to the union, health and safety representatives and workers when it became clear that their concerns were generating growing public conflict and attention. This also offered opportunities for the more enlightened line managers within INCO to pursue some of the new management strategies being promoted at the time by business schools and governments (e.g. the QWL movement).

The evidence thus far suggests that corporate and state efforts to control workplace hazards and worker responses to those hazards were pulled and pushed in contradictory ways, reproducing consent and compliance overall but also sustaining, and in some contexts even accentuating conflict and resistance. To demonstrate more fully the contradictory ways in which production and management restructuring and management control strategies can combine to both reproduce and undermine consent and resistance, I end this chapter by briefly comparing the health and safety politics in three mines.

A comparison of consent and resistance in three mines

The three mines were McCreedy, South and Levack. In terms of working conditions and labour process, South and McCreedy were very similar with high levels of mechanization, a system of ramps heavily reliant on diesel trucks and scoops, and entirely VRM production.[1] In terms of health and safety risks, there is no clear evidence that one mine was substantially worse than the other. Levack was an older mine with a mix of conventional and VRM. Although the reported injury rates were somewhat higher at South, serious fall incidents, the reporting of which are less influenced by workplace politics, were much higher at McCreedy and even higher at Levack (Table 5.1). There was also some evidence that McCreedy and Levack had higher dust levels (Inco//Local 6500, 1982–84). Neither mine had a fatality during

Table 5.1 Mine accident comparison

	Lost Time Injuries per 100 Employees/Yr[a]				Ground Fall Incidents[b]	Fatalities
	1982	1983	1984	1985	1983–86	1980–87
South	20.9	10.3	6.5	7.2	2	0
McCreedy	15.8	12.5	10.9	4.9	10	0
Levack	18.1	10.0	0.3	5.6	56	6

a Inco, Ontario Division, Mine Injury Statistics, Provided on request, August 11, 1986.
b Ground Fall Occurrence Reports, 1983–86.

Table 5.2 Proportion of miners concerned about risk levels

	South (%)	McCreedy (%)	Levack (%)
Dust	68	52	40
Diesel	72	55	34
Oil Mist	40	17	21
Ground	72	48	41
General	68	62	39
Proportion Judging Frequent Exposure to High Risks			
Danger	48	28	40
Unhealthy	56	48	50

McCreedy N = 29; South N = 25; Levack=71.

this period but Levack Mine had six fatalities in three separate events. My observations and interviews with miners, health and safety representatives, shift bosses, company safety and ventilation engineers, some of whom had worked in both mines, tend to reinforce the conclusion that McCreedy and South were largely equivalent in the risks they presented to miners, while Levack was the more dangerous of the three.

However, although South and McCreedy Mines were relatively equivalent in terms of the labour process and risks, miners in McCreedy were somewhat *less* likely to judge their general and specific working conditions as presenting significant risk to their health and safety than South miners (see Table 5.2). Levack's miners were the least concerned about specific hazards but were similar to South in their characterization of their work as dangerous. As compared to the conventional mine (Levack), miners in South and McCreedy were more likely than Levack miners to contest their conditions by complaining to supervisors, managers, and union representatives. However, South miners were more likely to exercise their legal right under OHSA to refuse unsafe work (see Table 5.3). If we look at work

Table 5.3 A mine comparison of actions by miners

Actions	Proportion of Miners Reporting Action		
	South (%)	McCreedy (%)	Levack (%)
Complaints	68	66	46
Complaints about Dust	52	35	23
Complaints about Diesel	44	34	10
Arguments	48	40	40
OSHE Rep.	76	76	36
Work Refusal	92	72	43

refusals as the end product of a series of complaints and actions, which is the way it is usually described by the miners, the differences in work refusals are particularly indicative of a greater degree of sustained conflict at South Mine than McCreedy. In 1984–85, there were only 25 refusals in the entire Ontario mining industry, representing almost 47,000 workers. Yet, over 25% of those refusals were at South Mine alone (Ministry of Labour Annual Report, 1985). During that year, South had six refusals and four formal ventilation complaints requiring Ministry intervention. McCreedy had none.[2] In 1985–86, South Mine increased its refusals and ventilation complaints to 15 while McCreedy had only one. According to interviews with union representatives and supervisory personnel in these mines, individual complaints and demands regarding working conditions from miners at South Mine were submitted at a substantially greater rate than any other mine, approximately twice the average rate per month.

Within South Mine, the union OHS representatives were also considerably more militant both within and outside committee meetings, ideologically and in terms of their conduct. Confrontations within the committees were considerably more frequent and intense at South Mine. Union activists at South were also more likely to engage in organized campaigns using a variety of tactics aimed at broader problems in the mine as a whole (e.g. campaigns for improved scoop maintenance schedules and safety booths for remote operators). Representatives were actively encouraging and organizing workers to use their refusal rights to pressure management. Less than coincidently, when there was a new round of transfers in 1986, the union OSH representatives at South were split up and isolated across all the other mines in the Sudbury complex.

There were also other indications of more direct job action by South Mine crews and individuals, including "illegal" work stoppages, slowdowns and sabotage. This was also the only mine from which I received frequent reports of sabotage in interviews with management, supervision, union representatives and miners. While this evidence is harder to track down given the lack of records, there were two acknowledged labour conflict "hot spots" according to senior union and company management – South Mine and the nickel refinery – which was further confirmed by a review of grievance records showing that South mine had twice the rate of grievances in Mc-Creedy or Levack, many of which were addressing disciplinary actions by management (USWA, Local 6500, 1982–86). By 1987, relations and production shortfalls had gotten so bad that corporate management replaced the mine manager and superintendent following the receipt of an outside consultant's report.

This does not mean that McCreedy had no significant conflicts or resistance. Indeed, McCreedy exhibited some of the best examples of large-scale collective protests in 1983 and 1985 when workers organized two slowdown campaigns to protest concerns about diesel fumes and dust. These reactions offer further support for the argument that the mechanized and VRM processes werecreating more grist for significant labour conflict. However, as acknowledged by union officials and managers, compromises at McCreedy were more readily worked out at the informal level and through the health and safety committee than was the case at South, and rarely reached the crisis point of outside intervention (i.e. senior union officials, government inspectors, outside consultants). Levack, on the other hand, exhibited even fewer signs of resistance, again with its miners tending to resolve concerns directly on their own and informally with their supervisor.

Explaining the difference

I would argue that the different OHS politics in these mines can be explained as follows: First, as mechanized VRM mines, management, supervisors and miners at South and McCreedy were under considerably more corporate pressure to produce at higher rates of productivity and efficiency than the more conventional mines such as Levack. Second, as newer more productive mines, South and McCreedy were not under the same threat of closure as the older conventional mines, which meant somewhat less insecurity shaping consent and compliance. Third, as mechanized and VRM miners, the workers in McCreedy and South were less positioned within the labour process than conventional miners at Levack to negotiate limits to management demands and the risks they were being expected to take. Fourth, with less control and power within the transformed labour process, this also meant a greater tendency among miners in the VRM mines to externalize responsibility for judging and controlling risk. Accordingly, consent and conflict then turned more on whether miners perceived and trusted other means of acceptable judgement and control.

The crucial difference between South and McCreedy was that Mc-Creedy had the more elaborate team-based management system which gave workers relatively more power and control even within the context of the more limited skill requirements of VRM. This not only limited the powers of mine managers to impose hazardous conditions but also gave miners a greater sense of knowledge, control and responsibility. They had more room to negotiate compromises at McCreedy, even in the face of intensifying production pressures, narrowing skills, new technologies and corporate expectations. By reorganizing the political structure of the mine into semiautonomous work groups with collective decision-making and bonus systems and open-ended job descriptions, McCreedy was more able to maintain the sense of worker involvement and responsibility for working conditions and to establish stronger informal work-group pressures and obligations for meeting production and cost objectives (Boyd, 2003). These experiences of collaboration also reinforced management's claims that the labour process and labour relations changes were linked in a positive way to increased control and security. This further strengthened and legitimized the claim that management and workers had common interests which could be dealt with in a cooperative and non-confrontational manner (Belanger, Edwards and Wright, 2003). Paradoxically, the team system at McCreedy also enabled workers to move to a collective slowdown response when they were unable to work out OHS or other production issues through persuasion or other measures.

At South Mine, however, with its traditional top-down management and supervision structure, and no apparent manager commitment to group or team participation and input, the labour process empowerment of mine management was accentuated, while shift bosses lost room to manoeuver when trying to negotiate miner commitment to production targets. Two foremen from South and McCreedy provide a good contrasting description of the differences in supervision and planning:

> A Shiftboss at South: The day to day planning is being done by the engineering department which is directly controlled by the superintendent. It is extremely frustrating and very dangerous situation. To me there's a direct relationship between the frustration, the bad attitudes and the accident record here. Workers are not being left alone to do their job...I'm getting pressure all the time from [superintendent's name] to make my men to do this or that without proper planning – he threatens me.

> A Shiftboss at McCreedy: We are discussion oriented not order oriented. The way we do things is not to tell people what to do. We discuss things with them, ask them what they are doing and how they plan on doing it. They feel better and get the job done. This is my beat and it's up to me and my crew to decide how we are going to do the work.

Indeed, at McCreedy, many of these issues were resolved among the leaders and workers without involving supervisors or managers.

As Littler and Salaman (1984) argued, the power of any controlling ideology lies in relations which at least partially confirm management's claims (p. 65). However, if day-to-day relations fail to provide confirmation and consistently contradict the ideological discourse, control ideologies and discourses are not only weakened, they can become significant sources of conflict. Within the context of more advanced technologies and intense production pressures, the McCreedy system provided a much stronger relational basis for the development of consent-based production politics and ideologies, and thus was more effective in mediating and moderating the effects of mechanization on worker control and power, and more consistent in supporting INCO's discourses on cooperation, responsibility and teamwork. In Bourdieu's terms, the rules of the game were more readily confirmed. While the maintenance of traditional structures of power and authority at South Mine did not guarantee major conflicts and confrontations, as was clear in the Levack case, it did increase the potential for disrupting risk habitus and breaking down ideological consent around issues of knowledge, control and responsibility. Moreover, in the context of production and cost pressures, the lack of participative constraints on management increased the chances that managers and supervisors would use their authority to try to impose new conditions of risk-taking using coercion and threats, which is exactly what happened.

Summary and conclusions

The chapter has demonstrated that management reproduced consent in the context of labour process restructuring through explicit human resources and OHS management strategies, many of which emphasized worker involvement and management communication. At the same time, the company sought to exercise increasing control over production and the costs and production disruptions of accidents through a variety of safety technologies, programmes and discourses, some of which were aimed quite explicitly at controlling or influencing worker thinking and behaviour. This suggests that a dual effort to exert more management control *and* to reconstruct worker responsibility within a more sub-divided, deskilled and technologically controlled labour process operated in concert to reproduce worker consent by communicating to workers the substantive management commitment to health and safety, while also giving the workers some sense of control and of being informed, enhancing their willingness to accept management claims of safety or risk-taking necessity. At the same time, these communication and participation mechanisms became important sources of information and intelligence for management in monitoring and centralizing its efforts to control disruptions to the production process.

However, as I've tried to show, restructuring efforts to reduce costs and intensify production were also critical sources of tension undermining central management efforts to reproduce consent and control risks. Moreover, the HRM/OHS strategies, especially if implemented poorly or inconsistently, often introduced significant tensions which undermine consent and contribute to resistance, which then fueled additional adaptations or changes in management discourses and strategies. As we saw in the mine comparison and other data, managers also came with different orientations or habitus regarding employment relations which played big roles in whether participative programs yielded consent or conflict. Objective conditions for management and workers were also critical. Consent was often contingent on whether managers and supervisors were able to negotiate new terms and rules of the game with workers within the new labour process (Bourdieu and Waquant, 1992). As the comparison of South and McCreedy Mines indicate, what may be particularly critical is whether power and authority is sufficiently restructured by the corporation, and to some extent the state, to limit management powers and give workers some real control and influence (i.e. capital) with which to negotiate the new rules of the game Hegemony and a consensual-based risk habitus were accordingly dependent on buffering workers from the arbitrary and coercive actions of managers or supervisors who were responding to production and cost reduction pressures (Boyd, 2003).

The chapter also shows that both the union and the state contributed in somewhat contradictory ways to the politics of health and safety. As limited as they are, the regulations, workers' rights and Ministry enforcement often provided a basis for resistance, and limited management's power, moderating extreme conditions and/or coercive actions that ultimately hurt the company by leading to major incidents or a complete breakdown in consent and compliance. The actions of senior corporate management suggest that they understood that the best way of limiting the state's involvement and public attention more generally was to make the internal responsibility system work at least at some level. In creating a technocratic process that was predictable enough to control and moderate health and safety costs and disputes, and still providing flexibility for risk allowances, some managers and supervisors began to realize the value of substantive worker participation, encouraging elements within the company to incorporate participation principles much more broadly than just the joint committees. It was in these contexts of greater management commitment to worker involvement where the emphasis on the internal responsibility system was more effective in depoliticizing health and safety within and outside the workplace. However, the fact remains that the law gave workers certain standards and rights which they could and did use to challenge management again within the limits of management interests in overall productivity and cost control. Some joint committees and politically active representatives continued to be major thorns in management's side with some very positive impacts on working conditions. And, in the context of all the company cutbacks and worker insecurity, the role of the state, even a neoliberal one, in mediating this conflict was arguably enhanced quite significantly, at least temporarily, while the various changes were being implemented.

The contributions of the union were also quite complex and contradictory (Walters, 1996). By the mid-1980s, the local union was often a vehicle for spreading the management discourse on the crisis and the need for greater cooperation and collaboration. There were indications that the union, or at least some union and OHS worker representatives, were actively discouraging activist and worker complaints and challenges on health and safety issues. More importantly perhaps, as an organization, the union leadership was doing very little to prevent the depoliticization of health and safety issues within the workplace. Most mine health and safety committees were effectively controlled by management, with representatives operating more as extensions of the corporate health and safety programme rather than as politically active worker representatives. Workers were as inclined to hide things from these representatives as they were from management, and would rarely consider using them for advice, for information or to challenge management. This clearly had a significant negative impact on workers' power and their willingness to resist.

Yet the union and some of its worker representatives and stewards were often forces for positive change and important sources of resistance to OHS hazards. Some union stewards and health and safety representatives, most notably in South Mine, were extremely active and effective in politicizing and pushing health and safety issues. Some were also quite knowledgeable through self-education and union training, and pushed by these activists, the local union was sometimes an important source of challenges to existing definitions and management assessments of risk, often at considerable cost to the activists in question. Government inspectors also acknowledged in interviews that they paid more attention to inspections and complaints in unionized mines simply because they knew that the union would demand action if something happened. Most managers and supervisors also understood that it was almost always better to avoid union involvement than to provoke it, which was part of the reason why managers were often willing to negotiate and compromise on conditions, and/or seek worker input on alternative suggestions for doing something. In other words, like state regulations and the structuring of management authority, the union was important because it placed substantive limitations on management power, provided a level of employment security and gave workers access to independent knowledge and negotiating power. Once again, however, I stress that a habitus of overall worker consent and compliance to considerable risk was still being sustained in the majority of its mines, suggesting that these sources of worker empowerment and resistance were simultaneously operating as channels for continued management control.

Notes

1 Note that this meant that miners in VRM miners were either classified as VRM miners or as mechanized if they were involved in drift and stope development work.
2 These statistics were compiled by me using the joint committee minutes for INCO's South, McCreedy and Levack Mine, 1984–86).

Chapter 6

The transformation and fragmentation of Canadian agriculture

A "farm crisis"

In the decades following World War II, Canadian grain and livestock production became increasingly capital intensive with larger, more mechanized and more specialized farms entirely dependent on expensive chemicals and hybrid seed and feed inputs (Basran and Hay, 1988; Winson, 1992). Keynesian state intervention during the 1950s and 1960s in commodity markets along with farm support programmes had institutionalized continued growth in productivity and output by supporting increased chemical inputs, mechanization and intensive exploitation of soils. In the sociology of agriculture literature, this emphasis on government support for high levels of productivity is often called "Fordist productivism" (Pelucha and Kveton, 2017). With the increased production and the increased concentration of food processors, seed and agrichemical companies and retailers, both in Canada and internationally, the prices paid to family farmers for their production did not keep pace with rising input costs, which in combination meant a sharp falling rate of income/profit for farmers (Winson, 1992, p. 90). This concentration also yielded extremely close connections between agribusiness interests, farmer organizations, and state agri-service and policy providers (Hall, 1998b; 2003).

In the 1970s, the growing contradictions of this Fordist system were partly relieved by expanding export markets. Farmers were encouraged by initial recessionary rises in food prices, low interest rates, inflating land values and state export policy to invest heavily in land purchases and new equipment. As a consequence, there was a major increase in their debt load as many farmers sought to enlarge, mechanize and modernize (Ashmead, 1986; Keating, 1987). These developments continued to fester until the 1980s when Canadian agriculture entered a period of much more serious crisis (Basran and Hay, 1988; Lind, 1995). Overproduction and price instability became a problem when developing world debt undermined export markets and prices. Interest rates also increased dramatically further worsening the debt problem and undermining land prices. Net farm incomes as a share of gross farm receipts declined from 36.3% in 1961 to 21.5% in 1981 and then dropped to 14.5% in 1983, while debt ratios in Canada increased from 3.7 in 1971 to 5.4 in 1982 (Gertler and Murphy, 1987, pp. 246–247). Many lost their farms or were forced to seek off-farm employment to survive (Ashmead, 1986; Lind, 1995).

This farm crisis created substantial political and fiscal problems for the Canadian government. On the political front, there was increased militance among farmers including the rise of the farm survivalist movement and membership growth in the increasingly ecologically militant and social justice-oriented National Farmers Union (NFU). From both conservative and militant ends of the political spectrum, there were also demands for major increases in government financial assistance programmes and price supports (Gertler and Murphy, 1987). These problems all represented significant challenges to the long-standing federal policy emphasis on increased production or "productivism," based on the expansion of export markets to offset price declines. These pressures fueled a policy shift in several Western countries including Canada from Fordist productivism to what some have called a post-Fordist neoliberal market productivism or "neoproductivism" which emphasized further trade liberalization, adoption of risk and flexible farm management practices and reduced government supports (Tilzey and Potter, 2008).

As recognized by several scholars in Europe and elsewhere, the restructuring of agriculture and agricultural policy from the 1980s has also been strongly affected by environmental concerns, both from within and outside agriculture, leading some to argue that agriculture and rural policy was moving more to a form of "post-productivism" (Wilson, 2001), where there was a greater focus on land conservation, labour issues, food safety and environmental protection. Sustainable and organic agriculture were supposedly the key examples of this new policy direction. In this chapter, I take the position that in Canada at least, state policy and agribusiness sought to integrate a vision of sustainable agriculture and organic farming consistent with neoliberal or market neo-productivism (Hall, 1998b, 2003; Marr, Howley and Burns, 2016; Wilson and Burton, 2015), with significant contradictory effects on the development of farm health and safety and OHS governance within the industry.

The challenge of sustainable agriculture

During the 1970s and 1980s, alternative visions of agriculture emerged to challenge many of the productivist assumptions underlying conventional agriculture (Macrae, Henning and Hill, 1993; Manning, 1986). Rather than the emphasis on capital-intensive, large-scale, highly mechanized, high chemical input farms, proponents of alternative or sustainable agriculture (SA) advocated the elimination or reduced use of synthetic farm chemicals, smaller farm units and appropriate technology; the conservation of finite resources; greater farm and regional self-sufficiency with more direct sales to consumers; and minimally processed foodstuffs (Beus and Dunlap, 1990). Although there were important variations in the definitions of alternative and sustainable agriculture from the outset, these initial formulations were widely understood and presented as substantial moves away from the dominant conventional productivist regime in agriculture (Buttell, 1993; Manning, 1986). As such, while the SA movement included supportive agricultural scientists and agro-ecologists, it was initially a grassroots movement emerging at the level of the farmer and consumer communities.

In Canada, the Ecological Farmers Association of Ontario (EFAO) and the Canadian Organic Growers (COG) were both substantial grower- and consumer-based organizations which started work in the 1970s and 1980s with a primary emphasis on organic gardening and farming. Although the Ontario Federation of Agriculture (OFA), the main farmer organization in Ontario, steadfastly resisted organic farming, the NFU, which was particularly powerful in western grain-growing provinces, adopted a number of policies supporting the development of alternative and organic agriculture in the 1980s and 1990s (more on this later). The development of an urban-based environmental movement, severe pollution problems in the Great Lakes arising from heavy fertilizer use, and the increasingly politicized recognition of the health and ecological effects of pesticide and fertilizer use were also central forces in pushing the development of these alternative ideas (Hall, 1998b).

While concerns about the environmental effects of conventional agricultural chemicals and cropping and tillage practices were central emphases of sustainable agriculture from the outset, there was also a social justice component which carried over from the family farm activism of the 1970s which was also decidedly anti-corporatist (Hall, 1998b, 2007; Hall and Mogyorody, 2001). This linked sustainable agriculture and organic farming to the maintenance of the family farm and rural communities, and cast sustainable agriculture as a challenge to corporate industrial agriculture and the globalization of agricultural markets (Buttell, 1993, p. 22). This included a greater concern with the plights of farm workers, general labour practices and occupational health and safety. Within the Canadian context, a good illustration of this challenge was evident in Vision 2000, a 1990 policy document on sustainable agriculture sponsored by the NFU and the Catholic Rural Life Conference (CRLC). There were two major sets of sustainable objectives established in the document; one was "environmental," and the other was the protection of the "family farm and economic security." To achieve these objectives, the document called for an end to farmers' dependency on chemical-based production, endorsing a major state and research investment in organic and low-input farming methods. The NFU/CRLC policy document also directly challenged the central elements of neoliberal

agricultural policy popular at both the federal and provincial levels, by calling for an end to the cheap food policy and the export dependency policy,
and an expansion of supply management systems and price controls on all
commodities.

> The Canadian economy must be a protected economy or be aligned
> with a trading bloc with the same aspirations. The Canada-U.S. Free
> Trade Agreement promoted liberalized access...which is in conflict
> with our policy objectives of agricultural self-reliance and domestic
> priority... Economic security of the family farm must be ensured since
> it is only through the family farm that the land and its people can be
> maintained. There can be no greater indictment of our system than
> the loss of many thousands of farm families who were demonstrably
> the most efficient producers. While this has shown the failure of the
> 'free market' system, it has also exposed the economic system as being
> void of any social dimension...Supply management must become more
> prominent.
>
> (NFU/CLRC, 1990, pp. 2–3)

Solving the crises: "neoliberal market productivism"

During the 1980s, Western governments including Canada began to widely
advocate neoliberal governance policies as solutions to the agricultural crisis sectors (AgCare, 2001; Agriculture Canada, 1992; OMAF, 1991; Ontario
Roundtable on the Economy and Environment, 1991). Although export
market promotion had been part of the state agricultural policy for some
time, this broader policy shift to neoliberalism and globalization greatly
intensified the push towards the deregulation and de-subsidization of agriculture (e.g. 1995 Uruguay Round of GATT negotiations on farm trade).
Meeting global competition especially from lower cost Third World producers was key goal to this strategy but the means for achieving that goal
revolved around a call for new approaches to farming that could simultaneously meet the environmental and economic challenges and solutions within
a single state policy framework. As documented elsewhere (Hall, 1998b), in
Canada, the central elements of this strategy were outlined in a key 1989
federal agriculture policy statement:

> The vision which the federal government has for the future of agricul-
> ture rests on four pillars – more market responsiveness, greater self-
> reliance in the agri-food sector, a national policy which recognizes
> regional diversity, and increased environmental sustainability.
>
> (Agriculture Canada, 1989, p. 34)

The first two pillars in the government's 1989 policy statement were particularly crucial in signaling the major neoliberal shifts in both production and marketing practices. Similar to the shifts in thinking within the mining sector, market responsiveness spoke specifically to the need for more flexibility in production – the capacity to rapidly adapt to *global* changes in the marketplace (Hall, 2003). As stated in the government policy paper: "More market responsiveness means moving quickly in adapting our production and marketing systems to respond to new market opportunities" (Agriculture Canada, 1989, p. 35). Concretely, this meant first that farmers (again like miners) were encouraged to become more *flexible* producers, capable of shifting their production from year to year to different crop combinations and moving into new (usually global) markets as they developed (Hall, 2003). Similar to the push within INCO to exploit rare-metal opportunities, agricultural policy emphasized expanding into certain smaller but higher value-added crop markets, such as speciality kinds of soybeans for the Japanese market. Although the government did not initially encourage organic farming as one possible avenue for doing this, this eventually became an element of state policy, especially in certain provinces such as British Columbia (BC) and Quebec (Hall and Mogyorody, 2001). As part of this orientation to global markets, farmers were also encouraged to develop more comprehensive quality controls in order to meet global and export requirements, which translated into proposals for developing and adopting ISO standards and auditing processes for agriculture, and eventually stimulated state involvement in organic standards and the setting of health and safety requirements such as protective clothing (ISO, 2017).

Second, farmers were encouraged to find new ways of lowering costs on the farm rather than simply emphasizing increased production which has been the emphasis in the old productivist model (Hall, 2003). A joint federal/provincial document, endorsed by the various components of agribusiness, was released entitled "A National Agriculture Strategy." (Agriculture Canada, 1987) put it this way:

> In Canada, the dominant 1970s pattern has been for production to exceed demand, and Canadian agriculture has been beset by the high cost of production relative to farm income...Greater production efficiency together with the development and production of products to meet specific consumer demands will be essential.
>
> (Agriculture Canada, 1987, p. 2)

"Smart farming" was redefined as farming with less – in effect, an agricultural version of lean production where farmers sought to maximize cost-efficiency rather than just crop yields (OMAF, 1993). Along with government encouragement, there was a major shift within the farm media towards the promotion of lean production ideas and methods (e.g Ontario Farmer's Western Edition, 1993; Voice of the Essex Farmer, 1992; Ontario Soybean Marketing Board Newsletter, 1988–93; Essex County Soil and Crop Improvement Association Newsletter, 1989–93; Voice of Essex Farmer, 1990–92).

Another aspect of the risk management language that developed within this governance model was firmly in the managerialist camp in the sense that farmers were told that more knowledge-intensive *management* practices would be the foundation of their competitiveness. Several farm-specific programmes were created and promoted with specific cost-saving strategies in the areas of fertilizer management, pest management, human resources management and so on using a more explicit cost-efficiency and cost-benefit logic (Hall, 1998b; 2003).

A third key emphasis of the new policy was classic neoliberal thinking – the removal of inter-provincial and international trade barriers (Agriculture Canada, 1989, p. 8). This focus on trade barriers in turn introduced a central concern with deregulation, but this focus on deregulation was developed substantially around a discourse of "self-reliance" (Hall, 1998b; 2003). However, the issue of deregulation as constructed within the concept of self-reliance was not just about eliminating traditional trade barriers such as tariffs, but also the elimination of government farm support programmes which were presented as discouraging farmers from being cost-competitive and from seeking out new markets (Agriculture Canada, 1989, p. 10). Farmers were encouraged to understand their ability to achieved success without government supports as revolving around tighter m*anagement* control over their farm's production and marketing practices:

> The importance of our export markets to the strength of our industry, and a continuation of strong global competitiveness means that we will have to ensure that market signals are transmitted to individual entrepreneurs. Programs, policies and traditional industry practices cannot distort these signals which are critical if the industry is to respond aggressively. We also need to work at ensuring that Canadian farmers have the *highest level of management skills*...to allow them to be as efficient and competitive as possible.
>
> (Agriculture Canada, 1989, p. 27)

As happened in Europe and elsewhere, the move to deregulate agriculture in Canada was limited in part by farmer and consumer resistance, and some subsidies and import restrictions remained an integral part of the system. This was especially the case in the dairy sector, where a supply management system is maintained to this day in Canada. The sale of wheat also remained under the control of the federal government–established Wheat Control Board for some time, although this monopoly was ended in 2012.

Soybeans, a key export product grown widely in south-west Ontario where my research was done, was not regulated. Yet, it was also clear within the context of Canada's negotiation of the North American Free Trade Agreement and its participation in the 1990–92 GATT negotiations that the Canadian government was continuing to press towards a freer trade regime in agriculture into and through the 1990s. As a government press release put it:

> Public assistance to the economy has been reduced. Under WTO Agreements and strong internal budgetary constraints, expenditure on agriculture has been reduced by 20 per cent since 1994, mainly because of the elimination of grain transport subsidies. The impact of such cuts has been offset by strong export growth [within a] ...more market-oriented environment for grain and major livestock industries.
>
> (Canada, 1996)

Reconstructing sustainable agriculture: neo(liberal)-productivism

The early grassroots' versions of SA as described earlier can be characterized as a direct challenge to the neoliberal globalization discourse in the sense that they were explicitly criticizing the economic, social and environmental impacts of capitalist agriculture (Beus and Dunlap, 1990). Again, as I've documented in greater detail elsewhere, (Hall, 1998a,b; 2003), both the state and corporate agribusiness were very active in responding to the threat posed by the early sustainable agriculture movement. One of their most effective strategies was to appropriate and redefine the concept of sustainable agriculture in ways that integrated neoliberal agricultural policy and agribusiness interests. In short, similar to what we saw in the mining case of health and safety, the state and agribusiness recast and reintegrated the solutions to the environmental and economic problems of agriculture so that economic growth (i.e. market responsiveness, cost-efficiency, self-reliance) appeared to be the simultaneous **means** of achieving environmental sustainability.

> The goal of the sustainable movement has always been shared by members of the agricultural community...However it has come as a bit of a surprise to many that our daily activities can become both environmentally sensitive and more economically viable...the economy and the environment are not eternal enemies but in fact natural allies.
>
> (Report of the Government Roundtable on the
> Environment and the Economy, 1991)

This government and corporate version of sustainability was, as noted earlier, the fourth pillar of the federal government's 1989 policy paper (Agriculture Canada, 1989, p. 34). As documented in Hall (1998b), this version of "sustainable agriculture" was promoted in virtually every major government agricultural policy statement dealing with restructuring through the 1990s, while this same perspective was widely promoted through various corporate and most farm organization channels as well (AgCare, 1992a,b). Just as we saw with the corporate and government discourse on health and safety in mining, the discourse increasingly claimed that efficiency and environmental protection went "hand in hand" (Hall, 1998b).

Again, as demonstrated in greater detail elsewhere (Hall, 1998b; Hall, 2003), a critical message within the government, and indeed, within the establishment farm media and agribusiness corporate discourse was that "sustainable agriculture" could be achieved **without** the need for radical changes in capitalist agricultural production or in the corporate structure of the agri-food system (see AgCare, 1992a). Accordingly, only certain kinds of changes were being promoted by the government and agribusiness (Agriculture Canada, 1995), and alternative models of mixed organic farming were certainly not at the top of the list. Sustainable practices were reduced to improvements in chemical technologies, tillage practices and management practices, while chemicals and eventually bioengineering were front and centre as the core elements of sustainable discourses. As such, neither the specific changes being promoted in farming practices nor the rationales behind them challenged in any way the basic productivist assumptions of government policy on the restructuring of farm production and marketing.

One good example of this, again as documented fully elsewhere (Hall, 1998b) was the government and industry focus on conservation tillage, especially a version called no-till, offered as a dual purpose strategy aimed reducing grower costs while at the same time reducing fertilizer pollution in the Great Lakes. As Lighthall (1995) documents in the US case, these kinds of sustainable methods were strongly supported by agribusiness and the larger heavily capitalized farm operators precisely because they sustained a reliance on chemical inputs and new technologies to achieve ever greater economies of scale. Other widely promoted "sustainable" approaches such as the promotion of integrated pest management for fruit farmers, although potentially beneficial in encouraging a reduction of pesticides, were also presented in efficiency terms (Agriculture Canada, 1992, pp. 51–52).

When pushing these solutions, the government agricultural ministries (OMAF and Agriculture Canada), conventional farm organizations such as the OFA, and agribusiness more generally emphasized *voluntary* compliance, education and research over a more tightly regulated regime. In Ontario, an illustration of this was a voluntary program called "Food Systems 2002: A Program to Reduce Pesticides in Food Production" which was introduced in 1988. Its stated goal was to reduce pesticide use by 50% over a 15 year period (Food Systems 2002, 1992; see also Surgeoner and Roberts, 1993) which was to be achieved largely through education and extension assistance programmes. One of the few new regulated requirements regarding agricultural pesticide use was a new accreditation requirement for farmers which involved taking and passing a one-day pesticide safety course (OMAF and OME, 1993) but this was again consistent with the government's claim the environmental change could be achieved largely through education rather than actual enforced regulations on practices (see Hall, 1998a). In practice, the course and the certification exam were less than demanding (OMAF, 1993b). Very few farmers, if any, failed the exam and some farmers in their interviews argued it was not a serious effort to educate farmers. As one farmer described it: "The Pesticide Safety course is just a public relations exercise to keep the public off the farmers' backs. But you know it costs me too. I have to spend a day there and send my workers" (fruit farmer). I took this course myself at the very beginning of my research when I knew effectively nothing about pesticides and pesticide application practices, yet I passed the exam with a perfect score. While my case-study interviews and observations suggest that this course may have helped farmers to recognize and confront at least some of the issues regarding pesticide safety, whether it substantially changed their practices was hard to gauge.

What we saw then within the government and agri-business sustainability and environmental discourses was a similar effort to construct the control of risks arising from production, in this case risks to the environment as well as health, as being achieved through production and management restructuring. Thus, the same transformative efforts and line of thinking needed to achieve greater efficiency and productivity were presented as yielding reductions in environmental damage. But, if environmental risk was being constructed in this way, what about farmer health and safety?

A push for occupational health and safety

While the gazes of the state, the public and the farming community were principally directed on the economic and environmental crises in agriculture, the mid-1980s to mid-1990s was also a period of mounting recognition of its occupational health and safety problems (Denis, 1988; OFSA, 1993b; OMOL, 1991; OMOL, 1985c Thu, 1998). As pessimistically noted in a 1991 Ministry of Labour discussion paper:

> Historically, farm work has been considered hazardous, associated with a large number of fatalities and lost time injuries. Based on current information, its status has remained virtually unchanged and the fatality/ injury rate continues to be high, comparable to mining and construction [the two other worst industries in the province].
>
> (OMOL, 1991, p. 1)

Despite this acknowledgment at least by one ministry in the government, the damning statistics (OFSA, 1993b) and the report's recommendation to proceed with legislation, the government persisted throughout the 1990s in resisting calls to include farm workers under the Occupational Health and Safety Act (Denis, 1988). Although the Ontario government actually changed stripe three times during the late 1980s and 1990s, from Liberal to NDP and then Conservative, the government position remained the same – essentially that as small businesses in a volatile industry, farmers could not afford to be included in OHSA.

Instead, corporate agriculture and dominant farmer organizations such as the OFA insisted that technological improvements, better management and better education and training would resolve OHS problems without the need for extensive regulative control (OFA, 1994, 1995). As in the mining case, and in the environmental farm discourse, farmers were encouraged to shift from the "safety first" framing to thinking of safety, productivity and profitability as commonly linked objectives (Canada Safety Council, 1987; Canadian Agricultural Safety Association, 2000). Much like the corporate and government discourses on the environment, and one we also saw in the mine case study, the claim was made that the health and safety problem was being fixed by farm owners and farm organizations *voluntarily* as they sought to transform themselves into more productive and efficient operations. Almost all the case-study farmers agreed with this position – that is, with one exception, they insisted that health and safety regulation was not required to reverse the high injury rates or health problems among farmers. Apparently, in this respect at least, these farmers were looking at the call for regulation as owners rather than as workers, taking positions reminiscent of INCO's opposition to union calls for more regulation, worker powers and enforcement in the 1980s. As one farmer put it, "we need more education not more regulation."

This was also reflected in what I heard in interviews with several local and provincial farm leaders representing the OFA and the major soybean and corn crop associations. A key position expressed by these groups was health and safety legislation in the context of the 1990s restructuring and globalization would undermine their efforts to address the farm crisis (see also OFA, 1994, 1995). While advocating for inclusion under the health and safety act, the only major labour union pushing for both unionization and OHSA rights for farm workers in Ontario during this period was the United Food and Commercial Workers (UFCW) which began its advocacy in 1980. Although the NFU, again stronger in the western Canadian provinces, and a small organization of migrant farm workers in BC called the Canadian Farm Workers Association, were also advocating inclusion (Wall, 1996), these actions had relatively little impact in Ontario.

In as much as it wasn't until 2005 that farm workers were finally included under the protection of the Ontario Health and Safety Act (Ontario, 2005;

Ontario Ministry of Labour, 2006), farmers and the larger Agricultural industry were obviously quite successful in staving off demands for OHS regulation – that is, for well over two decades. I might also add that the agriculture industry was quite effective historically in resisting the unionization of farm workers (Wall, 1996). While the legislative outcome in health and safety was quite similar to the environmental portfolio in the sense that the government promoted voluntary over legislative and regulative measures, the Ontario government's focus on health and safety was considerably less robust in terms of funding and programming as compared to the environmental area (Denis, 1988). This weak policy and funding effort in health and safety, even relative to the environmental portfolio, reflected at least in part the very weak power base of farm workers as opposed to the politics surrounding the environment (Adkin, 1992; Denis, 1988; Wall, 1996).Certainly as opposed to the politics of workplace safety in most other industries including the mining sector, which as noted were quite volatile and substantial in the late 1970s and early 1980s, the farm sector advocacy was very limited. Although a social democratic NDP government legislated some unionization rights in 1994, albeit still without the crucial right the strike, even those restricted rights were rescinded by a neoconservative PC provincial government in 1995. Although restored in 2001 after a Supreme Court decision, the resulting legislation again provided only weak association rights. Moreover, many farm workers were (and still are) temporary and/or migrant workers, and as such, had (have) extremely limited political power both in the workplace and in the public arena (Arcury et al., 2005; Basok, 2004; Basok, Hall and Rivas, 2014; Grieshop, Stiles and Villanueva, 1996; Halfacre-Hitchcock et al., 2006). Rural women and teenagers also were and still are a good proportion of the seasonal and part-time work positions. And certainly, working on family farms geographically dispersed has made it extremely difficult for farm workers to organize (Denis, 1988; OMOL, 1991, p. 4; Wall, 1996).

At the same time, however, it is important to note that family farms were experiencing very high rates of *recorded* injuries and fatalities, and sadly, often including younger children during the 1980s and 1990s (OFSA, 1993b; OMOL, 1991; Pickett et al., 1999). Occupational disease especially among grain farmers was also being increasingly recognized in the occupational health literature over this time period with calls for more monitoring and research (Dosman et al., 1987; Zejda et al., 1991). Whether individual farmers were entirely unwilling to consider the prospect of regulations, inspections and the need for training and equipment expenditures is somewhat difficult to discern given the lack of substantial historical research in this area, but as noted, my case studies and interviews with other farmers and farm organization representatives suggest that most farmers were opposed to OHS regulation on the grounds that it would introduce too many costs and limit their capacity to organize production as they saw fit. However, the long road to OHS regulation in farming may be less a function of the power of farmers and more a consequence of the political status and power of the industry as a whole – or what Winson (1992) called the "Canadian agro-industrial complex" (for similar arguments about US, see Hendrickson et al., 2017). At the same time, the lack of any significant counterbalance to that power in terms of worker unions and a strong oppositional farm-owners' movement at least within Ontario, helps to explain how a policy of exclusions for farm workers could have lasted for such a long time (Denis, 1988).

Still, there was enough public pressure generated by the Canadian labour movement, in particular the UFCW, and the NFU, to at least force the farm community and agr-business to make some effort to present the industry as attending to the problem. As the pressure grew in the 1990s, the Ontario Farm Safety Association (OFSA), the main provincial accident prevention organization funded by the Workers' Safety Insurance Board via charges to the farmers, was especially important in sustaining the claim that agriculture could deal with the problem through education and voluntary programs. As part of this effort, the OFSA and the agriculture arm of the provincial government (OMAF) intensified their efforts to promote safety awareness programmes, best management practices and audits (Reed, 1990). Reflecting this development, almost all the case-study farmers and their spouses observed that health and safety was being pushed much more by their farm organizations and the government. Initially, most of the OFSA publications preached very basic prevention articles about identifying, controlling or eliminating various hazards such as tractor rollovers or "farmers' lung," much in the same way as INCO did in its safety communications to miners, outlining specific work procedures and responsibilities (OFSA Factsheets, various 1985–1991; OFSA, 1993c). However, similar to what we saw in the government's environmental and later OHS discourse, the key emerging argument encouraged by the government was that farm safety risks could be integrated within a comprehensive and intensive management approach that calculated, assessed and controlled safety risks, as part of its planning and productivity improvements (Canadian Agricultural Safety Association, 2000; OMOL, 2006).

These ideas were just beginning to take hold in the early 1990s with respect to farm safety but my observations of farm organizations and pre-case study interviews with farm leaders and farmers suggests that these underlying governance notions about being able to audit, calculate, and control health and safety risks within a more systematic management framework that simultaneously addressed economic and environmental risks, were gaining increasing acceptance among many farmers (see Chapter 7). But, there was also evidence of resistance as other farmers insisted that their practices were as safe as they could be, which meant in practice they were not participating in the educational programs or adopting the auditing and other safety technologies (see next chapter for the specific case studies illustrating these two responses). Of course, at the time of the farm studies, farming was still not covered under OHSA or health and safety regulations which meant that farm workers had none of the OHSA rights available to other workers in terms of right to information, right to refuse and right to representation, and farm owners had relatively limited responsibilities as employers.

This brings us then to a key question – in the restructuring context of a push for more flexible and leaner forms of "sustainable" farm production, to what extent were safety management ideas and discourses shaping farmers' approaches and practices farm health and safety in the absence of regulation, and with what consequences in terms of OHS outcomes? As before, we are asking these questions with reference to the broader restructuring context; that is, how were the significant pressures to adopt intensive management and sustainable farming practices on both environmental and economic grounds affecting the farmers' understanding and approaches to health and safety? What I want to show in the remainder of this chapter is that all three discourses – the restructuring discourses on farm management, production and marketing, the sustainability and environmental discourses, and the health and safety discourses – *combined* in complex ways to influence the risk perceptions and practices of farmers. Later in Chapter 7, I show that different farmers respond differently to these discourses, again with interesting and often conflicting results in terms of their approaches to health and safety.

The restructuring of conventional grain production

To stay true to the labour process aspects of my analysis, something needs to be said at this point about the actual changes in farm management and production practices and relations. As implied in the above discussion on the crisis, conventional farmers were encouraged to focus on increased efficiencies through the adoption of leaner, *less* labour-intensive methods such as reduced tillage or no-till techniques or rotational grazing. By reducing or eliminating tillage, these techniques reduced time and labour and use of fuel while also limiting the need for costly chemical fertilizers and preventing nitrogen and phosphate run-off, as well as helping to enhance soil quality.

Both reduced and no-till methods also involved some changes in technology, including the use of different planters called no-till drilling planters, but most significant perhaps from a health and safety perspective, was that they required much greater use of herbicides (Lighthall, 1995). The main labour process differences between conventional tillage and no-till field crop farming is that when a corn or soybean crop was harvested, the farmer did not turn the corn or soy bean stalks back into the ground, nor did he/she till to control the weeds in the spring. Instead, the farmer sprayed the fields with a kill-all herbicide spray such as Gramaxone, Dicamba or 2,4D before planting, and then planted directly through the debris from the previous crop using the special no-till "drill" planter. Any further problems with weeds during the growing season were again controlled with chemicals only, and the crop grew in the untilled soil until it was harvested and the process was repeated. Additional herbicide applications were usually necessary during the growing season given the avoidance of mechanical tilling and farmers often switch chemicals around to prevent resistance. For soybeans, the lack of cover often meant planting a winter cover crop such as wheat or clover, which was then sprayed and killed off in the spring before the regular planting (OMAF, 1993a).

While adding substantially to the pesticide bill, there were substantial savings in time, fertilizer and fuel costs, since tilling is by its very nature a slow and fuel-intensive process. The farmers could also often dispense with the larger, more powerful tractors needed for plowing, although the planting drills were heavier and more expensive than regular planters so the cost savings on machinery and fuel balance out to some extent. However, there was also a need for more careful management of timing for planting to ensure proper seeding. A greater sensitivity to wet weather often complicated things considerably in terms of seed germination and mould problems, and consequently, accentuated production and timing pressures. On the other hand, no-till crops tended to weather drought conditions better (OMAF, 1993a). Although no-till could reduce the amount of time farmers and their workers spent in a given field area, in practice no-till was partly encouraged by the government and agri-food industry as a way of expanding the amount of land farmed by a given operator. Several case-study adopters claimed that they were working harder and longer hours as a result (see also Hall, 1998b)

Rotational grazing for cattle and dairy farmers was also widely encouraged in the same fashion, as an approach that was cheaper in terms of inputs, was better for the fields and soil, limited the accumulation of manure, and enhanced herd health. Although by no means a "new" approach, it was an effort to move away from the model of the captive grain-fed herd that spent its life in the barn. This meant getting the herd off grain feed and into the fields where they would feed more "naturally" on grass. Like no-till, this was seen as greatly reducing costs associated with feed and health management, which also spoke to reduced labour time, but it also called for more intensive management and planning of field production. The fields were systematically rotated over time through grasses and cash crops with much of the fertilization taking place naturally as the cows spread themselves across any given field. This was also supposed to reduce various animal health problems which meant less time spent treating sick animals and less time using pharmaceuticals such as antibiotics. There were some additional tasks that went with this approach including much greater attention to fencing, protecting the herd at night and, if dairy, bringing in the herd for milking each morning.

With respect to both no-till farming and rotational grazing, changes in crop and herd management were arguably more significant than changes to direct production practices. The accompanying emphasis on increasing land size, coupled with an emphasis on a greater variety of crops, both for rotation/soil enhancement and marketing reasons, were important aspects of this push for more emphasis on information-based risk management – with particular reference to planning, scheduling, monitoring and assessing. The importance of measuring and tracking inputs and outputs was also emphasized in the government literature as critical to making this system work for the farmers as they sought to balance shifting market and soil conditions, and the risks associated with those conditions (Innovative Farmers, ND; OMAF, 1990, 1992). Flexibility was also emphasized in planning crop rotations to meet market developments. Cost containment was a key focus and again, a careful recording and assessment of costs and a constant search for alternatives was encouraged. Similar to what happened in mining, this was also seen quite clearly as a move towards a more knowledge-intensive management process with a major emphasis on knowledge acquisition and innovation. Various training programmes were made available to farmers to develop these management and marketing skills (Agriculture Canada/ OMA, 1994a,b; Stonehouse and Bohl, 1993). Since no-till and rotational grazing were often adopted together, this meant that these farmers shared a greater overall exposure to chemicals. However, while the management of the environmental impacts became quite integrated into these risk assessment and management programmes, relatively little attention was devoted to the occupational health implications (Hall, 1998b, 2007).

Of course, once again there were no regulations forcing farmers to adopt these new state and corporate promoted sustainable methods or approaches, or for that matter the new risk management methods. As noted and as we'll see more when we move to the case studies, some conventional farmers resisted and continued to farm as they always had, while others moved more radically towards alternative forms of agriculture, rejecting the core neoliberal, globalization and productivist premises, reflected most notably by the growth in organic farming (Atari et al, 2009).

The rise of organic farming: neo to post-productivism?

Although some provincial governments, with particular reference to Quebec, BC and Saskatchewan, began to provide more support for organic farming in the 1990s, Ontario was one of the last provinces to devote meaningful state funding or resources to organic farming (Hall, 1998b; Hall and

Mogyorody, 2002; Macey, 2000; Sustainable Farming, 1994). The federal government was also very slow to act but started to provide more funding and other support in the late 1990s including the regulation of organic standards and certification in 2009. Even at its best, however, state support in Ontario and most of Canada remained well below western European countries throughout the period under study (Agriculture and Agri-Food Canada, 2001a,b,c; Hall and Mogorody, 2002; Lynggaard, 2001). Despite the lack of state, corporate and farm organization support, organic farming grew slowly but steadily through the 1990s (Archibald, 1999; Macey, 2000). While organic vegetable and fruit producers often began their operation as organic producers, organic grain producers and mixed-grain livestock producers tended to be conventional producers who had decided to shift to organic, as was the case in all my case studies (Hall and Mogorody, 2002). However, exactly what it meant to farm organically, and in particular what the marketplace and the state would accept as organic production standards, became an important political struggle during the late 1990s (Hall & Mogoyrody, 2001, 2002). Partly reflecting their lack of support for the industry, the Ontario and Canadian governments were initially reluctant to establish fixed national standards, but pushed by elements in the organic community and eventually grocers seeking to justify higher prices, the Canadian government relented in 2009 introducing national requirements. This established the need to be certified by certain authorized certification bodies with requirements that included a range of production requirements and restrictions including prohibitions on the use of synthetic chemical pesticides and fertilizers. Largely missing from these standards was reference to worker safety or labour conditions more generally. Although government regulation placed clear restrictions on the right of farmers to use the term "organic" to describe their produce, the actual standards did not change appreciably from the pre-existing private certification system that was already in effect prior to the government standards. In fact, the government standards largely adopted the existing frameworks, while several private certification organizations were registered to administer and enforce the certification process.

While there is flexibility in these standards, a key point to make here is that in terms of external controls over farm production practices, the production practices of certified organic farmers tend to be more "regulated" than conventional farmers, although the direct regulator (or auditor) has always been a private agency (e.g. Organic Crop Improvement Association Canada). To gain certification, farmers have to follow a strict set of rules and standards and submit to annual on-site inspections which are, in effect, compliance audits. After farmers complete detailed questionnaires on their practices, inspectors tour the operations and take soil and product samples (Agriculture and Agri-Food Canada, 2001c).

According to interviewed officials from Agriculture Canada and Ontario Ministry of Agriculture and Food, one of the central motives behind the push for government certification was to enhance the export potential of organic production, in particular to Europe and Japan. The farmers and agribusiness interests behind this push were accordingly careful to ensure that the standards did not impinge on their competitive capacity to achieve high levels of cost-efficient production, including a capacity to use some "approved" pesticides. It is again important to note that even after 2009 when the federal government introduced a Canada-wide regulation, the government still left the certification task in the hands of private certification organizations. This left enough flexibility in the standards to allow some very real differences within the organic community regarding ideological orientations and practices. As demonstrated in several studies in Canada and elsewhere, including my own research (Hall and Mogyorody, 2001, 2002), some organic farmers think and operate much like conventional farmers with the core difference being that they do not use standard pesticides or chemical fertilizers (Allen and Bernhardt, 1995; Clark, 2007; Guthman, 2004). Moreover, since organic farmers cannot use kill-all pesticides, they have to rely on plowing, precisely what SA farmers were being told was bad for the soil.

In a survey we did of organic farmers (N = 249) in Ontario, we found that about one-quarter of the farmers shifted to organic farming from conventional because they believed it would be more profitable without expensive chemicals and with the higher prices for organic produce (Hall and Mogyorody, 2001); but significantly, the newer converts and grain/livestock farmers were more likely to cite profitability as very important to their switch. In short, for many of these farmers, organic was a way to be leaner, more efficient, and cost-effective, mirroring the arguments made by farmers who switched to no-till and rotational grazing. As we shall see, their labour, management and marketing practices were also very similar to the conventional farms in as much as they relied heavily on mechanized production and information technologies, they purchased many or most of their inputs, they focused on export markets, had limited crop rotations, were more likely to use cheap migrant or child labour, and were focused substantially on success through economic and farm-size growth (Hall and Mogyorody, 2001; Guthman, 2004). Many of these grain farmers were also highly specialized producers with no livestock at all which meant they purchased rather than self-produced many farm inputs such as fertilizers with more limited rotations and increased tilling, while relying entirely on wholesale sales and global marketing systems.

However, as our research also showed, there was another whole segment of organic farmers who had started organic farming from the outset with different motivations and orientations (e.g. environmental, health, social justice), with very real differences in practices (Hall and Mogyorody, 2001). Most notably, these were the kinds of organic farmers who sought to produce their own seeds and their own composted manure; who limited their use of machinery, a few going as far as replacing their tractors with horses, and who intentionally limited the sizes of their farms, often planting trees in retired land; who sought to market most or all of their products locally often at the farm gate; and, who maximized the mix of different crops and livestock. In other words, these farmers reflected more of what might be called an alternative post-productivist orientation. While there were variations within the two extremes (neoliberal productivist and alternative post-productivist), the latter were moving in very different directions from the rest of agriculture – that is, towards more labour-intensive practices with less reliance on technology and less reliance on corporate-controlled markets and distribution networks (Hall and Mogyorody, 2001). However, as will be shown, unfortunately this alternative thinking was not necessarily extended to OHS.

Farm hazards and the impact of farm restructuring

Before moving to Chapter 7 on the links between safety practices and the adoption of these different patterns of production and management within agriculture, it is important to recognize some of the implications of these different production approaches for concrete health and safety conditions (Thu, 1998). As in the case of mining, there were both positive and negative impacts, although in many respects the overall picture with respect to no-till in particular was more decidedly on the negative side. Since rotational grazing was usually adopted in concert with no-till or reduced tillage, I'll focus here on the no-till shift. The shift to no-till and reduced tillage limited some of the time spent in the fields and assuming that the farmers did not substantially increase their land base, which most did, this could mean reduced exposure to the outside elements like the sun, less time spent driving a tractor and less time on the road with tractors However, as noted, no-till tended to increase exposure to pesticides given the increased dependence on chemicals for controlling weeds rather than using mechanical means. Moreover, to the extent that no-till encouraged many farmers to expand their land base, in part because its adoption was presented as a major selling point for no-till, the total effect here in terms of work time and stress was usually negative, especially with the use of even more chemicals and comparable amounts of time being spent planting, spraying and harvesting given the extra land size. This also often meant that the land being purchased or rented was also spread over a larger geographic area sometimes 30 or more kilometers from the home farm, so many farmers were actually spending more time on roads with their equipment, often at night, getting to and from their dispersed fields. The emphasis on marketing multiple varieties of crops also meant an increase in the variety of herbicides being used further adding to the risk. While the increased amount of spraying encouraged some smaller farmers to hire out the spraying job to other farmers or contracting businesses offering this service, this also meant that some farmers began hiring themselves out as contract sprayers or harvesters in order to get a full return for their new equipment. Some farmers could neither afford this technology nor contract out the work which meant using antiquated spraying equipment and open tractors. To the extent that some farms were able to expand their land base, some also began hiring more wage workers. The danger here was that given the lack of regulation, very few of these workers were given proper safety training or information (Halfacre-Hitchcock et al. 2006).

The greater range of crops planted, whether as a market or an environmental strategy, also meant having to change combine blades and planters more often, which, again depending on the condition and age of the equipment and time pressures, could present more hazards. The need to change blades often translated into more time spent on the roads with the combine or a truck pulling the blades on trailers, which once again added to the roadway hazards. Arguably, and certainly the case-study interviews support this, the effort to achieve greater efficiencies, flexibility and sustainability through more complex crop rotations also added more stress in as much as farmers had to juggle the different schedules and demands of these crops, including a more complex storing and marketing strategy. In an effort to reduce vulnerability to price fluctuations, farmers also increased storage capacity on their farms, which in turn increased the potential for grain bin incidents – falls, entrapment, suffocation, gas explosions, crushing and dust exposure (OFSA, 1985b,e).

For those farmers who had shifted fully to organic, there were also some mixed health and safety impacts. Clearly one of the main advantages was that they were no longer using most synthetic pesticides or volatile chemical fertilizers like anhydrous ammonia which is highly toxic and unstable. However, this did not mean no exposure as organic farmers are permitted to use some "natural" pesticides and fertilizers (e.g. toxins in manure) which can be harmful to farmer health. Organic farmers were also engaged in more physical labour over mechanized production, and this did vary quite substantially with the farmers, but this also had some mixed implications. More physical labour in the fields introduced the potential for various strains and sprains, cuts and scratches from the plants and tools etc. along with increased exposure to the sun and heat exhaustion, but it also meant more physical exercise for farmers, with health benefits not available to mechanized conventional farmers who spend virtually all their time driving machinery. Less use of machinery also reduced the risk of turnovers, road and impact accidents as well as presenting fewer machine hazards to children. On the other hand, most of the organic grain operations were still using large machinery such as combines and tractors, in some cases as much as conventional farmers.

However, organic farmers tended to have older machinery, often because they were unable to generate the income necessary to carry a higher debt load, but also because some organic farmers objected in principle to expensive technology and/or the debt required to finance it. And again, while many organic farmers sought to intentionally limit the size of their farms, organic grain farmers were often just as large as the conventional farmers (Hall and Mogyorody, 2001, 2002). The workload and the stress and production pressures associated with the larger size operations were often quite similar to conventional farm operations. Moreover, even among those who restricted their farm size, there was a tendency to have more mixed livestock and grain operations which introduced a number of hazards associated with animal production as well as enhanced the amount and physical nature of the workload, organization and production pressures. On the other hand, mixed organic farms tended to adopt rotational grazing and as such had less exposure to animal diseases and close encounters with potentially dangerous cattle, while as organic producers they were also less exposed to the chemicals used to control bacteria and disease problems that were more common when animals were penned.

On the whole, then, the differences between the organic and conventional grain production approaches in terms of safety hazards and risk were not as great as the similarities, even when comparing farmers within the different farming orientations across the spectrum of conventional and organic spheres. On the other hand, the very obvious advantage that organic farming enjoyed was the major reduction in chemical exposures almost entirely, while conventional sustainable farmers actually intensified their pesticide exposure risks.

Restructuring and labour control

A key argument in the mine study analysis was that the labour process changes altered worker control over key workplace hazards as well as their capacity to perceive and judge the resulting risks to health or safety. In the mining context, I argued that the restructuring nested more risk control and assessment in management, management systems and technology, while management strategies sought to get workers to accept more responsibility for following the rules and procedures set out by management. Although we are talking in this case about self-employed land-owning producers (as well as their workers), we can still ask the same question – to what extent did the reorganization of production and the introduction of new production management models alter the control and judgement nested in the farmers themselves, as opposed to the technologies, discourses and management models they were adopting? This is clearly a different question than asking what the farmers believed about the impact on control, and I'll get to that issue in Chapter 7 as it relates to the critical question of risk acceptance. Certainly, as I've already suggested, the dominant government and corporate discourse on restructuring claimed that these new approaches to farming would enhance farmer control over economic, environment and safety risks.

However, when we look more closely at the changes in the organization and content of particular work tasks within the transformations of conventional grain farming and from conventional to organic, the changes in the work itself appear much less substantial than the shift we saw from conventional to bulk mining. The changes in technologies were relatively minor from conventional to no-till, and of course, in the context of the more alternatively oriented organic farms, there was movement *away* from technological to human control. Nevertheless, even for those organic farmers (and indeed the conventional farmers) who continued to embrace technology and large machinery, on the whole these farmers were in relatively unchanged positions as far as their capacity to know, judge and control hazards in their work. The same argument applied to their workers. The main exception was the added potential for more production and workload pressures within the lean and flexible production model, which then limited the time for the recognition and judgement of hazards, and induced a higher level of risk-taking based on production and time-pressure rationales.

Still, as owners and managers of the farm, farmers were in structurally privileged positions as compared to miners – in theory, they organized their work, decided the pace and made decisions about inputs. As I've suggested, the new approach to sustainable conventional farming was presented to them by corporate food producers and the government as a more knowledge-intensive approach to risk management which could balance their needs in all three key areas – costs, the environment, and health and safety. By appearing to give farmers more planned control through enhanced predictability and calculability, farmers were told they could make the right choices, take the right actions, and keep their work paces sane enough to achieve growth and profitability without harming their environment or themselves. As will be shown in Chapter 7, farm operators and workers didn't always see these OHS benefits. Yet, as I will also demonstrate, a farmer does not have to fully embrace this neoliberal neo-productivist way of thinking to be influenced by it. Indeed, a key objective in Chapter 7 is to show the varied impacts of this discourse on farmers' risk-taking practices and in their management of their workers' practices.

Conclusion

This chapter suggests that there was a major thrust towards neoliberal and post-Fordist restructuring in agriculture during the 1990s which reflected many of the same elements we saw in the mining context – a crisis in security generated by global forces addressed through an emphasis on leaner, more efficient production methods, knowledge-intensive management, cost-benefit risk assessment, and the use of performance technologies such as audits. We also observed a similar effort to address and depoliticize environmental, health and safety concerns within an integrative discourse which presented production and management restructuring as the simultaneous coordinated solutions to those problems. However, this chapter also recognizes that unlike the mining context, there were significant external and internal challenges to the corporate/state-promoted neoliberal productivist models of restructuring which emerged in the 1980s and 1990s, both from within agriculture and from a growing environmental movement. Also unlike the mining contexts, I've suggested that the production and public politics of farm safety were relatively slow to generate regulations or reforms, failing to garner anything like the same public or state attention given to farm environmental issues or to the health and safety issues in mining. What we need to address in Chapter 7, then, is how were these contrasting developments reflected in the risk subjectivities and practices of farmers (as producers and as managers), and, even more specifically, did farmers with different orientations to restructuring define, perceive and respond to OHS risk in different ways?

Chapter 7

Health and safety in farming

Chapter 6 suggested that restructuring pressures and resistance to those pressures have fractured agriculture into different production and marketing orientations. As a basis for comparison, this is somewhat similar to INCO in as much as the company responded to different market pressures by using different production methods and management structures in its different mines, with what I argued, were varying consequences in terms of OHS risks, subjectivities and politics. Of course, INCO was a single multinational corporation and therefore able to coordinate its strategic efforts to reconstruct management control over workers and health and safety; although even there, important variations in design and implementation were evident (see Chapter 5). In this chapter, I compare the operation of 20 distinct farm businesses which took different paths in responding to the market, environmental concerns and political developments outlined in Chapter 6. This fracturing within farming offers an opportunity to address a somewhat similar question addressed in Chapters 4 and 5 – are these different methods of production and management reflected in substantially different health and safety subjectivities and practices? Like the mining chapters, part of this analysis is concerned with the impacts of labour process changes on risk subjectivities and politics; but my more significant concern here is to explore whether a neoliberal neo-productivist orientation, one promoted by both the state and the transnational food industry, translates into greater levels of risk-taking relative to the conventional productivist orientations or alternative post-productivist orientations of those farmers who saw themselves as directly challenging neoliberalism and globalization on environmental and/or social justice grounds. By focusing on self-employed farmers, this chapter also adds a new wrinkle to the analysis. If self-employed farm owners have more power to protect their health and safety than wage workers, were they still taking personal risks grounded in risk subjectivities similar to those shown among miners and front-line managers in Chapter 3?

Drawing the distinctions: categorizing and comparing case studies

The analysis in Chapter 6 pointed to two broad divisions in the farm community during the 1990s restructuring period. First, there was the division between chemical-based and organic farmers with the core distinction being whether or not synthetic chemical pesticides and fertilizers were being used as the principal methods of pest control. Second, there were divisions *within* the organic and conventional categories which roughly ran along a continuum from neoliberal neo-productivism (managerialist, lean production, highly mechanized, global and flexible market orientation) to alternative anti-neoliberal post-productivist orientations and practices (environmentalist, labour-intensive, smaller-scale family farm, input independence, local markets). Based on these two dimensions, I've divided the case studies into five groups – making three distinctions within the conventional chemical-based camp, and two within the organic (see Table 7.1).

Groups 1 and 4 represent one end of the neoliberal productivist spectrum as farmers principally interested in the profit, efficiency and productivity gains of sustainable or organic farming with global market orientation and a commitment to intensive management, mechanization, land expansion and other technologies to achieve these goals (Wilson and Burton, 2015). The main distinction between Group 1 and 4 is that Group 1 consists of conventional chemical producers who have fully adopted the government discourse on "sustainable agriculture," while Group 4 members have also adopted this discourse without using the full range of available synthetic chemicals (Hall and Mogyorody, 2002). As will become clear, Group 1 is also distinguished by a greater commitment to risk management strategies including the use of auditing and other performance technologies but both groups are characterized as being heavily dependent on similar forms of mechanization and global marketing.

Table 7.1 Categorizing the farm case studies

Conventional			Organic	
Group 1	Group 2	Group 3	Group 4	Group 5
Productivist adopt no-till lean and flexible principles	Productivist Limited Changes	Sustainable Environmental Health Focus	Productivist Market-Oriented	Alternative/ Environmental Health Focus
CS1	CS3	CS7	CS10	CS9
CS2	CS5	CS11	CS15	CS13
CS4	CS8	CS12	CS17	CS14
CS6			CS18	CS16
			CS20	CS19

Within the conventional area, I've also identified Group 2 as conventional farmers who have adopted relatively few changes in their production or management practices, with little or no commitment to the state or alternative visions of sustainability, but who largely remain committed to the conventional productivist model (high yields through the substantial use of fertilizer, pesticides and mechanization). Groups 3 and 5, on the other hand, represent a clear alternative orientation to the environment and health, with the former trying to achieve this within the context of conventional production, and the latter, organic (Clunies-Ross and Cox, 1994; Hall and Mogyorody, 2001). Both these groups are highly critical of the government discourse on sustainability which they view as a smokescreen aimed at sustaining a continued emphasis on high-energy, chemically and technologically based farm production.

As I go through the comparison, it is important to keep in mind that the labour process changes are split differently in that Groups 1 and 5 made the most changes in their production and management practices, followed by Groups 4, 3 and 2 in declining order. As we shall see, this makes some difference when trying to explain the links between farmer orientations to restructuring and health and safety. However, again my objective in this chapter is to go beyond the labour process analysis emphasized in Chapters 4 and 5 to focus more on the different "management" orientations and practices, both in health and safety and in general.

Knowing, judging and taking risks

Group 1

If we compare the awareness and knowledge of hazards within the five groups, the main difference was not that certain groups had a more complete list of farming hazards in their heads, but rather that some, in particular Group 1, were far more conscious of those hazards in their planning and thinking, both long-term and short-term. Part of being neo-productivist was that these farmers were "managerialist" in the sense that they believed wholeheartedly that an emphasis on information, knowledge, analysis and planning within a formal risk management system was crucial to business success (Barkemeyer et al., 2014). As this implies, they embraced the government and corporate line that economic, environmental *and* OHS goals could be integrated and coordinated as a holistic management system, which includes the adoption of pre-packaged planning and auditing programmes to ensure maximum predictability and management control of outcomes. As we saw in the mining context, this means that accident prevention was constructed within the management discourse as simply another aspect of efficiency and productivity (see Chapter 5 as well as Chapter 9 for more on this point within the context of the auto study). The notion that productivity, sustainability and safety go "hand in hand", was widely accepted as a truism.

> I'm not ready to go no till with the corn because it is not clear it will pay but soy beans, with the new [planter] drill and using different herbicides and containers I'm saving on labour time and physical labour, and that means I can do more acreage.
>
> (CS1)

As well, compared to the other groups, the four farmers in Group 1 exhibited very little uncertainty about their knowledge regarding the safety of their equipment or chemicals. They were confident that the use of the WHMIS label instructions and the basic procedures protected them, in part because they had done the research and had gone to the workshops and training sessions in the winter. As this implies, they invested considerable faith in technology and in science as providing the foundations of lean, flexible *and* safe farm production and management. These constructions also reflected their identity as "effective managers." As another Group 1 farmer put it, "I have the knowledge to control the different factors that impinge on farm profitability" (CS2). This included an emphasis on being able to quickly respond to changes in the marketplace by constantly monitoring commodity markets.

This emphasis on flexible knowledge-intensive management was associated with a conviction that they had full control over what happened on their farm, and as such, were responsible for its success or failure, as well as their own occupational accidents or health problems. This also translated into a strong neoliberal market-based voluntarism.

> We don't need more regulations. I know what I have to do to make this farm work, and getting hurt is not something I can afford. I prevent accidents by being as efficient and up to date as I can manage. It just makes sense to do it.
>
> (CS2)

Their commitment to reduced or no-till farming was a business strategy rather than an expression of their environmental concerns. As noted in Chapter 5, no-till deepened their dependency on and exposure to chemicals, but perhaps reflecting this commitment, they tended to express complete confidence that their pest management strategies and up-to-date production technologies presented minimal risk to themselves or their workers, and the environment. In two of these cases, where I was able to interview the FT worker(s) on each farm and in one case, an adult son, the workers were also convinced that their bosses were careful, competent and knowledgeable, and as such, expressed few concerns about their chemical exposure. One of them, a long term employee, was mixing and spraying a range of toxic chemicals throughout much of the summer because his employer was paying for his new sprayer by taking contract work for other farmers. Nevertheless, he expressed limited concerns about the health risks grounded in his belief that he had the knowledge and the technology to prevent exposure:

I. You are doing a lot of spraying aren't you? Any concerns about health?

CS1Worker. Not really. I've taken the [pesticide] course, it's a closed [air-filtered] cab. I know what I'm doing.

As the quote reflects, these farmers were also much more likely to seek and provide training and education for their farm workers or family members. Both owners and workers also seemed to share a certain confidence in the more advanced technologies and machines they were using which was also evident as underpinning their expressions of control over risk

All four of these owners/producers firmly rejected organic farming as un-economic and impractical, using the often-repeated argument that "without chemicals, there wouldn't be enough food to feed the world" (CS3). They also constructed their production practices as "sustainable" in line with the government and agribusiness construction of sustainable agriculture (see Chapter 6). For them, sustainability was demonstrated through the use of no-till and the adoption of several new "sustainable practices" such as planting grass buffers and wind blocks and installing better drainage systems. While none had moved fully to Integrated Pest Management (IPM) at the time of the study, they saw the profit value in attempting to reduce pesticide use through programmes like IPM, and in a follow-up one year later, I discovered that two (CS1 and 2) were using some IPM procedures such as band and targeted spraying (Baker, Green and Loker, 2020). However, again the economic benefits of chemical reductions were the central motivator for them rather than their own health If there was an "environmental" issue for them, it was not pollution but rather the degradation of the soil and the possible long-term effects on productivity (see also Atari et al., 2009).

They were also avid supporters and early adopters of the Environmental Farm Plan (EFP), a voluntary audit procedure which was being heavily promoted by the main farm organizations in Ontario at the time (AgCare, 1992b; Atari et al., 2009; Robinson, 2006). As well, these farmers adopted a simple OHS audit programme promoted by the Ontario Farm Safety Association which involved conducting regular inspections using a hazard and safeguard checklist (OFSA, 1993d). All four farmers adopted wholeheartedly the notion that the "environmental practices" and "safety" promoted through the audits were simultaneously making their operations more efficient, such as pointing to the multiple benefits of better storage facilities and procedures and improved spraying equipment. As two of the farmers stated:

Moving away from jugs of pesticide to storage tanks is saving me labour time and it's safer for the environment and the operator.

(CS1)

I've updated my storage of pesticides using bulk whenever I can. This really limits my exposure. It is so easy to get hurt, slipping with containers of chemicals when pouring it in the tank. It saves some labour time as well and its quicker.

(CS2)

They were also very conscious that these activities were not just good business with respect to their individual farms, but also understood them more generally as part of a political strategy to minimize any new regulations in the pesticide area or in health and safety more generally.

If we don't get going with the environmental farm plan they [the government] will impose it on us.

(CS4)

We can get where we need to be in safety by showing that we are following best management practices.

(CS1)

This awareness of what was happening in provincial politics reflected both their involvement in the main farmer lobbying organizations (e.g. Ontario Federation of Agriculture) and their general world view that collective farmer activism and lobbying in provincial politics was central to controlling the economic and regulative conditions facing farmers. While all four strongly shared the view that OHS regulation would limit their flexibility, they insisted that trade liberalization forces were positive forces pushing them to employ the best environmental and OHS management practices. As one observed: "Part of the push for EFP was the banks telling farmers they were going to foreclose on us if we don't deal with the environmental costs so this drove the EFP in a way."

Also reflecting their managerialist orientation, the farmers in Group 1 were also more likely to calculate and judge environmental, health and safety risks in cost-benefit terms.

When they banned Lasso (brand name of herbicide) that was what really got us going. I thought it was going to make us uncompetitive. There's risk involved in everything – you just have to look at the risk and benefit of using it. You have to weigh risk and benefit of taking the risk and to me there was more benefit than risk in using it. There's no clear proof it causes environmental or health problems for the user.

(CS1)

I've reduced my tillage in last several years because it was cheaper, and now I see it's better for the soil and environment but I've not gone no till yet because I still had some money I need to get out of my equipment. Now I'm looking for a no till planter and I'm going to make the shift. I'm sure it is going to pay.

(CS4)

Accordingly, risk was not understood as an absolute high or low, but rather as a relative judgement grounded in science, information and calculation. The various different *kinds* of risks, economic, environmental, health and safety were considered here as pieces of an integrated model of profitable sustainable farming. We saw this thinking within the context of INCO's Loss Control programme, although not surprisingly among the managers more so than the workers. Reflecting their managerialist orientation, these farmers strongly identified the "sustainable" labour process changes as accentuating the significance of their management role, one that resonates with many key features of the post-Fordist emphasis on quality and flexibility:

With zero tillage or conservation tillage I feel I'm using my head a lot more, how to organize the work. So it's brought back challenges and I can see possibilities now of succeeding where before the challenges of wealth was overwhelming... by lowering capitalization. Now it's all about management, it's quality not quantity, tightening the operation, being more flexible, more tailored herbicide application, being really careful about timing. It's about being on top of it – a change in management and organization.

(CS2)

Although economic interests underlie this way of thinking, this should not be confused with the farmer feeling forced to take risks because of explicit concerns about business failure or job loss. Risk itself is being judged or perceived in relative terms as part of a well-structured business model, much in the same way as mine managers talked about how they weighed and balanced costs, production and safety. However, while Inco's mine managers and miners were often under significant pressure from corporate management to produce or perish, three of these farmers were very secure with large profitable businesses and limited (carefully calculated) debt loads, while the younger farmer in this group was in no immediate financial danger given considerable family support and an off-farm job. As such, it was not immediate insecurity, or even a question of affordability, but rather a more generalized disposition or habitus for making risk-related decisions which sustained maximum profitability. Eerily similar to the statements of mine managers, CS1 put it this way: "if I can't make it pay in some way, I'm reluctant to do it [take the safer option]."

While this logic was often turned towards rationalizing the taking of higher levels of safety and health risks on economic grounds, these decisions were usually constructed by these farmers as rational management choices, even if it sometimes meant exposing themselves, their workers and indeed their family to increased exposure risks. It was this capacity to construct risk-taking as a free rational choice that linked cost-benefit thinking to neoliberal risk management theory, and why it was so powerful in terms of reproducing consent to risk-taking. As a ruling technology or discourse, cost-benefit and risk assessment thinking were not just rationales for accepting risk, they were ways of judging what risk itself means – that is, that risk cannot be judged or assessed *independent of cost*, while any actions must be taken with reference to the benefits that were accrued (O'Malley, 2004). Yet, as there was in the mining context, there were some OHS advantages to this managerialist orientation over other approaches, as will become more evident as the farm comparison proceeds. For example, as noted in the mine chapters, a cost-benefit logic also sometimes worked to reduce OHS risks. One farmer explained:

> I don't use ammonia nitrate any more. You get a mouthful of that vapour and it can knock you right out. There is no reason to put yourself at risk for the pennies you save – it's just not worth it. I weigh the risk against the cost [for everything I do] – it's not a formula or anything but it's how I judge what is safe. I look at the formula of the chemical, what I have to do to mix it, how much exposure I get, what handling is involved, and then at cost, and what I get from using it.
>
> (CS4)

Certainly, as happened when INCO developed its new OHS management system, Group 1 farmers were paying much more explicit attention as managers to health and safety in their planning and purchase decisions. However, because the economic parameters of the equation were still the deciding factors, these farmers were often funneled into making a variety of decisions which impinged negatively on their health and safety in significant ways. For example, when CS2 made the shift to no-till in an effort to be more cost efficient, he decided he needed to buy a new expensive spraying machine, again largely because he thought it was a more cost-efficient sprayer. However, part of his thinking was that the newer, more efficient machine would offer more protection than using the normal sprayer and tractor combination. As someone who had health issues, he had been advised by his doctor to limit his exposure as well as his workload. But then he realized that under his cost-benefit business model to make the machine pay for itself, it had to be used more often than was needed for his own fields, even though he already had 2,000 acres. And so, he began to do contract spraying for other farmers which meant he and his full-time worker were working substantially more hours around chemicals and working harder more generally. While he was careful in handling chemicals and followed all the recommended procedures, he felt frustrated and trapped by the way all his plans seemed to lead to more work not less. Yet, he continued to rely on this cost-benefit logic as he tried to plan his way to an improved situation.

Indeed, all four farmers were taking a number of significant health or safety risks because, in the final analysis, all their planning, flexible cropping and marketing strategies and their use of no-till, had translated into more work, tighter schedules and health and safety compromises. Accordingly, they (and often their workers and/or working family members) found themselves working at night, driving on the roads more with equipment, and spending more time in dusty grain bins. This is to say nothing about the additional emotional stress of managing all the competing demands on their time. As one stated:

> As the technology gets more and more sophisticated, and more expensive, there's more pressure. The decisions you have to make now are more important, the end results are more serious problems. Having more equity now makes it easier, I can haggle now at the banks. But the work always did it for me, love working outside. But 6 or 7 years ago, I realized it wasn't fun anymore.
>
> (CS2)

I would also say that in the end these four farmers talked much more about feeling driven to take certain risks than those farmers in much more difficult financial situations, pushing themselves to get the work done on schedule. All four farmers also said several things during the study which suggested their identities were grounded in high standards of success to which they held themselves when conducting their work, standards by which they judged other farmers and saw themselves as being judged. For example, fulfilling a planting or harvesting schedule regardless of the context were in themselves symbols of their control and success. As one stated: "It means a lot to me to get the fields done and the crops off. It just bothers me. [Why?] I don't know. It just does" (CS1). Not surprisingly, they didn't acknowledge that this often meant that they took risks unless I raised the issue.

We saw a similar kind of disposition or habitus among some conventional miners who saw themselves as independent stope operators where success meant not only the bonus they earned but also the sense of pride that they were excelling at their craft by working through safety and health risks. Framed differently, in Bourdieu terms, risk-taking was a source of "embodied cultural" and "symbolic capital," which in turn helped to form or support a risk taking habitus of these farmers (Bourdieu, 1977; 1990). There are contradictions here as well in that part of the individual habitus was formed around the assumption that certain safety and environmental practices and procedures demonstrated their effectiveness as cutting-edge farmers (Shortall, McKee and Sutherland, 2019). As I observed at farm training sessions and meetings, these and other farmers often exchanged accounts and stories of their various practices, constructing and symbolizing what it meant to be a 'good' farmer. This prestige or symbolic capital also seemed important to them in building social and institutional connections – social capital which they used to access knowledge, funding and other resources which were applied to further improvements in farm production and profitability (Atari et al., 2009).

However, as we saw with the miners and mine managers, these farmers were selective in what they recognized or attended to as OHS hazards or recommendations, even when they had knowledge of those recommendations. CS1, for example, reported that neither he nor his son used sunscreen and rarely used hearing protection.

> R. I find the stuff sticky. And I've never given much thought to UV rays really. But I'm not out in the sun that much. We don't use open tractors any more.
> I. What about hearing protection?
> R. I wear ear plugs occasionally. I just don't think of it and if it is a short time period, I just don't take the time to go get my ear muffs.
>
> (CS1)

Although limited attention to hearing protection was not unusual among farmers, the father in this family acknowledged he had a constant ringing in his ears as well as reduced hearing acuity. As we saw with mining, the difference between these kinds of risk practices and others where more significant changes were reported (e.g. the use of disposable masks when handling grain, installing PTO alarm indicators on tractors), often seemed to relate to whether or not they were seen as affecting and impeding short-term production output, rather than the question of long-term health effects. There was a certain level of health awareness among the farm workers as well following their employers' lead, they saw health protection as question of individual responsibility and preference. As one put it, "I use a dust mask sometimes but they are too hot to use them all the time." In other words, as was often the case in the mines, longer-term health issues or any large-scale safety problems didn't tend to fit into the cost-benefit management framework they'd learned, and thus tended to be ignored, dismissed or minimized with reference to the use of personal protective equipment. As CS4 put it,

> I'm more aware of things now so I take care, I use gloves when I'm mixing and handling, I tend not to spray on windy days, that sort of thing. But I don't pay too much attention to toxicology of chemicals I just look

for something that will work. I'm not too concerned about the health effects on me.

Yet, he was more attentive to safety with respect to equipment maintenance and operations.

Again, one key difference between the mining case and these farms is that management decisions in the mines ultimately had few consequences for the health and safety of the managers involved since they were rarely exposed. In contrast, the farmers were making decisions that were impacting their own health, and in some cases their family members'. And although these farmers were more careful than the other groups, even when they were aware of some effects and risks, they continued with some risky practices, *sometimes contradicting* their well-laid safety plans and their claims of a science-based approach to OHS. In other words, as they actually performed their work, these farmers often talked about risk taking in the same way many miners did – sometimes taking pride in their ability to take certain risks, sometimes denying other risks by insisting exposure was too low or infrequent, and sometimes coping by clearly avoiding any effort to think through these claims or seek more information (i.e. cognitive dissonance).

In as much as these farmers indicated that these were long-standing risk practices and rationales, some of the day-to-day dispositions toward OHS risk that they had developed earlier in their careers persisted despite the adoption of a neo-productivist model. This may simply reflect what Bourdieu (1977) called hysteresis, the persistence of habitus assumptions and practice despite significant structural change, as their adoption of these management ideas were relatively new, but I suggest that the persistence of these ideas also reflects the structural continuities in neo-productivism itself – that is, farmers adopting a neoliberal neo-productivist model were still faced with some of the same contradictions and limitations to their control as farmers in traditional productivist systems. They had more resources, tools and technologies to deal with these challenges than the other groups, but ultimately, they were often making OHS compromises to meet production objectives and images of the good farmer, which in turn reproduced understandings and rationalizations which supported continued risk taking.

Groups 2 and 4

Reflecting the overlap between productivist and neo-productivist orientations, there were some common tendencies among farmers in Groups 1, 2 and 4 including a strong emphasis on farm growth and profit. This included the productivist organic farmers:

Most organic farmers are idealists but I don't know how good they are at the business of farming. I think farming is a business like any other and it has to be run like a business. Organic farming makes good business sense because you don't need expensive equipment and limited inputs.

(CS17)

You have to keep growing bigger. If you are a little guy like me, it's the only way you can make it. I'm constantly looking for more land to farm.

(CS20)

Accordingly, like Group 1, they tended to see decisions about the environment or safety in financial terms. They also recognized to some extent the economic value of preventing accidents and acute health problems. However, in terms of actual management and production practices, specific health and safety issues or concerns played a much less explicit role in their planning or discourse on daily work practices or on various production decisions than in Group 1. Concretely, this meant they paid relatively little explicit attention to OHS management in the way they scheduled and planned their production activities, and did not approach health and safety as calculable risks that could be managed in cost-benefit terms.

For me a lot of decisions are not a sitting down planning thing. I don't lay it all out and I make a lot of decisions while I'm working on the tractor.

(CS4)

This is not to say that they did no planning, but rather that their planning was less detailed and extensive, and certainly not calculated numerically. There were also few explicit efforts to integrate environmental and OHS allowances into their production planning. There were certain ways of doing particular tasks which they had evolved over time and, from their perspective, these needed little or no apriori reflection to assess OHS or environmental risks, and any contingencies could be dealt with as they arose. Production planning decisions were largely economic in nature, focused on costs and expected profit returns, although the organic farmers also had to take into consideration organic production standards. For example, when I'd ask farmers from Group 2 or 4 to explain what they do to prevent accidents when fixing machinery out in the field, the typical response would be "I'm careful"; whereas from Group 1, it would be something like: "I make sure the equipment is properly maintained so it's less likely to break down," "I always block or lock the equipment," "I make certain I have lights and the right tools," "I watch how many hours I've been working so I don't get careless" etc. With respect to training or supervising their workers, Group 2 farmers would again say something like "I tell them to be careful" with little or no direction or training. Similar to what other researchers have found among many small businesses in other industries, they tended to "leave it to them" to be responsible in working safely (Eakin, 1992; Hasle et al., 2011). Group 1 also paid much more attention to formal training and observing and supervising their people in the fields. In other words, accident prevention in Group 1 had been thought out in much greater detail and more explicitly as part of the production plan, with concrete procedures not only to avoid disruptions but also to minimize the delays if such disruptions occurred.

For themselves and their workers, Group 2 were inclined to approach health and safety as something that "you just did," or as "common sense" rather than something you planned and prepared for in advance. In many ways, this orientation was also reminiscent of the conventional miners who saw themselves as being responsible and in control of their stopes without needing or wanting any "outside" direction, knowledge or judgements. Again, this is not to say that Group 2 farmers were unaware of safety and health issues, and certainly some reported changes in their safety practices resulting from the required pesticide course or via other inducements such as an insistent spouse; but as long-time conventional farmers who had not made many changes in their production methods – similar to conventional miners – they believed that they knew from past experience what to do to avoid problems and that any health and safety management programmes, expert advice and technologies were unnecessary if not a hindrance. And like the conventional miners, while they had indeed learned largely through hands-on experience how to assess and control many of the hazards in their workplace they also accepted certain significant risks and any possible safeguards as fixed quantities. As one put it: "I work a lot by myself and when moving stuff I use the tractor and rig it with chain. I should have someone to guide me and tell me if I'm slipping but I don't so I'm taking a risk there but that's just the way it is (CS8).

These farmers also paid relatively little attention to the emerging recommendations coming from the Ministry of Agriculture, the Ontario Farm Safety Association (OFSA) or the Workplace Safety Insurance Board (WSIB). This partly reflected a common suspicion or criticism of the government as untrustworthy or incompetent, but also a general lack of connection to farmer organizations or safety and environmental networks. There is arguably another parallel here with the conventional miners who were relatively distant from their union and the OHS representatives in their workplace and thought of themselves as rugged individualists running their own little business.

Similar to Group 1, health issues tended to get less attention unless a hazard was seen to be overtly affecting their capacity to work, and again with very limited awareness or attention to things like sun exposure, noise and dust. But there was also less attention in Group 2 than in Group 1 to the issues of chemical handling and spraying which seemed to reflect the fact that these farmers were less involved in and committed to the government's sustainability discourse. Much of this I would argue comes down to the tendency of farmers from Groups 2 and 4 to work within a common-sense framework grounded in an indigenous knowledge base and habitus. One stated his approach to pesticides this way: "I don't worry too much about the risk of handling and spraying especially given the volume of spraying I do 2–4 days a year. My understanding is that the length of exposure is what matters and it's not frequent enough" (CS8). When questioned about his knowledge, he acknowledged he had done no research and read very little on the subject.

Since Group 2 was quite similar to Group 1 in looking at their farms largely as a business, all production, safety and environmental decisions were understood as cost and production compromises. The calculations were less explicit, detailed and careful but cost or profit returns were still central to most decisions made by these farmers, while safety and environmental issues were either absent or afterthoughts rather than integrated objectives.

> I went to band spraying five years ago. It was a cost saving thing, cut my chemical costs by 2/3rds but it is good for the environment. Still I gave it up last year cause it was too much time for cultivation. With 700 acres, it was just too much and so I'm back to overall spraying.
>
> (CS3)

In shifting from conventional to organic farming, Group 4 had made more changes to their production in as much as they now had to rely much more substantially on plowing and tilling to work in the manure and control weeds, but there was little, if any, acknowledgment of the added hazards (e.g. noise, road accidents) involved in operating and maintaining large mechanized equipment. On the other hand, they were much less involved in spraying pesticides and fertilizers so this whole area of health risk was seen by them as having been eliminated by their shift to organic (although again the shift to organic was not based on health concerns). As noted, this claim of no risk was not entirely true as organic farmers are exposed to manure gas and toxins and may employ some approved pesticides such as copper sulphate, borax and borates which can cause health problems. However, since personal or family health tended to be secondary to these farmers as motives behind their shift in farm methods, there were few indications of any generalized commitment to healthy work practices. Dust masks, hearing protection and sun protection were all used infrequently or not at all with many of the same explanations found at conventional farms. As one of these farmers noted, "I wear a disposable mask if it is really bad but it's not usually worth the trouble" (CS5).

While Group 1 looked to an integrated management system and leaner "sustainable" production methods as the key to their success, the Group 4 organic farmers saw their profitability mainly in terms of growth, higher prices and lower input costs gained through organic production. For them, this meant focusing on making certain they got the production end working properly, again mainly in terms of manure and weed control, so that the losses in terms of yield did not eat away at those cost savings. This focus on cost reduction could have translated into an effort to minimize costly accidents or incidents as in Group 1, but OHS did not figure in their planning to any significant extent. Instead, the pressures of sustaining yields within the limits of the certified organic production standards and a continuing commitment to a mechanized productivity approach, seemed to translate into frequent safety compromises (e.g. working alone and at night) and very little effort to achieve higher levels of efficiencies and cost savings through more sustainable OHS management techniques. At the same time, since these organic farmers were keeping with many of the key elements of a productivist system, with particular reference to crop specialization and mechanization, the changes in the actual work tasks appeared relatively minor, making it easier for these farmers to rely on their past experience and knowledge as sufficient to protect themselves against work hazards even when that experience included some injuries and close calls. As a result, the added dangers of increased mechanized work, heavier workloads and time pressures were understated as health and safety issues. As one stated, "you can't stop doing stuff that needs doing cause you are worried about an accident."

Group 3

The farmers in Group 3, the ecologically oriented conventional farmers, were much more mixed in terms of knowledge and their orientation to health and safety. Although using chemicals, they saw themselves as changing their pesticide and fertilizer practices to address environmental or health issues. Accordingly, these farmers, with one notable exception, tended to know more and take more care in handling chemicals than Group 2. Of course, their commitment to chemicals as a necessary input for production was also much weaker than Groups 1 and 2. In fact, two of these farmers (CS11 and 12) were experimenting in minor ways with organic production and all expressed an interest in radically reducing their chemicals through IPM methods or by going fully organic. These two in particular claimed that they were being prevented from moving further mainly by the fact that their fathers still owned much of the land they were working on, although everyone in this group stated that they were not entirely confident that organic farming would work on their farms. Nevertheless, they were all quite critical and mistrustful of the chemical companies, their claims of safety, and the government emphasis on chemical based no-till versus other less-chemical-dependent approaches. As one stated, "you can't believe what those [chemical] companies are telling you, just look at their history." None of these farmers were connected to the conventional farmer organizations behind the promotion of the no-till version of sustainability and, were either not involved in farm organizations at all or were involved in alternative farm organizations such as the National Farmers Union (NFU), which as noted was critical of the neo-productivist approach and its emphasis on global marketization. However, the farmers in this group were research oriented and reported readings a wide range of organic and sustainable agriculture publications and literature. Two had connections to Ontario environmental groups and were beginning to reach out to farm organic organizations such as COG, organizations which tended to place little emphasis on OHS. However, the only farmer of the 20 case studies to be involved in or particularly attentive to the work of the OFSA was also in this group.

With the exception of this latter farmer, their attention to safety issues other than chemical handling and spraying was not extraordinary. Like Group 2, they tended to take a common-sense approach to preventing accidents with limited amounts of conscious planning or attention to health and safety in purchasing and scheduling decisions. This conflicted with their approach to planning production around environmental goals which were more systematic and carefully planned. Along with being mistrustful of the government and corporations when it came to claims of chemical safety, they were much less critical or aware of the origins of other health or safety risks, or the need for a more systematic approach to safety planning. As one stated when he was asked about safety concerns other than chemicals, "it is my understanding that safety is just a matter of education and experience" (CS12). Nevertheless, compared to Group 2, there was a greater expression or commitment to the need to attend to and improve health and safety practices and for some this also meant reducing time, financial and production pressures.

However, there were also two somewhat contrasting extremes within Group 3. One of the farmers (CS11) had demonstrated a relatively careless approach to chemicals to me several times in both his mixing and spraying behaviour – not using proper clothing, spraying in open tractors, sometimes in windy conditions etc., and was not all that careful in terms of safety practices more generally. Once I watched him clean his sprayer nozzles after using the herbicide "Roundup" without gloves and then proceed to another task without washing his hands. He was also quite cavalier about chemical safety with his two farm workers, who reported little or no instruction or protections when working with pesticides. When I asked CS11, an older man in his 50s, to explain why he wasn't more careful when using chemicals from a health perspective, he responded, "I figure, whatever damage has been done already so I guess I just say what the hell." This fatalism about one's own health among older men was not unique to this farmer, as noted earlier, but somewhat surprising in as much as he was quite hopeful and enthusiastic about prospects for reversing the environmental damage of industrial farming. But when I asked him about safety in general, he just shrugged and said he couldn't really explain it, which was also odd in as much as he was usually very expressive. I was never completely successful in getting him to elaborate more fully on this, but in the end, it seemed likely to me that he really had not thought that much about safety and was embarrassed when I started asking him questions. So perhaps like many of the organic farmers who seemed entirely preoccupied with meeting both the farm's survival and their organic production standards, this farmer, although not financially threatened, was so heavily immersed in his efforts to transform the farm and farm production into an ecologically sound model, that his own or even his farm workers' safety simply did not occur to him as a significant concern.

On the other hand, another farmer in this group (CS7) demonstrated perhaps the greatest knowledge and care of all 20 farmers in following safety and health procedures, and had closely integrated health and safety in his planning and thinking. While he didn't have any wage workers, he was also extremely careful in ensuring that he did not bring any contaminated clothing inside his house which he explained as protecting his young family. I ak As an active executive member of the local Farm Safety Association, as noted the only active OFSA member in the entire sample, I asked him to explain his concerns and attention to health and safety. He responded that he had developed this focus "through his father," who was a full-time autoworker and a part-time farmer. Although he had his own farm, he worked cooperatively with his father exchanging labour for the use of his father's machinery. As a new parent himself, he was adamant that he had a responsibility to protect his family from the OHS dangers of farming, and in staying healthy himself in order to make the farm a success. His father and ultimately his concerns about his family had prompted him to seek out information from the OFSA, and eventually he became involved as a local representative when the opportunity arose, which in turn according to him, "really opened my eyes to what could happen." He was unusually conscious of health and safety matters but he had not internalized the neo-productivist rhetoric or discourse about efficient and flexible production. While he was not active in any other farm or sustainable farming organizations, he had absorbed many of the farm safety association messages that safety is not an afterthought but an integral part of work planning and organization. He was meticulous in thinking through various risks and researching them in planning his practices. He was the only farmer in the study to put on a complete protective suit, hat and mask when using or mixing chemicals, and required anyone else working with him, myself included, to do likewise.

> I'm always very careful. I always have a jug of water and soap right there. I won't touch anything without gloves and always wash fully and

use an apron when mixing. But it took the pesticide course last year to really get me to be more careful – I never skimp on what I do. Spilled a little on my sneakers, I threw them out.

(CS7)

From his perspective, farmers relying on common sense and the common farmer knowledge were part of the problem because as he put it, "if common sense has been so great why have we had so many farm accidents and fatalities all this time." In his view, training and education were crucial to develop the knowledge and the practices needed to better prevent accidents. He also adopted key aspects of the logic found in Group 1, in particular that safety was a "management" issue in that he had to consciously consider and plan for health and safety as he planned his production, made his purchasing decisions and carried out the work. He also firmly believed in the OFSA-promoted idea that "safety pays" and that every production-related decision, including the purchase of equipment should involve safety and health considerations. Accordingly, he still tended to employ a cost-benefit logic in thinking about safety improvements or safety-related decisions, but not quite in the precise calculated manner of the Group 1 farmers. Moreover, although he tended to weigh the safety end of the calculation more heavily in his decisions, he saw many of his "choices" as no choices at all - risks forced by economic circumstances.

I had an accident with anhydrous ammonia. That really scared me. Last year I wanted to contract it out or use something else but it was too expensive.

(CS7)

Similarly, he was very concerned about sun, dust and chemical exposure because he was using an open tractor but, for the foreseeable future, the most he could do was wear protective equipment and sun block.

Group 5

As one might expect, Group 5, the alternative-oriented organic farmers, were all quite aware of the dangers of chemicals. Also, not surprisingly, they often cited health as a key reason for moving to organic farming. As one explained: "I had a neighbor who taught me how to spray and he died of cancer not long after. I think these chemicals are dangerous." However, as with Groups 3 and 4, this concern about chemicals did not tend to generalize to an awareness or concern about other health- or safety-related hazards in farming. Indeed, perhaps the most committed alternative farmers (CS9) in the group, in the sense that both the husband and wife were fully committed to small-scale mixed organic production marketed locally at the farm gate, were actually the least safety conscious of the entire sample. Core practices identified by the OFSA as highly risky, like riding with kids on the tractor, working on barn roofs with children under 14 and without any safety ropes, driving long distance at night on the road, never wearing dust masks or hearing protection, riding in open tractors without sunscreen, and allowing young children under 10 to ride on the hay wagon while loading bales, were all observed on this farm.

While the wife in this farm was much more critical of these practices and claimed a constant effort to get her husband to be more careful, her husband insisted that their operation was not dangerous because of "the way we farm" by which he meant organic. As he put it, "if you are going to farm in many areas and go, go go, then it's dangerous but, if you farm as I do, it's safe, safer than lots of professions." But he then went on to explain that his philosophy to health and safety reflected his philosophy of life "that life is about taking risks, and you can only learn by doing." As he saw it, he was living and teaching his children the view that life is all about facing risks, not being afraid to take chances and overcoming them.

However, a lack of knowledge also seemed to be at play here in that he appeared to be unaware of many widely acknowledged health dangers. For example, when asked about silo gases, a significant and widely publicized hazard that had killed many farmers (OFSA, 1985b), his response was "I never heard about that." In other respects, he reminded me of the "hay-wire" miners identified in Chapter 4, those hazard "sceptics" who seemed to cope with many workplace hazards by simply refusing to acknowledge or think about them. Still, like them, this farmer had certain safety rules which he had developed from experience.

> Any accidents I had, it just takes experience, a matter of experience. I've never really thought of what I've done or what limits to the risks I'll take but if I have close call I usually do something, I almost lost a finger years back in a corn picker and right after I got something to protect my fingers. I have two uncles that lost hands and fingers working on running machines so I never do that.
>
> (CS9)

Again, for him, much like Group 2 farmers, health and safety was achieved through experience- and common sense and did not require any additional study, planning or training. He consulted with few other farmers, was not involved in any organic farm organizations, and did relatively little reading or research in organic farming or health and safety. This was very unlike most of the other organic farmers in Group 5 and indeed most organic farmers surveyed in Study 3 (Hall and Mogyorody, 2002), whose orientation tended to be that successful organic farming required a complete retooling in their knowledge and orientation to the soil, to the weather, to crops, marketing, etc. This retooling and rethinking were understood as requiring a fair amount of research, networking and involvement in organic organizations, and careful planning and coordination. As such, the importance of knowledge was also internalized among most of these alternative organic farmers, but the difference was that their management focus was not on efficiency or cost-benefit decision-making, but rather on achieving their farm and environmental objectives; that is, succeeding without having to expand or sell to global markets. Again, however, the key difference relative to the managerialism of Group 1 is that health and safety was not an explicit part of their planning or management decisions, and if considered at all, was largely an afterthought.

I would argue this lack of attention to health and safety planning was reflected in and reinforced by the broader discourses of the organic movement as a whole, which ironically largely ignored health and safety as core or central issues or concerns. This was evident in both the larger survey that we did with organic farmers and several years of observations of organic organizations (Hall, 1998b; Hall and Mogyorody, 2001, 2002). Their principal focus was on survival and farming ecologically, and similar to the Group 4 farmers, this was presented by them as a significant challenge. Similarly, the case study organic farmers were concentrating on making their farms work, often struggling with various principles, production problems and market conditions, while largely taking the OHS view that they were "careful when they needed to be." In practice, however, this meant a limited and inconsistent approach to accident and disease prevention and a frequent reliance on established unsafe dispositions and practices.

The meaning of risk calculations, costs and benefits

In trying to draw together the key points on the construction and judgement of risk, it should be emphasized that while cost underlied the risk taking of all farmers to some extent, in as much as cost was rarely seen as irrelevant to any decisions that were made with respect to health and safety, an understanding of the differences between the groups lies with both the level and formality of calculation and the kinds of costs and benefits that were being emphasized. Group 1 farmers were clearly doing much more extensive and explicit risk assessments, calculations and comparisons of costs and benefits when making major production and investment decisions, followed by Groups 2 and 4, and then 3 and 5. But where these groups also differed was on the weight or significance given to different costs and benefits. For Groups 1, 2 and 4, the core benefit was almost always profitability and maximum profitability was highly valued, while the principal calculation was a question of how much something costs, so that in the final analysis environmental or health benefits were constantly quantified in financial terms. This was largely what INCO management was doing in its senior and local management decision processes. For Groups 3 and 5, health and environmental impacts formed the core of their thinking, and these were seen as either principal costs or benefits which dictated many of their decisions, while production, profitability and financial cost were bottom lines which had to be met in order to keep the farm alive. As such, Groups 3 and 5 were much more reluctant to talk about taking health or environmental risks with reference to cost or profit *needs*. They liked to think of health and environmental concerns as separate issues that needed to be addressed without reference to cost or profitability. Their decision-making was accordingly less influenced by the production pressures that were operating among the business-oriented farmers, especially Groups 1 and 4, as they tried to enhance their profitability through significant restructuring.

On the other hand, this also meant that when costs did enter into Group 3 and 5 decisions, it was constructed by them as *compromising* their principles, as a matter of having or not having the money in situations where they really didn't feel they had much of a choice. As CS7 put it:

I would really like to buy a new sprayer and tractor, easier to circumnavigate around the field, it would be more efficient, less drift, plus it would be safer for me. The one I have is really unsafe – no safety features at all. I know it's [the chemicals] affecting me.

(CS7)

He wanted to buy a tractor with a cab and filtered air conditioning largely, it turns out, because of health concerns rather than productivity issues. But one can also see the cost-benefit logic creeping into his discourse as he tried to justify to himself such an expenditure. Still, he did not buy a new sprayer for quite some time and when all was said and done, he didn't because he couldn't afford it. I heard senior and mine managers and indeed many other farmers make exactly the same arguments for their OHS decisions numerous times. However, the difference was that the farmer in CS7 literally did not have the money, while the mine managers and many of the better off farmers constructed affordability in a more complicated calculus which included economic and political costs and long-term payoff.

While the organic farmers in Group 5 were more inclined to speak to the importance of preserving certain environmental principles, these goals may have put even more pressure on them to compromise on safety or their own health in the bid to keep their farms alive. A key and ironic part of the problem then, for both Groups 3 and 5, was that despite their greater concerns about the links between farming and health, their less-secure finances forced them to compromise health and safety on cost or economic grounds. Indeed, those few farmers in these groups who were better off, such as CS11 in Group 3, were much less likely to acknowledge that they ever made these kinds of compromises, always suggesting there were options for making the farm more viable if some environmental or safety decisions imposed higher costs.

> I can use my environmental approach to improve my farm income in various ways. It doesn't have to intrude so it's a matter of choosing one or the other. I can use various strategies to make it viable, and those options come out of making this whole farm more complex biologically.
>
> (CS11)

The impact of restructuring on farmer control and risk subjectivity

A central argument that came out of the mine study was that the transformations of the labour process shaped risk subjectivities and politics by altering the knowledge and control exercised by management and workers over workplace conditions and hazards. Although I've already suggested that the control impacts of the labour process changes in agriculture were less substantial than in the mining case, the question still remains whether farmers, as owners, managers and direct producers, perceived changes in their control within the context of restructuring more generally, and if so, whether this was linked in any way to their responses to workplace hazards.

As might be expected, Group 1 farmers, that is the conventional farmers who had made the most changes in the production and management process (CS1, 2, 5), reported that they viewed the changes in farming methods as improving their control over their production process because it gave them more time to plan and manage the process, and the enhanced management tools and procedures they needed. Although the changes in production practices (e.g. the shift to no-till or rotational grazing) meant some uncertainty about what they were doing, and introduced problems and new risks in certain contexts (e.g. late springs or wet weather more generally represented a greater challenge for no-till), they still saw themselves as being more involved in *managing* the process based on more intensive use of information regarding the fields, the crops, the weather etc.

This sense of greater control was also related to their adoption of management-intensive programmes, audits, information and new spraying technologies, and crop and market diversification strategies, and extended as well to the much more active way in which they monitored commodity markets and "strategized" around marketing and sales. They were especially enthusiastic about the increased control that they felt they were exercising over their day-to-day work and their decisions through their new management tools, while emphasizing the knowledge-intensive nature of their current practices and the idea promulgated by the state and corporate advocates of these changes that this was a "smarter" approach to farming. Emerging technologies and machinery were also seen as critical sources of control to the extent that they provided predictable outcomes with flexible and secure operations. Although they recognized the persistence of production and cost pressures given the continuing problems with commodity prices, they saw themselves as managing these pressures through technology, constant research, training, and new management techniques or systems, which they interpreted as giving them more control over production-level decisions. Again, this was very similar to the way mine managers described their new technologies and management systems.

Those who had made fewer changes in their production and virtually none in their management methods, that is farmers in Group 2 (CS3, 6, 8), insisted that they had experienced very little change in terms of production control, and in general expressed confidence that they were able to produce with the same consistency and quality as ever. Although more concerned about their sense of declining control over the business and the marketplace as a whole, they denied paying major attention to the market developments because these seemed "beyond their control." They did not have the same confidence as Group 1 that they were able to manage the various risks of the business and two of the three were talking about winding down the business or selling if things didn't get better. But other than the stress of losing their farms, OHS was not a major concern for them, in part because they felt that they knew the hazards from experience and could control the risks accordingly. In practice, this often meant ignoring or taking significant risks with inadequate safeguards since they were relying on assumptions, rationales and understandings (i.e. habitus) which understated the risks and overstated their control. This suggests then that similar to conventional miners, it was the lack of change in the labour process that limited the adoption of new OHS understandings and practices.

Group 3 conventional farmers, again depending substantially on their financial situation, reported much more difficulty in responding to restructuring pressures which included for them their ecological and social justice commitments. They were also more likely to report concerns about their capacity to control production pressures given their adoption of sustainable farming methods and were more likely to express a lack of confidence about their control over the production process. Critically, they also saw cost constraints directly in relation to health and safety but were unable to see ways of managing those risks given their financial situations other than being careful; that is, in four of the five cases, the availability or lack of adequate machinery and equipment was seen as a crucial impediment to adequate control over the production process, increasing several health and safety risks.

While their perceived lack of control was sometimes expressed as doubt of their own knowledge or skill in ecological farming methods, and this was also evident in Group 5, more often than not the lack of control over risk was understood as a structural problem – a lack of adequate resources, insufficient state support for ecological farmers and poor prices. As one Group 5 farmer put it when asked about how much control he had over production decisions:

> "the problem is that it is often more the idea of knowing what should be done and seeing what cannot be done. Most of the time I'm just making it, because it's so crucial to do things when that window is open... because if you have to do two things at the same time then you need somebody else to do one of the things, and I don't have the people"
>
> (CS19).

For farmers in Groups 3 and 5, a key part of their efforts to change their approach was the belief that the government and corporate push to increase efficiencies and productivity within conventional farming was simply intensifying production pressures beyond the farmers' control, and that all the managerialist language, ideas and programmes promoted by the government did not make any difference because commodity prices were simply too low and input prices too high. The result, as expressed by one was "that farmers were working themselves sick or to the bone," risking their health and safety to keep their farms going in impossible situations. As another organic farmer recalls why he decided to make the shift:

> I was thinking in terms of self-sufficiency, I thought maybe for a while one of the best ways to be self-sufficient was to get bigger. If you had more land you could do more different things and that would be self-sufficiency... I decided that wasn't the way to go...Probably rightly so because he (his brother who continued to farm with chemicals) died quite young of cancer... I'm not sure that part of his problem wasn't the financial worries he had.
>
> (CS19)

However, the conventional farmers in Group 3 and the organic farmers in Group 5, both of whom rejected the neo-productivist line on farming, also tended to see their approaches as empowering them and giving them more independent control over decisions. Given their rejection of the corporate and government line, and for some the rejection of conventional global markets, they often constructed themselves as being less subject to the constraints of the marketplace, while seeing conventional farmers as being under the direct influence of corporations and governments through forward contracts, extension services, corporate seed and crop advisers, and patent requirements. At the same time, many of these farmers expressed doubts and uncertainty about whether their alternative orientation was giving them the control they expected, noting that yield and market expectations were often not realized, putting them under even more pressure to take risks than they'd experienced as conventional farmers.

Indeed, sometimes these conventional farmers seemed to fit the description of farmers with little control over their farms. For example, although CS3 had clearly stated that he controlled the way he worked, it became very clear through observations that he relied heavily on seed company and processor company recommendations and advice. I recall going with him in the spring to the local Cargill office, where he picked up not only his seed, fertilizer and pesticides, but also step-by-step instructions from Cargill's crop adviser, which he stuck to quite closely in the weeks and months that followed. Others, even organic farmers who were producing under advanced contracts, acknowledged that they had to follow buyer instructions very closely:

> They are pretty picky. You have to weed more carefully for sure. The year before I grew them [soybeans] on contract but I didn't worry about weeds and they were stained. They wouldn't take them so I was just feeding them to my cows and I sold a few. I had about twenty acres that year too and I sold some but they were transitional beans and there's no market for them. I took a beating on them.
>
> (CS19)

As such, farmers who avoided conventional markets tended to see themselves as gaining more control because they were less subject to agribusiness controls and market pressures but, as the evidence shows, they were not immune to some of these same market pressures.

While the knowledge-intensive nature of organic farming was emphasized by all organic farmers as key to their increased control, the more mechanized organic farmers (CS10, 17, 18) were very much like the large conventional farmers in that they also saw much of their control as coming from their machines, their technologies, and ultimately from their entrepreneurial and marketing strategies which allowed them to accumulate enough capital to expand their land and their technologies. The alternative organic farmers, on the other hand, tended to emphasize the more labour-intensive nature of their labour process as their source of greater control, ultimately more knowledge since they saw themselves as being closer and in greater touch with the land and their crops. As one stated:

> Well from what I see, organic farmers have control of their own lives and when all is said and done they don't make very much money but they made a little as opposed to the people who bought all of the inputs and did all of the big things that they were supposed to do that didn't make any, or lost it, and it was a whole lot less efficient. I saw as well that if you did things in a little less mechanized way or less

automated way that more people had jobs to do, and they were meaningful jobs.

(CS)

While critical of mechanization, most organic farmers did not romanticize
or equate physical labour with skill and control. Rather, they saw the working of the land as a part of limiting the impact of mechanized farming on
the soil:

> Yeah. Some growers have these transplanters, tractor-mounted trans-
> planters, and there's a few growers that are the same size we are who use
> that and some wonder why wouldn't we because we do a lot of trans-
> planting. Yes, it's a conscious decision to put the plants in my hands...
> I wouldn't want to make all our raised beds by hand, that's for sure.
> 'Cause I think that would be exploitive of human labour. (laughter). So
> there's a line there. Some things are just too much work to expect it to be
> just human labour but then there are some things that... you can easily
> become machine crazy.

(CS4)

This less-than-positive view of the skills involved in the physical aspects of
farming was particularly evident among Group 4 organic farmers, where
again the skill and the knowledge were seen in management and technological terms: As one put it: "Sure there is more management skills in organic
but not really more skill on the labour [side]. It's just physical labour and
more of it" (CS10).

Perhaps just as important is the finding that many alternative organic
farmers also acknowledged that the labour intensity of their approach, especially those who were trying to sustain mixed livestock and grain farms,
often translated into a crushing workload that was both exhausting and
stressful. As one farmer stated, "Our way of keeping control was to stop
chemicals and this gave us a sense of freedom, but it also locked us into having to do more work and so we had less freedom to do other things. We feel
better about it but it's hard"(CS9).

The lack of state regulation

The mining analysis suggested that despite the limitations of a neoliberal
OHS regime with its emphasis on self-regulation and internal responsibility, state regulations and legally defined workers' rights including the rights
to participate and to refuse unsafe work, were significant in pushing the
company to direct more attention to health and safety and in sometimes
empowering workers to challenge their conditions. However, the absence of
any OHS law covering agriculture at the time, and indeed the absence of the

right for farm workers to unionize were, I would argue, significant factors in limiting the pressures on these farmers to attend to health and safety issues in a more systematic way. The lack of legislation likely explains as well why so few farmers in the sample had integrated health and safety into their "management" planning and thinking, into their supervision and training of hired workers and family, and into their own daily practices. A concern about preventing regulation certainly figured in the thinking of some of these farmers when they moved to adopt some of the OHS recommendations coming from the OFSA, OMOL and OMAF. But this shift was not a primary motive even behind Group 1's adoption of OHS management systems. The politics of inclusion as pursued by the UFCW and the NFU during the 1990s, as discussed in Chapter 6, certainly helped to put some pressure on farm organizations and the state, as did the environmental movement's pressure on chemical exposures. Still, these movements never achieved the level of public controversy and attention that was attained around industrial and mine health and safety in the 1970s and 1980s (Adkin, 1992; Storey and Tucker, 2006).

However, it was important that in order to sustain the image of the farmer as a responsible user of pesticides and a responsible employer, the industry and government discourse moved to emphasize the importance of farmer knowledge of hazards (Hall, 1998a; 2007). While used as an argument against the inclusion of farms under OHSA, the push to educate did shape some farmers' attitudes and practices in positive ways. However, in the final analysis, at least until 2006 when the OHSA was finally extended to farm workers in larger farmers, farm owners across the spectrum of production orientations were not compelled to take health and safety as seriously as environmental issues (Hall, 1998a).

Conclusion

This chapter suggests that the neoliberal orientation to the calculation and management of OHS risks had a significant impact on the way health and safety risk was understood and approached by some self-employed farm producers and ultimately their workers. As we found in the mining case, some of those effects were positive, in as much as farmers who fully adopted these ideas were more active than other farmers in thinking about health and safety in their planning and purchase decisions, in seeking other sources of knowledge, and in anticipating and planning for high-risk circumstances. Operating within this framework, farmers sometimes consciously invested in safety procedures or equipment which reduced or eliminated hazards. Yet the same kinds of limitations underlying this governance approach in the mining case were also evident in these neo-productivist farming contexts – the rationalization of risk on cost-benefit grounds, the obscuring of hazards in numbers and calculations, the individuation of responsibility, and the intensification of production and efficiency pressures. Despite all the planning and calculating, and some significant improvements in production practices, these latter tendencies still translated into significant amounts of risk taking by self-employed farmers and ultimately their workers. As such, while the introduction of these new technologies and management systems likely disrupted aspects of the pre-existing productivist risk dispositions or habitus of farm owners and workers in some significant ways, the new forms that these dispositions and understanding took still followed an economic logic and a set of game rules and parameters which concealed or encouraged compromises in health and safety for production and profit motives. Moreover, as the farmers' accounts suggest, some of the changes in OHS practices were more apparent than real, in as much as these farmers continued to construct their farm identities around a willingness to continue to take certain risks (e.g. working late into the night 'until the job is done').

Somewhat disappointingly, the alternative farmers' rejection of neoliberalism and neo-productivism based on environmental and social justice principles did not yield a greater awareness or commitment to health and safety either for themselves or for their workers– in fact, quite the contrary. Most alternative farmers devoted very little time or attention to health and safety planning, relying largely on personal experience and common sense. This "common sense" approach meant the persistence of dispositions and practices which exposed farmers and their families and workers to significant OHS risks, while giving little attention to other sources of knowledge, the need for training for themselves and for their workers, or the importance of health and safety planning and investment in their management decisions. Our observations of organic farm organizations in Ontario over several years from 1998–2002 also revealed very few efforts to promote or draw attention to health and safety prevention as an integral part of their management and production practices (Hall, 1998; Hall and Mogyorody, 2001). The larger survey and additional case studies conducted with organic farmers in Study 3 also reinforced this conclusion that health and safety was not a priority issue. These findings are somewhat puzzling in as much as organic farmers, as an alternative movement, placed a strong emphasis on health, education, science, and the need for a tight farm management system (Hall and Mogyorody, 2001). It is also puzzling to the extent that many of the organic farmers in this research were relatively new to organic farming and, in that sense, were adapting to somewhat different labour processes with relatively limited experience, which if consistent with the mine study, should mean some changes in their experience of control and habitus.

I suggest that the persistence of a limited common sense approach to risk taking among alternative farmers has its origins in some of the same kinds of factors that historically shaped a similar habitus within conventional farming. First, while productivist and neo-productivist farm policies promised a golden future, family farms have generally suffered through decades of crisis and economic pressures routinely threatening their survival (Hall, 2003; Lind, 1995). With the likely exception of dairy farms covered by Canada's supply management system, few family farms in Canada, whether conventional or organic, have not faced significant financial peril at some point over the last three to four decades (Hall, 1998b). In that sense, the idea that survival and financial necessity requires compromises in health and safety was relatively well ingrained in the minds of most Canadian farmers, even to some extent the better off farmers in Group 1. This history has helped to shape a risk-taking habitus which was tied in significant ways to the broader culture or habitus of the farmer as a rugged individual who works hard and does what needs to be done to keep the farm in business. As conventional farmers moved into organic farming, the habitus persisted. While the neoliberal neo-productivist call for smart managerialist farming challenged this habitus to an extent, the case study evidence suggests that the culture of risk taking persisted even among farmers embracing this model.

Ironically, many alternative farmers expressed an effort to move away from the pressures of the conventional productivist model and the neoliberal version as their reasons for moving to organic farming (Hall and Mogyorody, 2001). However, organic farming introduced its own financial and production challenges, given the tendencies toward more limited capitalization, old equipment, unstable local markets and prices, and inadequate labour supply (Hall and Mogyorody, 2001). I should emphasize that these tendencies were not only evident among the case studies discussed here but also among most of 240 organic farmers surveyed in Study 3 (Hall and Mogyorody, 2001, 2002). Farm survival was a constant worry for most organic farmers. As such, alternative organic farmers often found themselves feeling forced to make unexpected compromises on economic grounds, not only in health and safety but also in terms of their organic principles (Hall and Mogyorody, 2001; Macey, 2000). By relying on common sense and their own skills and knowledge as their main approach to prevention, farmers could avoid dealing explicitly and psychologically with many of the costs and compromise issues that would have been raised had they taken a more systematic and scientific approach to prevention. Moreover, for some I suspect the prospect of worrying or planning about health and safety issues was simply one challenge or stressor too many, encouraging some to adopt the same strategies of denial or understatement of risks that we saw among miners worried about the insecurity and their lack of control.

I want to emphasize again that these 'strategies' were often not conscious decisions in as much as farmers often claimed that they simply never thought about health or safety or feeling under pressure to compromise, and sometime they acknowledged that they consciously didn't want to think about it. As such, the default way for dealing with this perceived lack of control, especially when it became clear that there would be injuries, was to fall back on common sense and personal experience as the basis for risk judgement and control. Again, of course, this meant a persistence of certain assumptions and understandings that over time proved ineffective in preventing injuries or disease. Thus, like all farmers and farmworkers to an extent, organic farmers constructed themselves as being able to control those risks by being careful, using their skill or by using protective equipment or procedures. But like the miners who felt trapped by management pressures, insecurity and family obligations, some farmers and workers in these situations, whether organic or not, partially coped by ignoring and understating some risks, and/or by overstating their capacity to control other risks.

As also suggested, a second factor explaining a persistence of weak OHS practices across farming is that the public and state discourse and politics around OHS in farming never approached the level of organization and controversy around mine and industrial health and safety in the 1980s. The lack of unions, the weak position of migrant and young farm workers, and the contradictory structural positions of farmers and their farm families as self-employed producers, translated into limited success in pushing farmer and public awareness and discourse on health and safety, which helps to explain not only the delay and weaknesses in state regulation, but also the failure of many farmers to incorporate a concern for health and safety into their own approach to restructuring. By not requiring farmers as "owners" and "employers" to fulfil certain requirements and recognize worker rights, even if in only a limited way, they were not pushed as employers to recognize OHS as a core business interest (see chapter 9). As will be argued in the next two chapters, many small firms in the auto industry generally did not move in a sustained way to pay more attention to health and safety planning and management until new state regulations forced them to set up joint health and safety committees, inspection schedules and procedures and reporting requirements. Even in the context of weak enforcement, regulations introduced uncertainties which owners recognized as requiring management attention. As well, the lack of a union representing the vast majority of farm workers, and the precarious employment of many farm workers, was likely quite crucial then as it is now in limiting the political pressures on farm owners to change their views and practices for themselves and their workers (Arcury et al., 2005; Basok, Hall and Rivas, 2014).

While these arguments apply to all farm types, the absence of regulation allowed organic farmers as individual operators and as a movement to essentially ignore health and safety as an integral feature of their management and production thinking. The focus was firmly on making organic farming work from a business and an environmental perspective, and the significant challenges of this were perhaps more than enough to push health and safety aside in the absence of any regulative requirements (Guthman, 1998). Reflecting this orientation, organic farm organizations like COG and EFAO placed all their educational efforts and energies towards helping farmers to be successful organic farmers, in effect serving to perpetuate the lack of health and safety awareness and planning among farm operators (Hall, 2007). It did not help matters that the OFSA was at the time largely unconnected to and inactive within the organic movement.

A third line of explanation for the more limited changes in OHS awareness and practices among farmers during this time period is that the changes in the labour processes and production conditions were not transformative in the sense that they were for bulk miners in the mine case study. Unlike the mines, both from the worker and the manager perspective, most conventional farmers were not experiencing substantial work and job restructuring at the time of the research and, in that sense, their conditions and their control over those conditions had not been substantially altered, allowing them to continue to rely on their own personal perception and experience-based knowledge in judging and responding to risk So, unlike the VRM and mechanized miners, there were no major disruptions to the risk habitus they had developed over their life course and farm career, and for them, the assumptions were that they could respond to production pressures as they always had – by working harder and accepting or dealing with risks as they came along.

However, the relevance of the labour process also varied to some extent with the different groups. This was especially true for Group 2 since they made very few changes in their farming practices. Group 3 as well had only shifted their practices in limited ways by moving away from chemicals but their commitment to high levels of production and mechanization was largely consistent with their past practices and the belief that they exercised control and responsibility for the work they did. From the perspectives of these farmers in particular, they saw no change in their control over their work and accordingly didn't see the need to enhance their knowledge or alter their practices or their thinking (Shortall, McKee and Sutherland, 2019). If they had any effect on these farmers, the restructuring discourses in agriculture largely fed into and reinforced these self-reliant meanings, habitus and identities, while the relative lack of strong messages on health and safety as part of that management-restructuring discourse ultimately meant enhanced self-reliance without all the other neoliberal baggage – knowledge intensification, formal risk assessment, measurement and calculation, audits, and standards. And of course, for Group 1, it was their full adoption of the neo-productivist model of governance that made the difference in their practices and habitus, not the labour process changes per se.

In contrast, the alternative organic discourse emphasized the greater control they would exercise over the labour process by simplifying their dependency on chemical inputs and machines (Hall and Mogyorody, 2001). Again, in practice, the changes in their day-to-day production may not have been that radical but there were new demands and hazards associated with the increase in physical labour. However, the discursive emphasis on simplification and individual physical control helped to reinforce the view that health and safety could be managed through common sense and care, rather than detailed planning and investment in OHS improvements.

A final point, one which segues into Chapters 8 and 9 on the auto sector is the extent to which the findings are linked to the small business status of most family farms. Some research suggests that small businesses in general tend to take less intensive and formal management approaches to health and safety prevention, while owners are more likely to dismiss or normalize hazards (Barrett, Mayson and Bahn, 2014; Eakin, 1992; Eakin and MacEachen, 1998; Frick and Walters, 1998; Hasle et al., 2011; Vickers et al., 2005). As Joan Eakin (1992) described it, they tend to "leave it up to the workers" (p. 689), a pattern we saw in many farms. But the evidence also suggests, there were important differences among farm owners, with some responding much more favourably to these governance trends than others, a finding also found in the small business literature (Hasle et al., 2011; Vickers et al., 2005). Similar small business patterns and differences will be relevant to the Chapter 8 analysis of the auto parts sector research in as much as many of these workers work in relatively small firms, while the case study in Chapter 9 examines a firm that adopts neoliberal strategies of OHS governance as it grows from a small to moderate size.

The transformation of production and health and safety in auto parts manufacturing

The main study employed in this chapter (Study 4) was aimed at understanding the links between several labour process and management characteristics and OHS politics in unionized auto parts firms. Plant union chairpersons completed a survey questionnaire about plant size; the use of lean production methods such as just-in-time (JIT) delivery systems; the levels of automation, skill, repetitiveness and direct supervision; job assignment flexibility; labour relations; worker solidarity; and, the use of worker participation programmes such as quality circles. Worker health and safety co-chairs of the joint OHS committees were also surveyed for their assessments of health and safety conditions, the functioning and effectiveness of the health and safety committee, their own effectiveness as reps. and management's commitment and cooperation. Follow-up qualitative interviews with 21 of the OHS co-chairs provided detailed accounts of the reps.' activities and efforts to gain improvements in conditions. As reported elsewhere (Hall et al., 2006), the qualitative analysis revealed three forms of representation which we called technical-legal (TL) representation, political activism (PA) and knowledge activism (KA). A cluster analysis of a later survey of 888 reps. across Ontario in multiple industries (Hall et al., 2016) offered further support for the argument that reps. approach their roles in different ways with different standards and expectations. This chapter first examines the relationships between lean production measures and health and safety concerns and politics and then moves to focus on how differences in rep. orientations affect OHS politics and outcomes.

The auto parts sector in Ontario

At the time of the study, several macroeconomic developments were affecting the auto industry as a whole including increased global and regional competition in the context of free trade agreements such as NAFTA in 1992 and the WTO in 1995, accelerating shifts to computerized production and automation, and a widening adoption of lean production management methods including JIT delivery systems (Babson, 1995; Lewchuk, Stewart and Yates, 2001; Lewchuk and Wells, 2006; Rinehart, Huxley and Rob-erston, 1997; Stanford, 2014). After a relatively strong period of industry heightened increasing competitive pressure as the Canadian dollar increased in value from 2002 to 2007 eventually reaching parity with the US dollar, although the full extent of the crisis did not happen until 2008. As well, by the mid-1990s virtually all parts firms were required by their assembly firm customers to be QS and/or ISO certified which imposed quality and production standards, pushing more effective record-keeping, auditing and management systems to insure implementation and compliance (Johnson, 2002).

A comparison of parts firms

Similar to the farming and mining industries, "leaner" production models and neoliberal risk management practices were being encouraged by government and industry organizations as solutions to management problems in these contexts. However, like the farm industry, this did not mean that all parts firms and managers adopted or applied these ideas and practices in the same way. The survey of the unionized firms sampled 36 different-sized manufacturing companies (20–800 employees) and revealed variations in the application of lean production principles, use of assembly processes, levels of automation, job skills and proportion of trades, use of participative or team approaches, and labour relations climate. Some of these differences reflected the particular parts being manufactured as well as the firm's history and corporate status. While there were also important differences in how different firms approached their management of health and safety, and in particular their approach to IRS (Vickers et al., 2005), all the companies were reportedly in basic compliance with IRS regulations in OHSA (1990), in the sense that they had committees and elected/union appointed worker representatives conducting inspections and meeting every three months with minutes and follow-up procedures.

Although based on a limited convenience sample of a provincial industry with 700+ firms, this description likely mirrored the industry as a whole in as much as the Ontario government had moved more forcefully in its 1990 OHSA and enforcement policy reforms to require IRS adoption (Hall, 1991; O'Grady, 2000; Ontario, 1989; Storey and Tucker, 2006; Tucker, 2003; Workplace Health and Safety Agency, 1994). If a functioning IRS was more the norm in 2003, this was in stark contrast to where the industry was in the 1980s, when firms with less than 50 employees were not required to have committees or representatives, and a significant majority of manufacturing firms with over 50 employees were found to be breaking IRS regulations (McKenzie and Laskin, 1987; Ontario, 1986). It is likely that some smaller non-union firms were still less compliant at this time, as this was evident in the larger surveys of workers (Study 7) and reps. (Study 6) completed in 2011 and 2013 (Hall et al., 2012, 2013). Certainly, several of the smaller unionized firms in the first rep. study reported that they had only recently moved into compliance in the previous five years. The case study firm examined in Chapter 9 also reinforces this argument as it didn't move into compliance until the mid-1990s as it grew from a small- to moderate-sized operation (<50 to >300 employees). Some support for this as a more general trend comes from a government agency study in 1994 which found that 80% of Ontario workplaces were in compliance with IRS regulations (Ontario, 1994), a major improvement from the results of the 1986 survey (Ontario, 1986). Although not assessed directly in the first rep. survey, the follow-up interviews suggested that most of the firms were also operating with formal health and safety management systems which included OHS staff, safety rules and procedure manuals, formal training, auditing systems, and organized safety communications, although again the smaller firms tended to have more limited systems (Eakin, 1992).

OHS conditions as viewed by reps.

Only a minority of Study 4 reps. assessed their overall health and safety (4%) conditions as "poor," while 50% rated them as "fair" and 46% as "good." Interviews suggest that what reps. often meant by good or fair included some significant exposures to risk. This was evident in the survey in that many reported that most workers in their plants were being frequently exposed to a range of hazards potentially dangerous to their health and safety (see Table 8.1). While noise and ergonomic exposures top the list, housekeeping such as oil spills or other trip hazards, dust exposures and machine hazards were also seen as frequently present by a substantial proportion of OHS reps., suggesting these were normalized at some level.

Only four of the respondents thought that *health* conditions had gotten "worse" over the last three to five years, and no one thought that *safety* conditions were worse. A majority (70%) thought that safety conditions had

Table 8.1 Frequency of exposure to hazards

N=36	Often/Very Often (%)	Sometimes (%)	Rarely/Never (%)
Chemical Exposure	36	28	36
Noise	84	13	3
Dust/Gas	49	38	14
Temperature Extremes	35	35	30
Biological	12	9	80
Ergonomic	68	30	2
Housekeeping	59	32	8
Machine	48	41	10
Stress	39	42	19
Work Fatigue	46	44	11

"improved," while a substantial but smaller proportion viewed health conditions as improving in "recent years" (44%). Overall, health conditions such as air quality were seen as the most difficult issues to get addressed, similar to what was found in the second rep. survey (Hall et al., 2013) and in the mining study (Hall, 1989).

OHS and the labour process

As noted, the initial objective of Study 4 was to determine if certain labour process and management characteristics were associated with health and safety problems and conflicts. Although the small sample size limited the quantitative analysis, the data suggests that worker representatives reported more concerns about health and safety conditions within certain labour process conditions. With respect to lean and flexible production measures, OHS concerns increased the more their firms were operating on JIT schedules ($r = -.555$ $p < .05$) and the more workers were being moved from job to job to meet production needs ($r = -.543$, $p < .05$). Qualitative data also supported this in as much as reps. were more likely to talk about production pressures as increasing safety risks when their plants were operating with limited buffers. As one put it, "you're trying to build that much quicker, trying to get the part out the door...people start getting hurt."

Consistent with the mine study, three measures of job skill were also related to OHS concerns. Less repetitive work in the plant ($r = -.422$, $p < .05$), overall assessment of job-skill requirements ($r = .497$, $p < .05$) and the proportional representation of trades to unskilled labour ($r = .571$, $p < .05$) within the workforce were significantly related to better overall assessments of health and safety and/or assessments of safety alone. Several hazard exposure measures – chemical exposure ($r = -414$, $p < .04$), dust ($r = .506$, $p < .01$), machine ($r = -567$, $p < .003$), and stress ($r = -463$, $p < .02$), as well as a combined scale of exposures to all nine hazards listed in Table 8.1 ($r = -631$, $p < .004$), were also strongly correlated with the number of trades in the plant suggesting that perceived hazard exposures decreased as the trades in the firm increased. On the other hand, most of the skill or lean production measures were not significantly related to most exposure measures with three exceptions: repetitive work and temperature hazard exposures ($r = .573$, $p < .002$), job movement and exposure to machine hazards ($r = -567$, $p < .003$), and job movement and emotional stress hazards ($r = -463$, $p < .02$).

While mechanization and new technology at INCO were critical in shaping miner and representative concerns about certain hazards (e.g. diesel fumes), in this study the level of automation (all production to none) within the auto firms, and change in the pace and level of automation and/or the use of robots were unrelated to any of the OHS assessment measures. Levels of automation were also only weakly related to the other labour process measures such as repetitive work, skill levels, job assignment flexibility and just-in-time.

This is perhaps not entirely surprising. In the mining case, the labour process was shifting from a high-skill multitask process to highly specialized and mechanized jobs with substantially reduced skill and control. Few of these parts firms were reporting the kinds of radical job shifts that were taking place in the mine study. Although not asked directly about the effect of automation on worker skills, only a minority of the worker co-chairs (17%) saw new technologies as having a negative effect on health and safety, whereas close to half (48%) saw technology as neutral in its impact; close to one-third (35%) saw technology as having a positive impact. In later interviews, these positive views were usually found to mean that representatives saw new machines and/or automation as eliminating repetitive, dirty or dangerous tasks.

While other interpretations of these findings are certainly possible, they give some support for the argument made in Chapter 5 that lower-skill labour processes, assembly lines and "leaner" production approaches tend to lead to more OHS concerns being raised at the joint committee level. This may have reflected higher risks within these jobs because managers had more control and power to impose more OHS risk on workers; or, as we also found in Chapters 4 and 5, the concerns may have reflected the greater reliance that lower-skill workers placed on external sources of regulation or governance as means of getting change, given their lack of control and power to resolve issues directly. As found in those chapters, miners with more skill-based control, power and status were more likely to address their OHS concerns by either directly modifying their conditions or practices or by negotiating informally with their supervisor or manager. What I also argued was that higher-skill miners were also often consenting to significant risks based on interest rationales and identities. Unfortunately, whether the latter was happening in the auto parts firms is impossible to say with this data.

Another possible interpretation is that the auto OHS reps. were paying closer attention to lower-skill workers in their plants. As may be recalled, in the mines, the higher-skill conventional miners and the trades people, such as the electricians and mechanics, tended to question the competence of OHS reps. to pass judgement on their practices; and whether the reps. agreed or not, they tended to be less likely to intervene or seek to discipline worker practices among the higher skilled trades and conventional miners. Further support for this argument comes from the survey finding that auto reps. in lower-skill plants reported receiving higher frequencies of worker complaints (never to very often; $r = .481$, $p < .05$). Of course, we would also expect lower-skill workers with limited control and power to be more reluctant to voice their concerns for fear of reprisals or other negative consequences but, if the same dynamics are operative in the auto plants as in

Table 8.2 Pearson correlations – labour process and OHS cooperation/conflict

Labour Process	Committee Productivity	Significance	Management Cooperation	Significance
Proportion Trades	.618	p < .002	.381	p < .07
Just in Time	.209	NS	−.058	NS
Repetitive Jobs	−.015	NS	−.172	NS
Flexible Job Assignments	−.150	NS	−.015	NS

the mines, concerns and complaints are still more likely to emerge despite those restraints when risks are seen as more serious.

With more safety issues, we might also see more committee conflict and/or more representative concerns about management cooperation. In the mine study, the more significant levels of OHS conflict were in the most intensive transformed VRM production mines. As the mine comparison in Chapter 5 also showed, the level of conflict depended in part on whether management restructured worker control in other ways, in that case through a more sub-stantial team-based system which gave workers and worker representatives more control over and confidence in health and safety related information and decisions. However, as Table 8.2 indicates, most of the relevant labour process and cooperation/conflict variables within the parts plants were unrelated – with the exception of the proportion of trades to semi-skilled production workers and joint committee productivity (r = 618, p < .002).

Again, the sample size was too small to do much more thorough quanti-tative analysis but it is worth noting that a firm's use of participative man-agement or team approaches was more common in firms with higher levels of repetitive jobs (r = .385, p < .05), which was suggestive at least that the findings in Table 8.2 reflected the moderating effect of participative manage-ment approaches.

However, employment security may also have affected whether workers and worker representatives acted on their concern through joint committees or other means. There was some variability as some firms were experienc-ing layoffs and were under threat of closure, while others were still growing and adding staff as the main crisis didn't hit the entire industry until 2008. Consistent with other research including the mine study that job insecu-rity discourages worker complaints (Basok et al., 2014; Hall, 2016; Lewchuk, Clarke and de Wolff, 2011), the number of complaints reported to the joint OHS committee co-chair did decline with the reported decline in employ-ment numbers (r = .558, p < .05). As well, where companies were introducing changes that negatively impacted on employment, worker representatives were more likely to report problems in management cooperation – that is, in those plants where employment increased, management cooperation tended to be perceived as better; where employment was decreased, cooperation declined (r =.568, p < .05). Perceived changes in health and safety were also linked to employment changes, improving with increased employment and declining with reduced employment during the same time period (r = 434, p < .05).

While only one OHS rep. rated the impact of job security on health and safety as "very negative" when asked directly in the survey, close to one-third of the reps. (31%) viewed the impact as negative, and in follow-up interviews, the political impact of insecurity was further acknowledged by a number of OHS representatives. As in the mine study, some saw their management as constantly pushing workers with this threat. As a representative in one of the more taylorized just-in-time plants stated:

> The mentality that runs that plant is, do more faster, harder, more, you know what I mean. More production, more production. And they [workers] get this drilled in their heads every minute of every day. And if I could tell you how many times that, if I got a quarter for every time I heard we're gonna close, we're gonna close, we're gonna close.

On the other hand, again like some of the miner representatives, some of the autoworker reps. accepted market and technological impacts on employment as legitimate pressures and demands outside management control, requiring some health and safety concessions. As one put it, "if you have resources, you can have your preventative maintenance…But when you cutback, it affects that area…You're not looking for those problems at that point."

While these findings were suggestive of important links between lean production, job skills, employment security and OHS politics, a number of the expected relationships were not found. This could be partly due to the small sample size and the lack of sufficient variability within the sample but, if we look at the qualitative interview data, it became clear that there were likely other reasons why the relationships may not be as straightforward as expected. Reps. often noted, for example, that while skilled machinists and trades people such as millwrights tended to keep to themselves and deal with OHS issues on their own, this depended on the issues specific to the plants and jobs in question. Ventilation was a critical problem affecting trades people as much or more than the less-skilled assembly workers, but this was not a hazard that workers could control directly and it was not something their supervisor had the authority to resolve. As such, those issues were more likely to come to the reps. as complaints from the trades and skilled machinists, just as they came from the trades people and the mechanized miners in the mines. Differences in the responses of skilled and less skilled labour to workplace hazards also reflected the use of bonus systems in some plants and not others, in the sense that bonus systems encouraged speed-ups and discouraged complaints, again findings reflected in the mine study. Other plants had quotas which allowed workers to limit their work time so that once they'd reached their quota they could either go home or take coffee breaks in the cafeteria, while others forced workers to work from "buzzer to buzzer." These differences in the structure of work time likely made a difference in whether reps. received complaints from workers about the pace and other issues regardless of skill.

Differences in representatives' orientation

The lack of predicted labour process relationships was also a function of the data itself in as much as we were relying on the reps.' accounts of conditions and concerns. These data limitations came into stark relief when we observed through the follow-up interviews that the representatives were often using different standards of OHS assessment, management cooperation and committee impact when answering these key survey questions. It was really this recognition that led us to an effort to differentiate the reps. according to different orientations and practices – which in line with this book's use of social field theory, can be conceptualized as different forms of "representative habitus"; that is, representatives performed their roles grounded in certain assumptions and beliefs about what they were doing and why, about their power relations with management and about what was possible and reasonable to achieve within the committee and the IRS as a whole. Concretely, this means different reps. engaged in different kinds of representation practices and strategies.

As noted, grounded in the 21 reps. interviewed in the first auto rep. study (Hall et al., 2006), we argued that OHS representation tended to take three forms: technical-legal (TL), political activism (PA) and knowledge activism (KA). The second rep. study (Hall et al., 2016) also found three distinct clusters of representations, using a survey instrument which asked 888 reps. to identify the relative amount of time they spent on various activities (e.g. reports, inspections, research, education, connecting with workers).[1] We identified two of these clusters as good representations of TL and KA forms of representation. The third was more of a middle range of the other two often related to the amount of time they'd been representatives, suggesting to us that many of these reps were in transition towards either a TL or KA habitus. The qualitative data revealed some differences among those within the KA cluster, enough to suggest to me at least that our PA categorization was worth retaining in representatives and, as such, I'm going to retain this category in the discussion that follows.

Representatives with a TL orientation or habitus internalized the apolitical IRS partnership model promoted by the state and many corporations, which meant that they largely assumed that their role was to monitor the workplace and workers principally through technical-based inspections (audits) and report to their management partners in the joint committee. Further, they tended to assume that management would then take responsibility to address the problem as conditions and circumstances permitted. As one put it, "we do the inspections, we (management rep.) do the inspections together, and then we sit in office, discuss some things, then write down everything, and then we leave it to them [management]." Delays in management action due to conflicting management or firm interests were either ignored or constructed as legitimate business concerns. As one put it, "Well there's some situations that we felt things are needed but they can't be justified because of cost and even the investigation of the concern. It would be something like some main ergonomic change." While reporting high levels of management cooperation and commitment, and high levels of committee impact, when asked to illustrate cooperation and impacts, smaller-scale issues and corrections were emphasized with limited attention to systemic or underlying causes. For example, TL reps. often talked about cleaning up oil spills as they were identified in inspections without ever identifying cuts to maintenance or production paces as the underlying causes of the spills. When asked about the changes they made, the focus was on relatively minor housekeeping matters as the following quote illustrates:

> Yeah, just little things like you know maybe putting up a mirror on the wall where the forklifts come around the corner so everybody can spot a forklift... just little things like that, make sure the floors are clean and you know no tripping hazards and stuff. (TL3).

While often aware of larger hazard issues, such as systematic air quality problems or systematic maintenance shortfalls, these representatives assumed that these conditions were untouchable aspects of production, based on technical or cost rationales presented by management. TL reps. then confined their role to protecting people within these fixed technical and economic conditions. As another TL stated:

> Every plant has back injuries [so] they [management] try to stop them but you are always going to get that. Ever since carpal tunnel syndrome came too, there's a lot of people with repetitive injuries and that's why we got the ergonomic committee to work on that.

These representatives also looked almost entirely to performance technologies such as audits to define, judge and control risk. As one TL rep. put it when asked what it meant to be an OHS rep.: "Every time we go down on the floor we're doing an audit. Every time we walk down on the floor you're doing a visual around". Mastering and employing these audit technologies was their source of power or embodied cultural capital. Such mastery also gave them some status or symbolic capital within the management hierarchy and among workers. As another TL rep stated: "The safety inspections are really good...you know, it's regimented so they know when I'm coming out there, so it's usually very clean. Sometimes you know they'll get a score, I score it out of ten...." TL representatives accordingly often defined their role in management terms as educating, monitoring, evaluating and disciplining the work force. As still another put it, "my job is to monitor things to find where people are not working safely and correct them."

For TL reps., then, management cooperation was defined by their mutual and coordinated efforts to ensure basic safety standards and procedures, while impact was largely gauged by the inspection and auditing metrics established by management, metrics which were weighted around whether workers were following rules and procedures. Cooperation also meant largely taking direction and assurances from management with limited debate or independent worker input, while defining their role in technical, legal and apolitical terms realized through formal meetings, inspections and audits. As another of these representatives responded, when asked how he differentiated his role from other union positions, "no, that's labour relations, it's not political, it's a different avenue – health and safety...I guess it's just about people's safety" (TL8).

For other representatives, referred to here as either "politically active" (PA) representatives or knowledge activists (KA), cooperation meant progress on some larger-scale hazards and a greater focus on health than just safety, plus a greater willingness on their part to meaningfully *contest* or change management decisions on significant OHS issues. These latter reps. also recognized the limitations of OHSA as providing relatively few strict standards or requirements leaving lots of "grey" area for disputes. As one put it:

> The stuff that's the hardest to get is the stuff that's grey in the occupational health and safety act – a manager would look at you and say that's a luxury...So sometimes you either have to figure out how to *force* their hand or they won't do it.
>
> (KA1)

Management cooperation *and* commitment on health and safety were not understood as givens, as often seemed to be the case with TLs, but rather were understood as *political* outcomes nested in their capacity as worker representatives to leverage the power they needed both within and outside the committee context to persuade and sometimes force management to work with them on addressing major as well as routine safety and health matters.

Ventilation is at the top of the list because it's an ongoing concern – all these fumes will cause health problems you get guys with bleeding noses...you can get quite serious about this and you just have to act a bit harder on the company... I've got the Ministry of Labour coming in and issue orders, getting testing done...Every day I'm on the production manager's mind, I'm in his face every day.

(PA3)

PAs and KAs were also more likely to recognize the importance of their relationships with workers in achieving changes which also meant consciously building worker knowledge and support.

Our ability as a membership to insist on safety corrections is far stronger than the company's ability to buy its way to the answer 'no'. So stalling became one of the patches in these grey areas. And I would go to back [at them] for any legitimate demand, but in some of them the company looked through me to the workers to see how serious is the anxiety. They'll actually ask the workers involved in those areas. If they get feedback that it's just bullshit, it's just a couple of guys complaining, they won't move on them.

(KA1)

Significantly, in terms of understanding the links between lean production and OHS conditions and politics, the politically oriented representatives were more likely to identify lean production, deskilling and job insecurity as systemic sources or causes of persistent safety or health problems, which may also further explain why the quantitative relationships were weaker than expected.

I would say the biggest issue today would be the lack of preventative maintenance of equipment and tooling. That would be the biggest one. Because when you downsize all your departments, something has to be impacted by that decision....And as a result, some things go unchecked for longer periods of time.

(KA6)

They also recognized these factors were making it more difficult for them to mobilize workers which is why they needed other strategies that didn't depend entirely on worker solidarity.

If we really got sticky with them on something, their answer would be that they're just gonna farm it out...Yep. The company would just say it's not feasible, it would cost too much money and they don't have the time to do it, they need the parts now, so we're just gonna ship out the work and let these other people do it. People, even senior people, just quit complaining.

(PA4)

To the extent they recognized them, the TL reps. in contrast tended to accept these "realities" as constraining what was possible, which may also explain how they came to construct the little things like cleaning up spills as their primary function.

Not surprisingly, autoworker representatives who took stronger political stances were more likely to report conflict and limited management cooperation (King et al., 2019; Walters et al., 2016). This may help to explain the weak quantitative relationships between many lean production measures and OHS conflict, since TLs and PAs were distributed across all kinds of workplaces and labour processes. However, as this also implies, the form that OHS politics took was not determined by the size of the workplace or the type of labour process (see also Hall et al, 2013).

Knowledge activists

As also outlined elsewhere (Hall et al., 2006), some of these politically oriented representatives (i.e. KAs) were more successful in establishing meaningful cooperative relationships with management within the joint committee and/or more broadly with management, leading to substantively better OHS conditions than other plants, while retaining their critical connections to and support from the workers and their union. This subset of politically oriented representatives we called KAs (Hall et al., 2006, 2016) because more so than the PAs, they *actively and autonomously* gathered specialized knowledge and information on hazards which they used to mobilize workers and position themselves politically to challenge management's monopoly on knowledge and risk assessment.

As was clear in both rep. studies, the level of cooperation and effectiveness in gaining changes varied with knowledge activists. Some of this variation reflected differences in the representatives themselves, in the sense that some were more active and more skilled than others in building and using their knowledge. For some, it appeared that the cooperation and effectiveness was a product of several months or years of effort on the part of the representative, in effect taking the time to build her/his knowledge, reputation and cooperative relationship with management (and the workers).

Our continuous complaints on issues I think did result in change...the more I squeak, the more oil they put on it because they don't like to hear me whining all the time...After a while, management, [union] leadership, the plant manager get sick and tired of hearing me complain and they start looking to their management personnel and subordinates, 'why can't you deal with this guy's problems, with the committee's problems. So they were putting pressure on the manager to do it...he had his budget constraints but where they could help they did...[To get there], in the 90s, we had to fill out lots of complaint forms and they would have to be addressed and in writing.. management understood we were not going to tolerate not getting these things done.

(KA8)

Other knowledge representatives were either at the earlier stages of this effort, or were facing a more difficult or reluctant management (see Hall et al., 2016). Many representatives also talked about the challenges of dealing with management changes, which in some auto firms, seemed to happen frequently, forcing them to have to start over to build their relationship. As one put it, "It took me a while to get them to come around. They didn't like it at first but once I showed them that I knew what I was doing, they started to listen" (KA8). Other KA representatives argued that their success was tied in an important way to changes in managers or owners, especially if they were more open to a knowledge-based cooperative approach.

R. If you went to the Essex aluminum plant for example, I know my counterpart J, has had a lot of problems with the management people. It all depends on your management, it really does...You have to have it at that plant manager level. When you get a plant manager who always just says 'where does it say that in the agreement or the law', and if it doesn't say it, 'I'm not doing it'. So again you're working with the minimum standard, you know, the occupational health and safety act, you're going to be in a lot of trouble.

(KA3)

Some of the TL representatives also saw changes in management or ownership as being central to their ability to get changes, but again, these representatives were using different standards about what kinds or levels of change were being achieved. Still, this reminds us that a TL orientation didn't just happen as a function of government and corporate promotion – it also reflected the strategic orientation of local owners and managers. As one TL rep. reported after experiencing a change in owner and HR manager:

I just simply go and talk to the human resource manager now and he takes care of it and it's no problem. Pretty much anything I wanted taken care of around there, as long it was, as long as it's legit and as long as it's not, you know, a huge ticket item that's gonna take some time, they pretty much, pretty much we're pretty, pretty good that way. The owner's son now runs the business too as well, that's been a lot of,

there's been a lot of changes going on in the last few years, and he seems
to be very self-conscious about health and safety.

(TL9)

However, as with other TLs, what this rep. meant by "anything" tended to
be limited to short term fixes and minor low-cost changes.

These findings also suggest that as with individual workers and worker
representatives, we can differentiate individual managers according to differences in "management" habitus, especially with reference to whether
they were disposed to accept worker input on production and OHS issues as
ways of achieving production and cost goals; as opposed to managers who
reportedly refused to allow worker input entirely or who paid lip service
to worker input as ways of delaying resolution and obscuring concerns. As
indicated, when TL reps. referred to their managers as cooperative, it meant
something quite different than what KAs or PAs meant. For TLs, cooperation ultimately meant accepting management direction, control and goals as
the prerequisites of continued goodwill, while KAs recognized cooperation
and their right to participate as political contingencies and opportunities to
make OHS gains for workers.

While the first rep. study admittedly did not involve interviews or observations of managers, these rep. characterizations of managers and supervisors were consistent with some of the variations found in the mining study
and what I will show in the case study presented in the next chapter. Moreover, while the habitus of different managers and owners regarding OHS and
HR management certainly affected how difficult or successful the KAs were
in terms of OHS outcomes (Shannon, 2000; Tuohy and Simard, 1993; Walters et al., 2016) my perspective is that the forms that worker representation
and committees took and their effectiveness are best viewed as the political
achievements of workers and worker reps rather than the concessions of
enlightened managers (Boyd, 2003). Managers were not all the same but
without some structural constraints on their power (as in the McCreedy
Mine, see Chapter 5) and/or substantive political opposition from workers,
"cooperation" meant more often than not concessions to management goals
and interests.

As was more evident in the farm research than the mine research, the
labour process was not the defining or determinative feature of OHS subjectivities and politics. With respect to the auto parts plants, KAs were somewhat more common in lower-skill lean production contexts, perhaps because
there were more OHS issues and/or perhaps because workers were looking
to the committee and representatives more so than in workplaces with more
powerful trades workers, However, there were TLs, PAs and KAs operating within all varieties of production contexts and management systems.
This was evident not just in comparing the 21 cases in the first auto study

(Study 4) but also in the large rep, study (Study 6) which included workers in a much wider variety of occupational and industrial settings (mining, transportation, construction, manufacturing, education, and health). In the latter research, construction, manufacturing and mining were more likely to have KA forms of representation than the other sectors, likely for a variety of complex reasons including the levels of risk within the different production processes, but KAs and indeed TLs were again present across the full spectrum of workplaces and labour processes (Hall et al, 2013).

It was also evident in interview accounts in both studies that the politics and/or form of representation in workplaces often changed within the same labour process or workplace context as either new managers or new worker representatives entered the workplace or the committee, or some other significant change occurred such as layoffs or hiring. Like the farms, labour process changes seemed to play a lesser role in some cases in shaping or encouraging a shift in OHS politics. However, similar to the mines, some KAs and PAs referred specifically to recent changes in the production systems including new technologies, speedups, or major job skill changes as provoking their decisions to get involved as worker reps. and/or changing their thinking about the need to challenge management more effectively. In some cases, it was also the reactions of workers to those changes. As one KA put it, "when they started to cut maintenance, and people started complaining, I knew what I was doing [as a rep.] was not enough." However, before dealing further with the complex question of how different forms of representation develop in different work contexts, let's first consider in social field terms why KAs were more effective in many workplaces.

Explaining the relative effectiveness of KA representation

Although admittedly based on self-reported impacts alone, KA forms of representation seemed to have better impacts. As compared to TLs and PAs, KA worker representatives in both studies (Hall et al., 2006, 2016) reported more frequent and substantial changes in working conditions, procedures and engineering design, and were more likely to identify and make progress on health issues such as air quality, stress and chemical exposure (Hall et al., 2016). As one KA put it:

> We now have a procedure we negotiated involving new equipment, we sign it off before it gets to the production floor – if they bring in a new press, we look at the PSR (regulated prestart safety review), we still *do not trust* the engineer who signed the PSR, we go look at the piece of equipment – the equipment [itself] may be safe but if there is no ventilation, lighting, anti-fatigue matting or support equipment or if it backs into a busy forklift area, the machine may be safe but not the worker.

As argued elsewhere (Hall et al., 2006), the key to this success was the emphasis that reps. placed on *actively and autonomously* seeking specialized knowledge and information on hazards. This worked because they positioned themselves politically to challenge management's monopoly on knowledge and risk assessment, especially relevant for health issues. As the same KA explained, this included an active use of social networks outside the workplace:

> Trial and error. Talking to other people within your local, other plants, magazines, internet...I sit on a committee at the Local level too, a health and safety committee where you talk to other reps from other plants and by networking you find somebody who's done the same thing before.... and call the Ministry of Labour inspector and ask them for advice.

This approach contrasted with the less successful PA representatives who tended to rely on their own personal knowledge, worker complaints or common sense to make their case. Perhaps more importantly, when rebuffed by management, they used political tactics which relied on traditional forms of worker mobilization (e.g. slowdowns), Ministry of Labour inspectors or official union support (e.g. grievances). Although potentially still important sources of power among KAs, the difference was how that power was mobilized. As PA reps. often reported, they were unable to mobilize or sustain the support and power they needed, in part because they did not have the necessary connections (i.e. social capital), knowledge (i.e. embodied cultural capital) and reputation (i.e. symbolic capital) within the workforce and the union for making and legitimizing their claims (Bourdieu, 1977). Without this capital, management was able to use their control of knowledge to obscure risks and importantly within a neoliberal context, make better use of employment insecurity to undermine worker and union support. As one KA representative put it, "if you're educated enough to know what you are talking about, they [management] don't talk down to you. They know it and they'll do it. Especially if it's in black and white. They like seeing things on paper" (KA1).

While again the level of the activism certainly varied among the KA representative, in that some were collecting information and building their knowledge base almost *continuously*, and others were more situational or issue-focused, they all shared the perspective that effective representation required them to seek information *autonomously* through non-management-controlled sources. Unlike TL reps., they understood they could not rely on management information or management's experts to give them the full story, while also recognizing both the technical and the political limits of their own common knowledge, particularly in more complex matters such as disease causation. As one representative acknowledged, "there were things in the [air assessment] report that were incorrect and I felt I needed some help from people that knew what they were talking about more so [than me] because like I'm still new to that..." (KA3)

As this again implies, they understood that to build an effective case they often needed to use scientific information and network sources as critical capital to validate and legitimize their arguments. More often than not, this involved not only consulting independent experts with credentials and relevant experience but also seeking more union or outside training so that they could build their own cultural capital and their credibility or symbolic capital with workers, managers, ministry inspectors and their fellow union officials.

> R: I've done a lot of work with OHCOW – they've always been there to help me interpret anything. Um, also to like help myself a bit, I took hygiene course. Actually, I took a whole certificate at St. Clair College. But it was all like introduction level courses so there was a hygiene course, and a toxicology course – the focus of it was not to come out and be an industrial hygienist. The focus was to know how to ask the right questions.
>
> (KA14)

In contrast, TL and PA representatives rarely reported going beyond the minimum education requirements, or seeking help outside the workplace (Hall et al, 2013).

The importance of being able to make a case for change using management's risk language (i.e. embodied cultural capital) cannot be underestimated. The key to their success according to many of these representatives was that after a time of building their capital, they were able to achieve significant changes principally through their interaction with managers with little or no need to mobilize workers, either within the context of the committee or through more direct one-to-one interactions, essentially relying entirely on evidence-based argumentation. Again, however, soliciting support from workers or the union, or using threats of government inspections, was sometimes integral to their strategies or, as one put it, "whatever works," but they generally contended that over time they were able to gain more changes through discussion and persuasion. Some also recognized that somewhat different tactics were needed with different managers. However, behind this capacity to persuade was again a history of building a reputation and support among workers. If significant costs were involved, ultimately reps. needed to be able to persuade managers that it would work for the company. As one rep. explained when asked to describe one of her successful efforts to get a change in working conditions:

> Doing my own research and seeing the people first who were complaining. I showed them [management] on paper...I went on internet and

got pictures of what I wanted and showed where the problems are, that the ventilation was behind the worker and the smoke goes past you...I showed it to the workers first and then we collectively started talking to the company and show them where the changes would be, that the guys will be on the job longer without having to go to first aid and report that there's itchy noses and coughing and spitting up black.

(KA14)

This emphasis on the strategic use of knowledge also contrasts with the accounts of PA mining-worker representatives who tended to complain about the capacity of the company to use science and engineering information and knowledge to obscure health and safety issues. While this obfuscation was still happening among the technically oriented auto reps. in their workplaces, the KAs have turned the tables on management by using knowledge and the standards of legitimacy as effective political tools or capital. In other words, they have adapted to and exploited the political-economic effects of increased insecurity and cultural conditions of neoliberal governance by shifting their political tactics and tools (Bourdieu, 1990).

While the dangers of getting trapped in the logic of cost-benefit analyses have been discussed in both the farm and the mine chapters, KA representatives recognized that costs were always an issue for management, and unless they dealt with them to some extent on management's terms, in part by constructing and calculating benefits that management could buy into, their ability to achieve any change through persuasive means alone would be limited. As in the case of the Group 1 farmers in Chapter 7, it was highly likely that many of these representatives were accepting some hazards because they couldn't "get the 'numbers' to work out"; but still, for many of these KA representatives, they had a clear strategic notion that there was a short and a long game that had to be played. While TLs also accepted that some changes would take longer than others, they had no strategy for achieving the long-term changes leaving it essentially to management.

As a final point in this section, I want to note that while the actions of some reps. were effectively limited by the technocratic regime of rules, legal regulations and procedures, KA reps. were more effective because they understood how to use the rules of the IRS game to achieve change. Although it often took considerable effort and sacrifice on their part, they played the politics of this game quite effectively with or without direct union or state involvement. Some cited the state and the law, or their union and the collective bargaining agreement, as significant sources of potential leverage and constraints on management. Part of this power was grounded in the IRS regulations which were also often elaborated in collective bargaining agreements including unilateral rep. rights to stop unsafe work, in the sense that they placed certain legal responsibilities and normative expectations on management. So, while it is accurate to argue that the politics of health and safety in most of these auto factories were somewhat less confrontational than we saw in some of the mines, this did not mean that workers and reps. did not need whatever protections the law and union could give them.

Explaining the origins of KA

The analysis presented thus far has suggested that a TL consensual approach to OHS representation and labour relations more generally was shaped by neoliberal approaches to economic policy, management and governance. As argued from the outset (see Chapter 1), the widespread adoption among corporations, unions and worker representatives of an IRS technocratic approach was achieved partly through Bill 208 OHSA reforms and subsequent changes in enforcement policy after 1990s (OMOL, 1989a). Along with expanding and enforcing joint committee requirements and procedures, while still limiting committees and reps. to advisory roles (Hall, 1991; Storey and Tucker, 2006; Tucker, 2003), the legislation created the Worker Health and Safety Agency (WHSA), which was a bipartite agency devoted to the development of required certification training for worker health and safety reps. (OHSA, 1990; OMOL, 1989). However, as a bipartite agency, corporations exercised considerable influence over the agency's training content, which, in effect, largely pushed the technocratic partnership model by emphasizing basic technical and legal knowledge and ignoring the political-economic aspects of health and safety. My interviews with labour activists and union officials active in the reform movement before and after 1990 (Study 8) suggest as well that the CAW, and indeed most major labour unions in Ontario, had largely adopted the technocratic partnership model to health and safety by 1990 and, in that sense, were receptive to developing a training regime that depoliticized health and safety and encouraged worker representatives to embrace the inspector or auditor role (Hall, 1992). As a senior CAW official put it,

> OHS in the CAW tended to be separated from our strategies against restructuring and cutbacks – it was seen as more technical and less political. We recognized that restructuring and cutbacks were affecting health and safety, in loss of power, but I don't think this then got expressed in OHS reform.
>
> (Interview, March 18, 1992)

This lack of connection also extended to the way in which the union leadership viewed and reacted to the work of health and safety committees and representatives. Thus, although CAW had a history of opposing production teams and quality circles in the auto industry (Gindin, 1995; Yates, 1993), health and safety committees were accepted as more collaborative sites where it was deemed appropriate for worker reps. to limit their activities to advising *management decisions* about health and safety (Gindin, 1995; Lewchuk, Stewart and Yates, 2001; Rinehart, Huxley and Robertson, 1997).

In effect then, the development of a technical-legal orientation among many reps. is not entirely surprising given this history. Yet, the PA and KA reps. got essentially the same initial training, and were likely subject to many of the same government and management efforts to capture them in a technocratic box. In this final section of the paper, I argue that paradoxically, neoliberalism has also helped to shape the conditions underlying the development of knowledge activism as an emerging and relatively effective political response. I have two main points I wish to make.

I. The contradictions of a participative knowledge-based governance model

Grounded in governance theory (Dean, 1999, 2007; O'Malley 2004), the same neoliberal governance technologies and discourses that shaped TL forms of representation can be understood as both creating the political space and shaping strategic and tactical insights that KA reps. are using to gain OHS improvements. The key observation, I argue, is that their participative rights offered them the political space, while the state and corporate emphasis on technical knowledge offered the strategic and tactical orientation which they adopted in a modified form. I suggest further that the power underlying knowledge-based activism reflects the broader cultural shift within a neoliberal information technology–based economy, where the persistent reference to the importance and significance of knowledge-based decision-making, with particular reference to actuarial risk assessment and calculation, is adapted by some worker representatives as strategic opportunities to empower themselves and workers in general (O'Malley, 2004; Rose, 1993). To the extent that they were successful, these workers were disrupting the dominant TL habitus within their committees and likely the worker habitus of risk acceptance operating in their workplaces.

Dean (1999) is particularly helpful in explaining how a participative ethic could create this political space of resistance when he argues that neoliberal theories of rule or governance are inclined "to guard against governing 'too much' by appealing to the rights and liberties of individuals" (p. 165), and I would hasten to add, individual responsibilities. By this Dean and others suggested that as neoliberal thinking gained dominance, there was a greater effort to assert that the full potential of organizations to compete in the marketplace could only be realized by ensuring the "freedom" of individuals to use their full potential, whether as workers or managers to act, to compete and to innovate as free subjects with choice and agency. The individual right to refuse within OHS law was the clearest expression of this impulse, but it was also expressed within human resource and OHS management more generally (Boyd, 2003). This individual freedom was then understood by employers and managers as essential to organizational success, an argument that has also been made to some extent within labour process theory (Belanger, Edwards and Wright, 2003; Lewchuk and Wells, 2006; Thompson, 2010). However, for Dean this was also part of larger cultural and political shift where the Keynesian adversarial divide between workers and managers was re-conceptualized as a "partnership" of free individuals who must be contracted, consulted and involved because it was only through this kind of collaborative orientation that agencies, firms and organizations of all sorts could reach their full growth potential within a free-market context.

If this kind of participative thinking was more than empty rhetoric as Dean and other governance analysts and labour process analysts have suggested, then the increasing adherence to this worker participation ethic had within it a capacity to "free" workers to exert more control over their work and to influence management decisions. Reading from this theory, we can understand how the idea of worker involvement or empowerment may have been resisted by many managers at INCO, in part at least because managers had still not internalized or routinized these notions in ways that gave them substance or consistency in everyday management practices (Piore and Sabel, 1984). Certainly, there is a large body of evidence on management participation more generally suggesting that the limitations placed on participation have been as much a function of management orientation to production as any structural or technological constraints (Vallas, 2003). Moreover, there is evidence that management attitudes towards participation improved since the 1980s and 1990s (Boyd, 2003). For example, a longitudinal study by Geldart, Shannon and Lohfeld (2005) found that managers had more positive attitudes towards the importance of worker participation in health and safety in 2001 as compared to 1990 (see also Geldart et al., 2010).

However, if there has been a more substantive shift in management thinking since then, why does research continue to find so many differences in corporate applications of participation principles as outlined in this chapter? Dean's work offers a line of argument here as well which brings us back towards the regulation perspective, when he reminds us that the "free subject" must also be subjected and disciplined, whether as a citizen (in schools) or as a worker (in training and supervision), and so we should anticipate significant limits and mechanisms constraining and shaping worker influence in significant ways (Dean, 1999, p. 118). However, as he went on to argue, subjection and subjectification were often laid upon another within neoliberalism so that they appeared identical at one point and quite disparate at others, while at the same time were understood as a condition of the other. It is in this context that we can have two distinct lines of political development occurring simultaneously. Just as we saw with the valuing and devaluing of scientific knowledge (Beck, 1992), workers were being subjected on the one hand to increasing management control through technological, technical and performance techniques, while at the same time, management was seeking *quite sincerely* to draw workers into planning and decision-making processes as part of an effort to enhance performance and prevent costly and disruptive injuries. We saw this clearly in the mining case when there were management efforts to standardize worker behaviour at the same time that we saw management designing or sustaining work processes where worker skill and knowledge could have an influence on production outcomes. This pattern was also evident in many of the auto parts plants, and I will document the same pattern in the case study firm presented in Chapter 9.

From this perspective, management control and worker participation work out in practice over time in specific workplace and market situations as products of various other structural and historical influences, including of course different forms of individual manager habitus. However, the notion that there are competing elements within neoliberalism helps to get us around the logjam that seemed to catch many of the early labour process analysts when they insisted that worker involvement and participation schemes were largely illusions of worker empowerment, and lean production principles like continuous improvement were simply modified forms of Taylorism (Lewchuk, Stewart and Yates, 2001; Rinehart, Huxley and Robertson, 1997). Having drawn those conclusions, labour process analysts have limited avenues to explain why many if not most workers in restructured contexts refuse to see in clear one-dimensional terms that they are, 'in fact' losing control and power through restructuring except to claim some kind of false consciousness (Belanger, Edwards and Wright, 2003). As I tried to show in both the mining and the farm study, and now in the auto parts context, restructuring within neoliberalism and post-Fordism is not a zero-sum game where management is always gaining and the direct producers are always losing. In part, this is because workers contest changes but it's also because, as regulation theorists have often observed, despite all the new technologies and control techniques, management does not exercise complete or even adequate control over production without worker motivation and commitment, thus requiring a constant process of both give and take (Burawoy, 1979; Littler and Salaman, 1984).

In combination, regulation (LPT), field and governance theory help us to understand that neoliberal corporate or production governance approaches like IRS contain within them a number of substantive contradictions in terms of management and worker control and power. Concretely, one of those contradictions is that modern HR and OHS management theories place emphasis on the value of *real* worker involvement and initiative at the same time that they are preoccupied with endowing managers with the extraordinary powers of numbers and prediction. Another contradiction which plays to the issue of empowerment is the simultaneous post-Fordist emphasis on flexibility and standardization (Beck, 1992; Piore and Sabel, 1984; Thompson, 2010). Again, how far different firms go in reflecting this relative emphasis on worker involvement, flexibility, standardization and management by numbers is accordingly quite variable. As we saw in the case of INCO, there were substantial differences in worker involvement - even between mines – as well as differences in worker security and rule enforcement. Accordingly it was a challenge to create management cultures, structures and decision-making processes which effectively integrated workers' qualitative experiences with management's quantitative reductionism while, at the same time, broadened worker responsibilities within a narrower frame of technological and organizational standards of performance.

Nevertheless, the core point remains: if a dominant cultural belief underlying modern neoliberal management theory is that firms succeed by interacting with and involving workers in decisions, not by ordering them around or beating the hell out of them, then this becomes a potential source of both worker consent *and* power to resist to hazardous conditions. Certainly, to the extent that workers' identities and values are also being reshaped both by this neoliberal discourse on individual rights and freedoms, and the management discourse on worker involvement, the work-participation discourse creates worker expectations in terms of influence and control which then underlie impacts on concrete social relations which are not going to be one-sided. As already suggested in Chapter 3, autoworkers were much less inclined to accept minor injuries and health symptoms than miners. While these differences may reflect the relative dangers of the two work contexts, I have tried to show in this chapter that it may also reflect larger continuing cultural shifts in the view of risk and risk responsibilities, and that these shifts are linked in contradictory ways to the further development of neoliberal discourses and ideologies.

II. The contributions of a knowledge-based union militance

In line with above argument, my second point is that by openly attacking labour union rights, neoliberalism also played a critical role in politicizing unions such as the CAW, not only fuelling its militancy but also shaping its strategic and tactical direction. This included a similarly placed emphasis on the power of knowledge and education in creating worker solidarity. While I've argued already that the CAW leadership had adopted a separate and mainly technocratic approach to health and safety, it is important to note that relative to most labour unions in Canada, including the USWA (the union representing INCO's workers), and certainly relative to the UAW in the US, the CAW took a militant position against neoliberal trade agreements, collective agreement concessions and team-based lean production systems. It also pushed political and social science–based education programmes for its leaders and representatives as ways of enhancing organizational capacity and militance (Gindin, 1995; White, 1987; Yates, 1993). I would further argue that CAW's adoption of a knowledge-based militancy was itself a reflection of these same broader cultural and political economic changes, including the emergence of the so-called knowledge economy (Piore and Sabel, 1984). It is worth noting, for example, that CAW's education programmes included courses on computer literacy as well as courses in history, political economy and economics.

Certainly, there were factors other than neoliberalism shaping the particular history of CAW militancy which have been dealt with in some detail elsewhere (Gindin, 1995; White, 1987; Yates, 1993). However, the central point here is that as CAW grew to become the largest private sector union in the country, it was promoting knowledge- and research-based political strategies and tactics in its community organizing, political lobbying, strike and collective bargaining activities, while encouraging an adversarial political-economic view of capitalism and labour–capital relations. While the central union leadership itself did not connect the dots in terms of applying these same principles or ideas to health and safety, local union leaders and representatives, the KAs often reported that the union was important in shaping their political orientations and research orientations. In short, these reps. adapted and applied some of the same political and research skills being encouraged in the fight against neoliberalism and lean production to health and safety. It helped as well that CAW saw aggressive collective bargaining as a crucial buffer against state deregulation in the mid-1990s in as much as PA and KA reps. often reported that they were able to use the collective bargaining process to gain them more time allowances, more powers and more protection for their OHS activities.

Additional influences were likely of critical importance in shaping the development of a KA representative habitus including the development of information technology. While doing much to undermine worker power in many workplaces, computers and internet offered tools which KA representatives could use to access the information and knowledge they needed. While Bill 208 in 1990 had pushed the OHSA emphasis on technocratic inspections and technical education, the legislation including the creation of the Health and Safety Agency, also reinforced the broader idea that prevention and change in health and safety was knowledge and education based. The creation of a certification training system delivered largely by experienced worker representatives was a disadvantage in some ways in as much as many of these teachers adhered to TL principles of representation, but over time this system also offered the opportunity for some KAs to develop and begin teaching their approach. In Study 6 for example, it became very clear that a significant number of instructors involved in certification training were promoting KA modes of thinking.

As well, although there were a couple of early clinics and information services available to workers in the early 1980s (i.e. the Windsor Occupational

Safety and Health Service in Windsor and a USWA-sponsored OHS clinic in Hamilton, Ontario), the creation of government-funded worker occupational health and safety clinic network (OHCOW) in 1991 gave reps. and workers access to new knowledge resources and experts in epidemiology, industrial hygiene, occupational medicine and ergonomics. Again, there were contradictions and limitations in the way some of the clinics operated but as a whole they offered workers and unions a much better funded independent network of OHS experts and health professionals and did much to promote the importance of scientific knowledge and research in prevention (see Chapter 10). With these cultural, technological and political developments, it is perhaps less than surprising that more and more worker reps. began to realize that scientific and technical knowledge were critical forms of cultural capital which could be used within the context of the internal responsibility system to steer owners, managers, government inspectors and fellow workers in desired directions.

Conclusion

This chapter further explored the relationships between the labour process, HR and OHS management and the politics of health and safety within the particular historical context of the auto parts industry in the early 2000s. Although the evidence suggests that the management or worker representative approach to health and safety is not fixed by any particular form of production orientation, there were some indications that an emphasis on low skill and/or JIT production schedules with limited buffers and flexible production and job structures yielded more OHS concerns at the joint committee level. At the same time, there was only limited support for the argument that worker involvement mechanisms mediated or moderated those effects. This may be a function of the weak forms that participation took in many workplaces. As we saw in similar situations in the mines, management opposition to meaningful worker participation weakened any capacity they might have had to give workers a sense of meaningful knowledge or control regarding the risks of the job.

However, consistent with the general argument, reduced conflict was linked to management responsiveness in and support for the joint health and safety committee And although the evidence suggests that management cooperation was often achieved by the political efforts of workers and worker OHS representatives, rather than freely given, it was also clear that managers brought different approaches, styles and levels of commitment to OHS committee, sometimes irrespective of what else was happening in the workplace (Lewchuk, Robb and Walters, 1996; Shannon, 2000). On the whole though, the evidence in this chapter further reinforced the argument that management OHS practices were not a simple function of a single uniform management theory such as participative HRM nor a necessary outgrowth of some labour process restructuring approach to lean and flexible production. Rather, they were also a function of a particular *history* of production politics and management discourses that had shaped the current regime, including the life histories shaping individual manager and worker habitus (Bourdieu, 1977).

Still, the fact that there were substantial levels of both cooperation and conflict within the context of the different plants returns us to some important questions which have formed the concern of much of this book. How and why do these politics take place within a neoliberal context if workers are supposedly so disempowered and workplace committees so encumbered with constraints, incompetence or corruption? Why would employers, given their strengthened leverage to threaten plant closure, continue to play this participation game if it leads in some cases to some significant and cost-heavy concessions? Some answers to these questions come from the LPT argument that workers were not as disempowered within automated or lean production as is often argued, and that there were areas of increased leverage in the jobs themselves that workers gained in these contexts (Belanger, Edwards and Wright, 2003; Thompson, 2010). However, another answer comes from a more nuanced line of argument that the state health and safety IRS policy and the corporate discourses on worker involvement were mutually reinforcing, albeit with contradictory effects in terms of management control and labour consent, yet still leading within those contradictions in important ways to meaningful OHS gains for some workers. This also leads to the argument that it was not just mutual contradictory interests that were generating support for worker participation in health and safety, but rather the power of the participative discourse itself, one that extended beyond health and safety to a broader management discourse. Understood in this way, the persistence of the IRS system was supported by the same neoliberal governance principles and logics that underwrote the management call for greater worker involvement in the areas of quality and efficiency, and it was this support which paradoxically afforded workers and worker representatives some room in which they could achieve and negotiate measures of influence over working conditions.

While LPT is certainly helpful in recognizing the dual potential here for worker resistance and management control based on interests and power struggles (Adler, Goldoftas and Levine, 1997; Burawoy, 1985), including the argument that managers strategically recognized the political value of providing spaces for worker challenges and controlled successes, I've suggested in this chapter that governmentality theory offers us a way of going beyond this argument to suggest the potential for more substantial empowerment contained within worker participation discourses more generally. In recognizing that there were structural and discursive spaces for worker empowerment within neoliberalism, post-Fordism and the knowledge economy, I am not abandoning my principal concern with the capacity of neoliberal discourse to reproduce labour consent to safety and health hazards. If workers were gaining some areas of control and power within the new management governance approaches, then the possibility of more worker control over hazards is implied. Just as clearly, I am not suggesting that neoliberal governance is all about empowering workers. My principal concern is still to understand how the reorganization of work, management by numbers, and worker involvement schemes combine to alter worker risk subjectivities and politics in ways that yield continued risk-taking and compliance, while also helping us to identify spaces and places where both cooperation **and** resistance lead to better conditions.

By recognizing the capacity of workers and worker representatives to adapt to neoliberal forms of governance, the contradictory impacts of the state and the union, and the larger institutional and public politics of health and safety, were also revealed. While the state's efforts to manage health and safety conflicts through the promotion and strengthening of the internal responsibility system contributed in significant ways to labour consent and individual responsibilization, often at a cost to worker health and safety, the IRS regime was also critical in shaping management attention to health and safety and their efforts to manage through joint health and safety committees and worker representatives. Those management efforts involved some very real concessions and compromises which made their workplaces safer and healthier places (Shannon, 2000). Although state and union support and responses were also often very limited, they played important roles in supporting the efforts of many reps. and contributed in that sense to meaningful prevention outcomes.

The farm study gave us some sense of what happens when there is no meaningful state regulation or union but we still haven't considered in any detail what happens in an industrial setting when there is law but no union. Some of the autoworker representatives in Study 4 talked about their previous experiences in non-union shops, as did non-union workers in the hazard and injury reporting study (Study 7), which were especially horrific. Some of the non-union reps. in the second rep. study similarly complained about owners and managers who refused to address even the most basic OHS problems like access to PPE, using threats and reprisals to gain compliance. In these contexts, there was a lot less consent and a lot more coerced compliance as the basis of management control (Basok, Hall and Rivas, 2014; Hall, 2016). But clearly, not all non-union auto parts shops were scab shops in this worst sense and certainly many had quite sophisticated HR and OHS management systems (see also Lewchuk and Wells, 2006). What actually happens in these more "sophisticated" contexts when there is no collective agreement and the workers and worker representatives are not backed up by a union, its grievance machinery and its other resources? Are worker involvement mechanisms other than joint committees more or less meaningful in these contexts, and is there politics of contestation that challenges management's presentation of knowledge, risk and rationales for risk-taking? The idea that workers' power may be more limited in a non-union context suggests yet another important question. What other dynamics are shaping management choices in terms of health and safety? As found clearly in this chapter, different firms, all unionized, often with similar products and positions in the supply chain, had quite distinct approaches to production, HR and OHS management. But why do firms develop a greater commitment to health and safety if there is no union and limited state pressure? Chapter 9 turns to these questions through an analysis of a non-union case study.

Note

1   As reported in Hall et al. (2006), there were 11 KAs in the sample of 21 politically active representatives, with the 10 remaining representatives classified as technical-legal. In the larger survey of 888 representatives (Hall et al., 2016), a cluster analysis found that just over half (57%) fit the KA criteria and 27% TL, with the remainder falling somewhere in between, principally in terms of their level activism and/or committee activity. In both studies, it is likely that the PA/KA representatives were highly over-represented in the given their greater interest in the research (see also Olle-Espluga et al., 2014).

Chapter 9

Participation and control in a non-union auto parts firm

A non-union auto parts firm: a case study of growth and transition

This chapter focuses on case study, LON Inc. (pseudonym), which was a privately owned firm which had grown from a small, single-plant operation with less than 30 employees to close to 400 employees working in 3 main manufacturing plants located within south-western Ontario. During this growth, the company went from one main GM contract to making a range of similar parts for all the main auto manufacturers operating in North America – Chrysler, Ford, Toyota, Nissan and Honda, shipping to both the US and Canada. In its early years, the company did not have a formal Human Resources department other than payroll and benefits, and no dedicated OHS staff. There was no formal training programme for safety and no formal system of safety procedure or rules. Sales, hiring and most management functions including human resources were handled by the owner and one manager. There was no quality control system to speak of and, as this implies, very little was tracked in terms of costs, productivity etc.

When the company began operations, the owner's orientation to employee relations was decidedly paternalistic (Lee, 2001). Elements of this orientation persisted throughout the company's growth and as the company shifted towards a more formal human resources management (HRM) system. At the time of the study, the owner was still handing out cheques in person to the workers on payday, while being revered among many workers for being able to address everyone by name even as the company grew to over 300 workers. He maintained an open-door policy for complaints and, according to the owner, provided loans to his workers on request. Workers also received Christmas bonuses and the employer met with workers in regular monthly full-company pizza lunches where company progress, markets conditions and other issues were discussed.

Also constant at the firm throughout the growth and transitions were relatively low wages, certainly as compared to the unionized workers examined in Chapter 8. The main labourer wage band was quite low and thin ($10.24–$14.99) covering sanders and cleaners, packaging crews, and assembly workers. The welder (1), machine operators and lift truck operators (material handlers) were generally divided into two levels 1 and 2, with pay bands of $11.31–$16.07 and $12.39–$18.22. Offsetting the low wage and the lack of a production bonus was a health-benefits plan which compared favourably with some of the union plants in study 4. Workers had a limited optional pension plan with an even split of worker and employer contribution to a registered retirement saving plan (RRSP).

Starting in the late 1990s, the firm moved to professionalize and formalize its human resources and health and safety systems, which included in both instances, significant developments in worker involvement programmes. By the time the study was being done in 2004–2005, most of these systems and programmes had been in place for three or more years.

According to the owner and senior managers, a central impetus for many of the management changes began in 1996 when the main client (GM) required that the company gain QS 9000 certification. QS 9000, a quality system developed specifically by a team of Ford, Chrysler and GM representatives, was essentially a compliance standard grounded in a continuous improvement philosophy. More concretely, the QS programme required companies to comply with a quality review process as defined in six distinct guides and manuals – a quality system assessment guide, a product quality planning manual, a potential failure mode and effects reference manual, a product part approval process manual, a measurement capability study manual and a statistical process control reference manual (Zuckerman, 1997; Johnson, 2002)). ISO 9001 certification was also obtained in 2002.

As the owner himself acknowledged in his interview, the firm in its early years paid very little attention to health and safety. Interviews with the owner, management and workers suggest that the single plant that it was operating from 1988 to 2002 was dirty, had no ventilation system and was increasingly dangerous as production expansion led to increased crowding within an inadequate working space. Up until 1996, there was no functioning health and safety committee, although one existed nominally, and there were no formal health and safety policies or procedures. Contrary to the law, worker health and safety committee reps. Were actually appointed by management rather than being elected by workers. WSIB claims and injury rates were not tracked nor addressed in any formal way by management and there were no formal company health and safety policies, procedures or rules.

Following the imposition of quality certification requirements, the company developed a more professionalized bureaucratic management structure with distinct office personnel and departments for record-keeping, design, engineering operations, human resources, sales and maintenance. Information technology specialists were taken on as well which reflected and fueled the development of various measurement, data collection and statistical procedures. A continuous improvement system was implemented in line with the QS requirements in 1996 as a single joint committee which was later expanded to allow all workers to participate in floor-level crew meetings.

With the creation of a separate professional HR department, the issues of compensation and OHS management were recognized as requiring attention, with particular reference to the question of compliance with the occupational health and safety law, which as noted was very spotty until this point. A part-time HR person was assigned initially to develop the health and safety strategy, which quickly led to the hiring of a full-time OHS coordinator. As the owner and other managers acknowledged in interviews, the initial results of these changes in OHS policy were quite minimal – aimed principally at passing MOL inspections without substantially changing much in the way of conditions.

Nevertheless, senior management claimed in interviews that during the initial transition period, 1996–98, they were increasingly aware of the need to become more than just minimally compliant with the law, in large part because as they began to identify and quantify the factors contributing to quality and cost, they "realized" the inter-related significance of accident and injury rates, leading to more significant change in OHS governance from 1998 to 2002. According to the owner and senior managers, this shift was at least partially prompted by an MOL safety inspection. Following a more serious injury to one of their workers, an inspector did a full audit and several significant orders resulted which went well beyond the injury issue, including orders to address ventilation and air quality issues. As the owner put it: "It was a big wake up call for me. I didn't like it at all and from then on we started to pay some real attention to health and safety."

The company finally expanded into a new plant in 2003 where engineering and organizational changes were made to the production line, storage and transportation space and ventilation systems. Although driven by unexpected growth including a high point in 2002–2003 when sales increased by 2 ½ times, managers, workers and OHS reps. uniformly saw this growth as allowing the company to make several OHS improvements. A worker recalls of the old plant:

> I'd say since we got from our old plant that we were at to this plant, I would say that it's fifty, sixty, seventy percent better for sure. It's a lot bigger, you can breathe a lot better than the other place and at the other place, in the wintertime, it was just cold...They'd let us stop and maybe take a few breaks to get warm but that was very terrible and here it's a lot better.
>
> (W1)

While growth and productivity were the main impetus for the change, both the owner and managers insist that a consideration of health and safety, especially a concern with ventilation but also a safer physical layout, was integral to the planning process for the new plant. In designing and organizing the new plant, management built on the negative experiences in the old plants, emphasizing a more orderly flow and separation of work areas and processes, and attending to key issues such as air quality by installing "expensive" heating and ventilation systems. A senior manager stated:

> It was a very dark, gloomy and compact building...And the building wasn't set up properly, because of our growth, we kind of piecemealed things together. So, where we found a hole, ok maybe that new machine can go there kind of thing...We moved here last year, we set this place up the way it should be set up where we have synchronous flow....So I would say it had a huge impact on making us a more productive company and certainly a safer company as well.
>
> (W4)

As this latter quote illustrates, the new plant was often described by management and workers as a perfect illustration of the way in which improved productivity and OHS could be accomplished simultaneously. As another manager put it when describing recent changes to the packaging process: "Health and safety has improved ten-fold but health and safety also means efficiency" (M5).

Most of the parts were produced in relatively small batches in distinct assembly stages by worker cells, with the company changing from one part to another from shift to shift or even within a given shift. As noted, when the company moved to its new plant in 2003, the production process was modified to allow for a more continuous flow of products through the different stages of production, and a better distribution of workstations and material-handling areas with more space. However, with the exception of the packaging process which was transformed substantially into a more continuous assembly line, the specific jobs within the production process changed very little. Much of the work remained labour-intensive and repetitive, although most of the batch production tasks in any given production cell involved multiple steps. The major technology investment was the introduction of automated robotic welding and metal-bending machines.

Although the company never moved to a full JIT production system, keeping three to four weeks supply of their major parts, this system became more difficult to manage as assembly customer orders shifted less predictably. There were many times, especially during the busy growth periods of 2000–2003, when a truck would arrive unexpectedly to pick up parts, and workers needed to switch and "crank up" the line to get the parts out "just in time." While we were doing our research, the market situation declined.

By 2005, the company was experiencing reduced demand for their products along with increased customer demands for cost reductions. For this period, workers were sometimes working well below capacity. This was also a period when workers began to experience and fear layoffs, although the owner (and many workers) insisted that the firm was doing everything it could to avoid layoffs (Sweeney and Mordue, 2017). Compared to the growth period, this decline also meant that workers were less able to move to different jobs within the plant.

The OHS situation in 2004–2005

By the time the study was being conducted, most of the changes in the firm's OHS and HR management systems had been in place for three or more years. Although the new main plant had by all accounts improved many aspects of health and safety, our observations of the joint health and safety committees along with a review of joint committee minutes and interviews with worker committee representatives and workers revealed several significant OHS concerns (aluminum dust and fumes, isocyanate exposure, periodic production speed-ups, lift truck accidents, RSIs and MSDs, noise, machine cuts and crushes). OHS committee minutes indicated that from 2002 to 2004, there was an average of eight medical-aid injuries in the two main plants which were mostly muscle and back strains, cuts, eye injuries and burns. There were no lost-time injuries reported in the committee minutes but this was likely because the firm actively encouraged "modified work" alternatives. We were unable to confirm management's claim of no-disease claims but there were a small number of skin reactions and dizziness episodes which may have been linked to fumes. Although the company did not provide any statistics, the workers claimed that so many employees were working on modified duty that the "healthy" workers were being overworked and put at risk, complaints which were also cited in the joint OHS committee minutes.

Despite workers' awareness of a range of hazards and some substantial concerns about some of the risks (e.g. aluminum dust and fume exposure), there were very few reported confrontations or conflicts of the sort that were visible in some of the mining committees and committees in the first auto study (see Chapters 5 and 8). All the worker OHS representatives interviewed (N = 4) reported high levels of satisfaction with management responsiveness and cooperation. While a few workers worried about possible management reprisals and coercion, they were unable to cite any concrete situations or examples where threats or disciplinary action were taken against workers for complaining about health or safety issues or reporting injuries. Most workers in interviews reported satisfaction with management's responsiveness to health and safety concerns and expressed no fear of reprisals for reporting injuries. Veteran workers and worker representatives with five or more years of experience all agreed that working conditions had much improved with the new plant and the new management systems. They also could not recall any major OHS conflicts, refusals or other confrontations prior to the new plant or management systems. In short, this was a non-union workplace context where the history and the current politics of health and safety seemed to exhibit limited overt conflict or coercion, and substantial labour consent to OHS conditions despite some significant risks (Lewchuk and Wells, 2006).

For the remainder of this chapter, then, the objectives are to first understand the links between these patterns of labour consent and the firm's approach to health and safety, and then to consider how and why this approach developed in this non-union context. In so doing, I further examine the links between the labour process and OHS politics, and elaborate on the argument that broader cultural, technological and political developments nested in neoliberalism combine with firm- and individual-specific histories to play critical roles in shaping localized management and worker actions and strategies, including the emphasis on rules and on worker participation.

Explaining labour consent

As compared to the mining case, one central difference was that with a couple of job exceptions, the firm did not make major changes to its labour process. As such, for most workers, machine operators, welders, sanders, painters and cleaners and hand assembly workers, the work and the control they exercised had changed very little by all accounts over the ten years. Moreover, given the firm's growth, many of the workers were relatively new with less than five years, again in contrast to the miners who were all long-term veterans. Up until 2003, the growth also meant that workers were able to move jobs within the internal labour market based on a seniority system set up by the firm, despite not being unionized. The sanding and cleaning jobs were viewed by management and workers as the worst jobs – worst in terms of physical demands, repetition, and exposure to dust and noise – which accordingly were assigned to the workers with the least seniority. Hand assembly was the next level in terms of skill. The workers and managers claimed that hand assembly could be mastered in two to three days, and indeed the work was repetitive. However, workers assembled a variety of different parts weighing as much as 40–50 lbs. which required somewhat different levels of strength, ability and skill. Managers and workers all agreed that some workers were much quicker than others and better at meeting quality requirements, again suggesting some skill nested in the workers.

Within both plants, most of the OHS concerns, and to some extent complaints going to supervisors and OHS management, seemed to be coming from these assembly workers revolving mainly around RSI and MSD concerns and cuts from sharp edges, but the split between types of jobs or plant location was much less clear than it had been in the mining context, perhaps because the skill variations and the safety risks were not as substantial. The sanding and cleaning workers in particular were concerned about aluminum dust, metal filings and fumes, noise, RSIs and MSDs. Assembly workers were more inclined to report substantial improvements in recent years citing safety mats, lift assists, PPE such as gloves, shin and other guards, and changes in racking and bench systems. These improvements were seen as coming through the actions of various sources – supervisors, the OHS coordinator and continuous improvement committees. The OHS committee was not cited as often by workers but committee minutes and interviews with OHS reps. (management and workers) suggest that many of those issues and changes were addressed in the committee context. Those on the cleaning and sanding tables were much less satisfied with attention to their issues as expressed in interviews, but this dissatisfaction was not visible in any significant conflict nor were their issues being pursued by health and safety committees at least observed or as recorded in Plant 1 and 2 OHS Committee Minutes (2002–2005). Machine operators and welders, higher paid and supposedly more skilled, seemed to express fewer concerns about the machine hazards in their work, although interestingly the OHS committee had moved on its own to identify several problems through inspections with machine guards not functioning and improper set-up. Although machine operators reported awareness of and sometimes concerns about welding fumes, faulty machine guards and burns, like assembly workers, they expressed confidence that they could prevent injuries by following procedures and employing proper guards. They also put considerable faith in their supervisor's attention to safety. As one machinist noted:

> There's one saw, the safety mat doesn't work. I brought that up to my supervisor, he's already said there's some safety guards going up.... And the supervisor will try and get things done [when they can] Meantime, I make sure before I go behind there, I make sure that the machine is stopped. Before I take the part out I make sure the blades are all locked.
> (W10)

Exercising control via procedures and to some extent personal skills and experience was also linked to an internalization of responsibility, although of course even these workers were subject to production pressures and recognized the constraints on their control. As one welder put it when asked about how much control he exercised:

> Yeah, I decide on whatever pattern. Welding is kind of a pattern, you just go in one pattern all the way around. So, if you can find your set pattern where you're doing good, there's only a couple of parts where they just give you directions on which way to weld it. [I: Do you have a

lot of flexibility in terms of how fast you can do your job] Some parts no, cause they got their quota or whatever. Some parts I got all the time in the world whereas others, I just got to keep my pace, always keep it going.

(W7)

This seems to provide some additional evidence that workers in low-skilled jobs were more likely to raise concerns with supervisors and through other reporting channels if they were not concerned about reprisals or job security, as was generally the case in this firm; but again, actions seemed to depend on whether the risks were understood as being linked to areas of worker or management control. Workers in the high-dust jobs who worried about their exposures saw some protection from their use of masks, but to the extent they still recognized effects such as coughing or sore throats, or black in their spit, they projected more concerns to management about their need for action (i.e.. engineering controls). As a cleaning table worker put it: I wear a mask but I get home I'm spitting black. Those cleaning tables – the ventilation system is not set up to draw the dust away...management needs to fix this." A welder in the other plant made a similar argument about welding fumes: "They have the ventilation tubing set up but it doesn't reach the welding tables" (W16).

There were also some differences among workers in the lesser-skilled jobs. Assembly workers were more likely than workers on the cleaning and sanding tables to argue that they could control some of the safety risks in their jobs by being careful, following the blueprint instructions and safety rules, using their skills and/or seeking rotations in their jobs. However, perceptions and judgments were also tied to the kinds of hazards. Assembly workers were concerned about lifting, cuts and scrapes and RSIs with relatively few concerns about dust or fumes, which were paramount concerns for workers in the cleaning and sanding areas. As one assembly worker, who was also an OHS rep. stated when asked about whether she was concerned about any serious injuries: "You just have to follow the safety rules and you should be fine." But as other workers noted, "it was hard to follow the rules" when they were being rushed. Still, assembly workers saw themselves as having skills and some control at least as compared to some other plant jobs. As one put it: "assembly, building the part takes skill. You have to know what you are going to build – lot of people don't build because they can't grasp it, too slow. As far as [assembly line] packaging goes, no skill there." As another stated:

You have to go by blueprints that tell you how the job has to be done... but you can come up with your own little like, and if you're building something or assembling something, everybody has their own little trick. So if you do it this way it's easier, if you do it this way it's quicker. When you're building you pretty much have the power, you just have to use the tools that they're telling you to use. But you can do it any way you want to do it. But for packaging, no, you have to do it, like there's a process you have to follow, you can't change it up.

(W4)

Whether these workers felt they could raise concerns also varied with the supervisors in as much as some were seen as much more responsive and competent than others. A perceived capacity to raise concerns also varied with the context. In the rush order situations, for example, few workers bothered to complain expecting if they did, they would be told by even the more responsive supervisors that the job "had to be done."

Some workers on the cleaning and sanding tables also thought they could exercise some control over the safety risks through care and skill, and over the health hazards through the use of PPE, but they were less confident that this meant adequate levels of safety, again looking to management to offer assurances and/or improvements. It was interesting as well that these workers reported feeling more driven by supervisors "standing over them" even during normal production periods, but again this seemed to vary with the supervisor. As one put it, "there's no consistency between supervisors... one will be saying 'oh no don't lift that by yourself', and the other is just like nothing."

The two locations where significant job changes were introduced – the bending machines which had been automated and the packaging line which had moved to a more subdivided assembly system, involved relatively small groups of workers. The bending machine operators became essentially technicians who tended to the robots and the computerized operation systems. This change was seen by them and management as requiring increased technical skills, although more control was nested in the technology than had been the case in the manually operated machines. However, this job only involved a couple of workers each shift (Belanger, Wright and Stewart, 2003). The size of the packaging crew varied from two to ten but the packaging crew expressed mixed views of the change. On the one hand, most seemed to agree with management that along with being more efficient, the process allowed workers to avoid certain risks most notably around bending and lifting.On the other hand, many noticed that they had less control over the pace, and when important orders were being rushed, they had to work much more quickly than they felt was safe.

The only time anybody really gets hurt is if they rush, if they're rushing us because the majority of the time if you're picking up a bar, you're careful how you pick it up, you're careful how you put it on. But when they're rushing us, you can't turn around, so when we're rushed, it's nuts. It's not safe at all.

(W13)

As should be clear by now, this concern about being rushed at certain times was a common complaint among workers across the plant. Although some workers expressed more concerns, workers from any of the range of jobs seemed to feel the effect. For example, as a welder reported:

> A lot of times they come in and they say they need so many parts by the end of the night. And then you'll be working on them and you'll see the supervisor eyeballing you and just keeping an eye on you, making sure you are working the whole time. So really, that's when you really feel... the most pushed.

> (W7)

Still, if most workers perceived some increased and substantial risk when the production pace was increased for "rush" orders, why were they accepting these situations even as they seemed to be increasing in frequency? Again, most were not worried about reprisals nor did they report feeling coerced to accept the risks. Interviews with workers suggest that their consent to rushed conditions first and foremost was based on their assumption that an increase in the production pace was necessary to meet the firm's financial needs, which in turn preserved their jobs. Similar to what we saw in the mine case study, some of this understanding was grounded in statements and communications by managers, supervisors and the owner, who explicitly indicated the importance of getting orders out the door on time and stressed the vulnerability of the firm to failure even when it was growing and certainly as the firm started to struggle.

Underlying these motivations was the fact that many workers were strongly committed to maintaining their employment with the firm despite its low wages and the availability of comparable and even better wages at the time in other local parts plants (at least until 2004). Some of this commitment revolved around other employment benefits:

> That's the most complaints you have here on the floor, people want more money, they're not happy, there's too much work for what we get paid... I would like to make more money, I'm not lying saying I wouldn't like to make more. But I'm perfectly happy with how it is because our benefits are amazing. We have good benefits there.

> (W4)

But more broadly, workers believed that the owner was exceptional in his treatment of and commitment to workers. As one worker put it, "I've never worked anywhere where the owner is so involved...three hundred workers...I've never had that anywhere where he would actually come around and say how are you doing." As perceived by workers, this commitment extended to employment security as both the owner and several workers noted a sustained effort to keep people employed even as the firm orders declined in 2004. When asked whether he was looking for another job, one worker responded:

> Well, is a very good question because I got reason for that point. It's very good because everybody loves money, but I got my job, there is no laid off. I got my benefits, I got everything because there is no big income for me. But I feel good. What happens if I change, if I switch from my job to the other job better? But what happen three month later, if I got laid off? It's no good for me. So I feel happy there and I keep my job....
>
> (W7)

This notion that the owner cared about them and provided a level of security, cemented a certain level of devotion to the firm which justified for some workers the taking of risks in order to maintain the employment relationship. For some, it wasn't just their interest in maintaining the employment, it was also a sense of obligation that the owner had done their best for them so they had to do their best for him. As another worker put it, "I think he's got enough to keep us going. He's been working his butt off just trying to get us jobs. I got a lot of trust in [the owner]...." The trust they had in him also helped to reassure or moderate the concerns that many workers had about OHS – that the risks were either not a great as they feared, or that they would be addressed at some time in the future.

Paternalism may have formed the foundation of this trust and commitment to firm but as the firm grew, workers also realized that much of the power over their employment and OHS security was increasingly held by others in management. Some workers acknowledged that they began to doubt whether the company was adhering to the same values as it expanded in size, especially as some supervisors and managers were seen as getting their jobs through patronage and favouritism, but also it seemed to them more generally that the rules, including safety rules, applied to some and not others. As one put it when asked about whether workers were being treated well:

> R: No. Cause even though it might not be [the owner] himself that takes part, there's a lot of people under [the owner that I don't necessarily trust, the management crew in the office, at all... He [the owner] seems to be a half decent person. [But] I don't believe he knows everything that goes on in there.
>
> (W11)

However, the firm's development of an IRS and audit-based OHS management system, and its new worker involvement mechanisms may have helped to reconstruct some confidence and trust, as well as give workers a sense that they could still influence management decisions. This also depended on the extent to which the more formalized rule- and knowledge-based HR and OHS management systems were seen by many workers as offering reliable control over risk to job and health security. A more organized OHS committee with elected worker representatives doing regular inspections, a participative employment standards committee hearing worker grievances on labour relations issues, and a participative continuous improvement process which allowed every worker to "have input," seemed to make an important difference for some workers in gauging their capacity to control immediate risks and to protect against coercion. These were also means of judging the effectiveness and integrity of management and technical controls. One worker stated:

> I've found in the shop floor meetings, most of that I would say has been health and safety related than just finding ways to save money or something like that. I think the first thing people look at is, I have a hard time when I have to close this door, can we look at closing this door or having a better system? That's the things I've found that they've brought up in the meetings. So I think their first impression of continuous improvement is let's find me a better way to do my job so it's not so much of a struggle day in and day out.
>
> (W13)

However, as we also saw in the mining study, the continued effectiveness of these participative and management systems in reproducing management control and worker consent to risk was partly dependent on whether workers perceived meaningful changes or improvements in conditions that concerned them. In that regard, most of the workers interviewed gave accounts of OHS changes which they saw as being direct management responses to worker concerns, and many described relatively rapid responses to some of their own concerns about back or muscle strains, machine guards or PPE. These changes were often small in scale and low in cost, often imposing responsibility on workers (i.e. requiring PPE) without addressing the underlying causes. Nevertheless, workers tended to interpret them as offering added protection, meeting their health and security needs as workers without endangering their employment interests. Given their direct participation in this process, workers perceived gains in their control and power to the extent that they saw management as listening to them. For example, several workers cited management's response to complaints that they were bruising their shins on a new conveyor. The fix was relatively low-cost and company quickly responded by installing foam pads and promised a continuing effort to address the problem in other ways (JOHC Plant 2 06/07/04). Reports of muscle and back strain led to a fairly quick approval for the funding of the purchase of back braces for any worker who could produce a medical letter. Wrist braces were also provided on the same basis. While low-cost fixes, these were real gains for workers, while also helping to reduce injuries and people on modified work, the latter of which was a major point of tension since "healthy" workers had to "take up the slack."

Some workers recognized that while there seemed to be action on certain basic safety issues like lifting or PPE, there was relatively little action on health issues such as air-quality ventilation and RSIs. As one worker put it:

> On some matters they work pretty fast but on other important matters, they're pretty slow...for example, the frets in the back that suck up all the dust from the cutting machines...the plastic bag is full, weighs maybe forty pounds, you got to lift thing and you got this dust coming off the bag and it's all over you...highly toxic...just a dust mask. I told [my supervisor], how they could do this.

Machine workers in the older plant with its more limited ventilation system claimed a long-standing wait for action on their air-quality concerns, while the sanders and cleaners made similar complaints about the failures of the ventilation system in the new plant.

As we saw in the mining context, the practices of individual managers also mattered in reproducing consent. One person at LON who was especially important in reproducing consent was the management OHS coordinator. Although she had no prior OHS experience or training, or even shop-floor production experience, she was frequently cited by workers as a person they trusted to address OHS issues. This was again substantially a reputation built by her responsiveness to worker complaints, even if those responses tended to be limited to low-cost fixes such as PPE improvements. But her trustworthiness was also built on her management style. Emulating the owner, she had an open-door policy and a tendency to walk the floor on a regular basis interacting and communicating with workers. As such,, she had earned a reputation among many workers as someone who "cared about them" and who was responsive. Other key plant managers were not seen in this same light and accordingly were not trusted nor used to address OHS concerns (Jeffcott et al., 2006). Varying perceptions of supervisors had similar effects in limiting acceptance of management claims and consent, which workers dealt with using various strategies, including the use of the OHS coordinator or an alternate supervisor on a swing shift, making a complaint to an OHS rep. or the committee, or speaking directly via the worker involvement mechanisms. However, if those strategies didn't work, or if they felt cynical or worried about possible pushback from the original supervisor or manager, they tended to comply while often seeking a transfer or other employment. But none of the workers seemed to consider the possibility of any other kind of political action such as seeking to unify work support for a work action, supporting unionization, refusing to work or requesting a ministry inspection.

The owner's paternalism coupled with the OHS systems and worker involvement mechanisms were not just shaping the responses to recognized risks, they also influenced the perception and judgement of risk. For example, the trust in the owner also meant to some extent that workers gave the firm the benefit of the doubt when accepting certain risks or certain claims about the risks being controlled or minor. As one worker rep. accepted without question, "there have been complaints about dust but those levels are monitored [by OHS department], so there's not a problem." Workers and worker reps. were often quite willing to take management's promises of some future change at face value while accepting the risk at least in the short term. A machine operator noted, "I've told my supervisor about the safety curtain problem and he tells me they'll get to it when they can, which is fine." Again, other workers would argue that some significant issues such as a lack of machine guards or dust exposures are around for quite a long time with little or no apparent action.

Management's responsiveness via participative mechanisms was important in giving the impression of a new stronger management commitment and competency around worker health and safety but, as we saw in the mine study, managers also communicated via these processes a variety of economic or technical reasons for why changes *could not be* made, which workers on the whole seemed to accept. In this context, the company was widely viewed by workers as doing its best, within the limits of technology and the financial situation, to protect workers.

Managers certainly saw participation in clear strategic terms as helping them to direct workers in ways that made them feel involved. As one manager put it:

> I think it's good to have employee input, it makes them feel more like they're part of the team too. If it's always 'here's what we're going to do all the time', I think when you bring it forward it makes it look like they're just being directed. Whereas this way, they're being involved.
>
> (ACSM-15)

Indeed, feeling that they could express their concerns within the context of the firm's worker participation approach, even if nothing substantial was changed, seemed to help in reducing both their anger and their concern: As one worker stated when talking about a recent rush order of parts: "Like we have a shift meeting every Tuesday… we get to say what we want to say. And I let him know at the meeting and while we're working."

Along with giving workers an opportunity to voice and address OHS concerns, the worker involvement mechanisms gave managers an opportunity to communicate and reinforce claims of management control over risks as well as areas of worker responsibility. When introducing changes to the production or packaging systems, managers routinely touted the OHS benefits of those changes, which was often reflected in how workers in turn presented those changes in interviews. As one put it:

> The whole packaging set up, like the conveyor belt is great cause we used to pack on the floor and that was killer on your back. Now this whole system is great, it seems like they're constantly looking for ways to make it easier.
>
> (W4)

It is also significant that many of these issues were addressed outside the formal IRS or OHS system in the context of the company's continuous improvement groups or employee relations committees, since those contexts reinforced the idea that productivity and health and safety were being achieved simultaneously and seamlessly. Interestingly, the owner expressed concerns that this mixing of OHS issues within continuous improvement groups was undermining the relevance of the OHS committee. In contrast, a number of his managers insisted that the lack of boundaries between the continuous improvement and OHS committees was a strength because it involved more people in decisions and corrective actions.

Further, these group interactions provided managers with opportunities to promote and reinforce messages to workers about their responsibility to follow safety procedures and rules. This was also partly achieved through the firm's greater attention to safety talks and training. Most workers expressed confidence that they knew about the main hazards in their workplace and had the training and information provided by management that they needed to avoid or prevent injuries.

> Everybody knows pretty much how to lift. They [management] really emphasized, when they give you your portfolio when you start, they got a big thing there about back lifting. You use your legs not your back, straighten out, it shows you right in there too. You would have to be pretty goofy not to know, if you know what I mean.
>
> (W3)

As the quote again indicates, rather than seeing management incompetence or intransigence as the source of risk, as was common in some of the mines and some of the other auto plants, workers and worker representatives tended to see worker carelessness as the greater problem. When asked if injuries were likely in her workplace, one worker replied, "the rules are there, you follow them, you are fine." This internalization of responsibility was not accidental in as much as management communications and training emphasized greater rules and procedures which placed responsibility on workers.

One final finding to explain is the relative absence of coercion in management's approach to health and safety. As noted with a few exceptions, workers felt comfortable making OHS complaints and reporting injuries without fear of reprisals. Others qualified their position by saying they felt they could selectively complain to certain people like the OHS coordinator and several supervisors and managers who they trusted as either competent or as adhering to the owner's philosophy of treating workers with respect. But the owner's paternalism and his perceived integrity, backed by an open-door policy which he continued to practice, not only underpinned workers confidence, it also acted as a fetter on those supervisors and managers who were more inclined to use coercion. Workers understood that they could not complain often or they would use up what capital they had as good and loyal workers, but they did believe that when something was seriously wrong, they could express their concerns with relative confidence that they would not be punished. Whether they would get what they wanted in terms of OHS protection was another matter, and certainly the owner and management more generally were skilled in providing economic and technical rationales for why certain changes were not possible making certain risks necessary. However, importantly, some of those complaints to supervisors, the OHS coordinator and the continuous improvement, employment standards and OHS committees were leading to substantive improvements in conditions, reinforcing the worker view that they could exercise some influence or control through the management system.

The role of the health and safety committee and representatives

Reflecting what we saw in many auto firms in Chapter 8, this firm's two joint committees and its worker representatives exhibited all the hallmarks of a technical-legal collaboration. This TL orientation was shaped by the company's move to a rule- and audit-based approach as well as the company's emphasis on participation and consensus building. As observed in both committees over a one-year period, the company was by this time following most of the basic requirements of the law in terms of elected representatives, rotated co-chairs, proper recording of minutes, regular inspections, follow-up procedures with required responses from management etc. While moving to some extent to establish more visible distance from management control through worker elections, the company was still influencing the selection process by explicitly encouraging certain workers to put their names forward. A review of minutes from 2003–2005 and observations in 2004–2005 suggested that although committee attendance by managers and supervisors was quite erratic in one of the committees, on the whole the committees dealt with health and safety issues in an orderly and systematic fashion, with formal reviews and follow-up. However, as observed, discussions in committees were very cursory often lasting less than 20 minutes with no evidence of disagreement or conflict. Although some immediate decisions were made on the smaller items, management was left with the responsibility to research or resolve any problem involving significant cost or production implications. Discourse within the committee and interviews with worker reps. reflected the view that health and safety prevention was a cooperative-technocratic process which operated in sync with production and efficiency improvements.

There was no evidence of "political" or "knowledge activism" among the six worker representatives as described in Chapter 8. Some representatives appeared to be largely inactive with little knowledge of the law or conditions in the workplace. In interviews, all the reps. defined their roles narrowly in terms of patrolling the workplace looking for rule and standard violations which were dutifully reported to the committee or to supervisors for resolution. Yet, unlike the mining context where many miners saw their OHS reps. as either incompetent or "management stooges," not a single worker disparaged their reps. and most seemed to accept the company line that the committee offered "another set of eyes and ears" to prevent unsafe conditions and practices.

As noted, as of 2003, the company had appointed a full-time OHS co-ordinator who was responsible for ensuring that committee procedures and regulations were followed. Worker OHS representatives tended to rely quite heavily on this OHS coordinator for information as she was the main conduit for communicating concerns to management outside the committee. None reported doing any independent research to make cases for change. Indeed, none reported ever making independent cases for change in as much as they simply reported the hazard and left it to management to resolve. Only the two worker co-chairs were certified as required by law and none had sought additional training or education other than CPR or First Aid.

The contradictions and limits of management control

As the evidence suggests, there were several signs that variations in management practices were yielding contradictory effects on OHS politics and conditions. For example, although the firm had moved to a professional HRM model, it is clear from many workers' comments that the owner's personal approach to his relationship with workers continued to play an important role in shaping both worker commitment to the firm *and* worker dissatisfaction with the firm. That is, while the paternalism often translated into a level of trust in the owner, several workers also complained about favouritism and the lack of consistent treatment across the firm, which undermined some of their commitment and trust. As the following quote indicates, this can lead to quite contradictory views even within a single worker:

> R. I never worked for a company before where the owner himself comes around day before payday and hands out the pay stubs. And he knows every single person by name...[But] You hear things and you know they're (owner and manager) friends outside of work. It's not just owner, whatever and I know he's known people cause I remember when BS got started. He had no experience at all, never worked in a plant and you could tell...Like I said, I've bypassed him totally. If I ever have an issue, I bypass him totally and I go right to human resources. The less I deal with that man the better I am.
>
> (W4)

These concerns about nepotism and line management competence and trustworthiness also extended to perceived inconsistencies between managers in their commitment to both the job and health and safety, and a sense that these inconsistencies were allowed to persist because of nepotism.

> R: Yeah, they all need to be on the same page because like I said, one week I'll have one supervisor and the next week I'll have another supervisor. That's how everybody has it but I've got one person for two weeks and one person for two weeks because I'm on steady days. And it's like, you see the different changes and how did one supervisor is like, oh no, don't be lifting that by yourself, you'll actually get hurt and miss work. ...Well, the other supervisor, you don't hardly see him all day, where the heck is he?
>
> (W10)

Another contradiction, which we also saw in management's OHS strategies in the mining case, was that while management was pushing job flexibility and worker participation as a way of giving workers more input, they were simultaneously pushing more rules and standards as ways of gaining consistent levels of production output and quality as workers moved from job to job. With those rules and standards often came more disciplinary practices and penalties, some of which were being exercised through the OHS committees and worker representatives. While encouraging workers to internalize responsibility, the use of a disciplinary approach can, as we saw in the mining case, backfire on management, undermining trust and creating resentment. As one worker complained:

> And now they've brought out the stupidest thing in the world. Now if you do something wrong [in health and safety], they're putting all these people back on probation. Like when we first started there. Jesus, what are they doing? I know of at least three people now. And one person works her ass off.
>
> (W15)

For the owner and for management in general, there was no easy route out of this conundrum between pushing for flexibility in job assignments and standardization of those tasks. Similar to the mining case, as cost, quality and price pressures continued to mount, the pressure grew on front-line managers to overwork the "good" workers, coerce the perceived "bad" ones, and ignore some of the rules and worker input in the process, while the persistent impulse to play favourites as part of the politics of control sometimes threatened to unravel much of the goodwill and trust that had been built up. Although one of the significant pressure valves during the firm's growth was that workers were able to bid out of the worst jobs such as the cleaning and sanding tables fairly quickly as assembly and other jobs became available, as noted these opportunities started to disappear in 2004–2005 when the market shifted and positions within the firm and across the industry started to shrink.

Certainly, for some workers, the emphasis on rules and procedures in some contexts and not others also brought into quick relief the tension between safety and production. As noted, periodic increases in the pace of work in response to rush orders and unexpected delivery requests were the most commonly expressed concerns – precisely because it was here that workers often felt that they no longer had control and capacity to prevent injuries since the rules that were supposed ensure their safety were often being bent and broken. Virtually every worker understood these situations as greatly enhancing the risk of injury, and again especially among machine operators, these situations were recognized as presenting some risk of serious injury (i.e. loss of limb). As noted, most people still accepted these rush orders as necessary, but for some workers, this also revealed the lie in the management's claim that safety was never compromised for production.

> It's a problem. Working at a higher speed, not so much me but the welders, they kill these guys. All of a sudden the standard is twenty parts per hour and then ten o'clock comes around, oh, the truck is leaving at eleven. Oh come on guys, we need you to so sixty an hour and we're

busting our ass going crazy, get the parts out, get the parts out. And then all of a sudden by the end of the day they're killing them and they're tired and the next day they walk in it's the same thing.

(W15)

Thus, as we saw in the mining study, the limits of management's capacities to reproduce labour consent were also nested in market pressures, the contradictions of management control strategies and mechanisms, and the failure of individual managers and supervisors to consistently apply those strategies and mechanisms.

Why this approach?

The final question we want to address in this chapter is what shaped this company's approach to restructuring, management and health and safety? Chapter 8 focused the analysis on the impact of larger external developments in neoliberal governance and the post-Fordist knowledge economy. While these also played a role in shaping this firm's orientation, here I want to look more specifically at the combination of economic and industry factors shaping this particular firm's actions.

The analysis thus far indicates that the impacts of restructuring on working conditions and OHS subjectivities and politics control were limited in part because the firm made only gradual changes using a continuous improvement process that limited the perceived impact on the workers' capacity to know and judge the hazards in the workplace, and thus moderating its potentially negative impacts on worker consent and habitus. However, I've also suggested that the professionalization of human resources and the 'actuarialization' of risk assessment and management increased management control over health and safety as well as the appearance of control in ways that encouraged workers to have confidence in management commitment to health and safety and their claims about current working conditions (O'Malley, 2004). I further suggested that the participative continuous improvement process and other involvement mechanisms did much to internalize responsibility for risk production and control in workers, while also reinforcing the idea that efficiency and safety were two common goals that were being achieved in a partnership-coordinated way. In as much as many workers and all worker reps. accepted this claim, labour process changes, safety and the politics of safety were not viewed by workers or managers as fetters to restructuring. Finally, I have argued that the combination of paternalism and worker involvement mechanism has generally reinforced workers' perception of control and influence along with their trust in management. But if there were no dramatic changes in the labour process, and no substantial worker resistance to the firm's transformation, no recent changes in legislation or health and safety politics, government inquiries, or unionization, what prompted the firm to take this new approach to health and safety when it did?

Growth and industry demands

As implicit in my earlier presentation of the firm's history, I would argue that the push towards a management-intensive health and safety programme along with a strong participative component at LON was shaped by two main sets of circumstances: First, as we saw in farming, there was the increasing client demand for quality, which in this context meant the imposition of a requirement for QS 9000 certification. This forced an emphasis on new managerial controls and systems to monitor quality and production standards, shaping the company's thinking about risk assessment and management and pushing management to recognize and integrate health and safety as key factors in its cost controls and risk assessment processes. For example, when asked if QS 9000 had had any impact on the company's approach to health and safety, a senior manager put it this way:

> I think it may have put the mind set in place. Like here's a system. When I started we had nothing here in terms of a quality system. Nothing. From even as simple as identifying the parts with a tag and we built it up to having a full system in place to address very many different facets of business. QS 9000 directly impacts health and safety mainly I would say there's a specific training component, but that brought about that mind set, well this is how we should run things, not just from a quality or production standpoint, but from a health and safety standpoint. And I know since that we've created many specific policies or procedures, health and safety ones.
>
> (M7)

Given the audit's emphasis on training, communication and worker involvement, the company was also encouraged to move from its paternalism to a professional HRM approach, which further reinforced the importance of managing health and safety *as* an important human resource issue and an associated recognition of the need to insure regulative compliance. And as noted earlier, a single event where a OMOL inspector cited the company for a number of infractions played into the greater weight placed on the need for corporate health and safety policy programmes as an important part of that shift. As a senior manager recalls the health and safety inspection:

> R: Well, I'll tell you. It had a huge influence because, well it was an eye-opener for me. I was new in this position...I had a lot of learning to do myself about health and safety and due diligence and the business risks involved and not doing the right thing, and all of that. So it was quite

the eye opener for me personally and it really changed my approach to that area and definitely put a lot more emphasis on it as a result of that... And like I said, it's something we track now weekly, monthly in our meetings and it's right out there in the open and a priority along with production.

(M4)

Second, the firm's strategy of meeting client price demands through low labour costs within the local context of highly unionized, relatively high-wage industry and a strong regional union environment, coupled with the dependency on workers given weaknesses in capitalization, further enhanced the need to develop a human resources approach that would build some worker commitment to the firm despite the low wages. Paternalism and a good benefit plan were integral to this non-union strategy in its early years, and as noted, these strategies continued to play a role to some extent in the transition, but as the company grew and the demands for quality intensified, and as the efforts of the CAW to certify the workers persisted, the company realized the need for other measures to keep workers non-unionized and committed to their jobs and the firm. Many of the human resource policies, including the use of seniority for layoffs and hiring and internal bidding process, along with the joint committee for dealing with work complaints, were important aspects of that effort; but, as argued, a more effective health and safety programme was also recognized by the owner and management generally as a key means of resisting unionization, in no small part because the organizing strategy of the CAW placed a heavy emphasis on health and safety.

When the firm's main clients began demanding quality assurances, in part through QS 9000 and ISO 14000 certification, the company was doubly forced to reassess its approach to management. Whereas much of the pre-existing paternalistic management model was grounded in personal relations and experiences, descriptive and impressionistic data, and personal judgement and common-sense knowledge, much more emphasis was now directed towards quantification, measurement and objective analysis of performance indicators and risk. Concretely in terms of management, this led to management procedures and systems including the development of quality assessment measures and tools, data collection, recording and analysis procedures, training etc. in an effort to reassure the firm's clients that the parts would be consistently produced on time and to specifications. This also created a substantive management interest in achieving a higher level of consistency and standardization in parts production and output.

The auto industry emphasis on continuous improvement as the best way to achieve productivity improvements was also important in shaping the firm's approach (Lewchuk and Robertson, 1997; Rinehart, Huxley and Roberston, 1997). However, as a single-owner firm, this company was also not in a financial position to develop expensive new technologies to respond to its growth and so the owner sought to increase its productivity and efficiency through smaller-scale changes, which also pushed the emphasis on a continuous improvement process which relied heavily on worker input and cooperation. The greater attention to the worker involvement aspect of quality improvement in turn reflected the fact that the labour process was labour intensive. Gaining worker commitment to quality was the central goal but in as much as health and safety was understood as being linked to quality assurance, OHS became a critical feature of the continuous improvement and quality assurance objectives.

The added initial dynamic was that when company started to grow in leaps and bounds as its parts' sales took off in the mid to late 1990s, the company had relatively few competitors. These market conditions increased the pressure in the sense that there were now numerous different versions of the two basic parts the company was producing. Like most parts suppliers, the company was increasingly squeezed by its customers for lower-cost parts especially as more competitors appeared on the scene. Initially, the firm was able to deal with these pressures by expanding its employee numbers and meeting the basic requirements of the audit but, over time as the clients demands continued to increase, and the client base and the variety of orders grew, the increasing complexity forced more substantial changes in management structures, procedures and technologies. And of course, as the company grew in size, it attracted even more attention from its arch nemesis CAW.

With formalization, codification and professionalization of the HR function, greater attention was drawn to issues of compensation management (as a cost), modified work requirements, and basic compliance with health and safety legislation (since inspections and accidents could also add unexpected costs), with particular reference to the health and safety committee. Perhaps more significantly, as the emerging management approach to human resources developed, company management also realized for the first time that health and safety was important to *employee relations* and *worker motivation* – that is, if the company wanted to develop an effective anti-union human resources approach aimed at motivating and sustaining worker commitment to the firm, without having to pay their workers a lot more money, it was important to have an effective approach to health and safety. And once the company started quantifying everything in a true managerialist and actuarial tradition, and health and safety in terms of incidents and dressings became measurable, it became even more important to firm up the procedures and policies, with particular reference to the joint health and safety committee.

The role of the state and the absence of a union

The evidence thus far seems to suggest a relatively minor role on the part of the state and the law in shaping this company's approach and in the day-to-day production politics of health and safety more generally. As indicated earlier, the 1990s through well into the first decade of the new millennium were relatively quiet years in terms of health and safety reforms. While a newly elected Ontario Liberal government in 2003 made some noises about the need for increased enforcement, relatively little was happening at the time of the study. According to management, the company had experienced few inspections. Workers had never used their right to refuse (or apparently threatened to use it) or their right to request a ministry investigation, nor had the worker representatives made any use of the ministry or the law to force changes. If we accept what the owner and managers said at face value, a single injury incident and the resulting inspection had a profound effect on management thinking, sending a signal that compliance and due diligence had to be addressed much more formally and seriously than was the case when it was a small operation producing a couple of parts for one customer. More than likely, the effect was not profound on its own. However, in combination with the moves being made towards a more professional HRM function, the threat of ministry involvement was likely quite important in further provoking the firm's interest in establishing a functioning IRS and OHS management system.

What also seems important here that the ministry's orders had not only reinforced the owners' understanding of his management responsibilities but also pointed to the need to *manage* the internal responsibility system as a whole, with particular reference to giving more attention to the health and safety committee as an important buffer from government oversight and involvement. It is perhaps less than surprising to find that the combination of a well-controlled joint committee and other participative mechanisms such as continuous improvement committees also meant that workers never thought about OHS in terms of their legal rights. For example, while a number of workers were concerned about aluminum dust and fumes, at no time did it seem to occur to any of these workers to request information under WHMIS or their legislated right to information, or to refuse unsafe work, or to make a direct complaint to the Ministry of Labour, as per their legislated rights under OHSA (1990).

In this context, it is difficult to avoid the suggestion that the lack of a political challenge from worker representatives, and for that matter the workers, is related to the absence of a union (Reilly, Paci and Holl, 1995). Certainly, the other studies reported in this book demonstrate that the presence of a union does not mean that unionized representatives are necessarily more politicized, never mind effective political actors. Nevertheless, in this case,

CAW's failed organizing efforts likely did play a dual role in pushing the company to improve actual conditions while also encouraging more sophisticated methods of controlling worker responses to risk. Still, it is difficult not to think that the almost complete absence of overt challenges by workers or worker representatives, in the face of some significant concerns about health and safety, was more than a result of some complete ideological hegemony exercised through management control strategies. Indeed, there were several workers, some of whom had come from unionized plants, who saw through the management veneer of commitment and competence and who felt that there were some significant problems with health and safety that were not being effectively addressed.

> At LON you're constantly burning your hands and from here to here. And I told TL [the OHS Coordinator], why don't we get some arm guards that just protect this area, the big heavy ones. I used to have one at TM for a press operator [union plant]. They go here to here and it just protects your arms, that's all we need. And she said, well she'd look into it. Well that was a year ago... Oh my god, you should see the burns they get on their arms. Oh my god, those big burns, they got to do something for those guys.
>
> (W15)

However, the critical point is that none of these concerned workers were joining or challenging the joint OHS committee. Although we saw little evidence of overt coercion or reprisals, a few workers expressed the concern that management had a black list. As one put it, "you get off on management and you end up on the 'list', they will find a way to get rid of you, they just bide their time." These workers also recognized the conflicting interests and the limits of what the firm's IRS and OHS management systems offered them as a means of preventing injuries and illness. As one stated when talking about the much-loved OHS coordinator: "Let's face it, she is part of the management team, it could put her in a bad position if she started to squawk too much and she realizes exactly what danger everybody out there is in over time." (W11). However, without a union, they saw no options in terms of action including appeals to the government, except to start looking for work elsewhere. Being unionized might have made some difference in creating a more robust OHS politics, or perhaps some opposition at the joint committee level on the more serious issues that workers had, not only because a collective agreement offers more protection against reprisals but also because the negotiation of higher wages might have made the job worth the effort. However, as the mine comparison and the second rep. study suggested (Study 6), the latter of which it may be recalled involved both union and non-union representatives, it is not just a question of being unionized, it also depends on the resources and political orientation that the union brings to the table (Hall et al, 2013; Walters et al. 2016).

Conclusion

The case study presented in the chapter addressed four main arguments. First, it reinforced the point made in Chapter 5 that management was able to encourage an internalization of responsibility for risk in contexts of restricted labour process control through worker involvement mechanisms, to the extent they were perceived as giving workers some meaningful security and influence over some of their conditions. At the same time, management also reassured workers (and themselves) through the various elements of its health and safety management system, including the joint health and safety committee, that management's responsibilities for identifying and controlling hazards were being competently addressed, in part by directing workers on how to work safely through rules, procedures and disciplinary supervision. However, as in the mine study, the impact of these management strategies on labour consent was constrained by weaknesses and contradictions in management orientation and/or implementation, which were often aggravated by market demands and production and cost pressures.

Second, consistent with the book as a whole, the chapter also showed how key elements of neoliberal governance within the auto sector, in the form of quality audits and continuous improvement in the auto industry and the IRS in the legal sphere, played critical roles in shaping the firms' actions and strategies (Lewchuk and Wells, 2006). Third, as also contended in this chapter, the lack of unionization played an important role in the limiting the opportunities or capital needed for political challenges (Basok, Hall and Rivas, 2014; Walters et al., 1996). On the other hand, the firm's efforts to avoid unionization often underwrote concessions by the firm as well as shaping its overall "consensual" approach to HR and OHS.

Finally, as the chapter also suggests, skill and control over work still played a role in shaping consent and resistance to conditions, although often in quite complex ways. Even in relatively limited skill contexts, workers exercised an influence over productivity and quality in ways that gave them some capital in relations with management in negotiating conditions and, at the same time, some sense of perceived capacity and responsibility for controlling some of the hazards.

What did all this mean for OHS conditions at the end of the day? This firm's official injury rate according to joint committee minutes was around 2.8 medical aid injury claims per 100 workers in both 2002 and 2003, with no fatalities and very few to no LST claims or disease claims. These results were generally better than current official rates of manufacturing injuries but, as noted, it is unwise to put too much faith in official injury statistics

(Barnetson, Foster and Matsunaga-Turnbull, 2018; WSIB, 2018). Nevertheless, like the neo-productivist farmers in Chapter 7, when we take into account all the evidence, my assessment is that this firm was doing a better job in 2004 with its OHS management and HRM program than it was doing in 1996 in preventing injuries and disease. It was also apparently more responsive to some of the OHS concerns that workers had (Frick, 2011; Vickers et al., 2005). Nevertheless, and again as was evident in both the mine and farmer case studies, many significant hazards and risks remained that workers were routinely accepting, and part of that acceptance was grounded in the very management systems that were supposed to be protecting workers.

Chapter 10

Conclusions and implications
for change

In this book, I have sought to demonstrate the origins and the contradictory effects of neoliberal OHS management and regulatory systems, including mechanisms of workers participation, on health and safety practices, injuries and disease exposures (Dekker, 2020; Frick, 2011; Frick and Walters, 1998; Walters et al., 2016). I have suggested through the three industry case studies that industrial, management and labour process restructuring shift the locus of hazard control and responsibility with attendant effects on power relations and the risk subjectivities or habitus underlying labour consent and compliance. As demonstrated, in these contexts of change both managers and workers seek to reconstruct their assumptions and understandings regarding control and responsibility. Guided by neoliberal notions of governance coming from various government and non-governmental sources including industry accident prevention associations and schools of management (Boyd, 2003), the neoliberal forms of OHS governance evident in all three industries seek to integrate and coordinate health and safety and productivity through technologies, procedures, rules and other technical means such as audits which promise to simultaneously reduce risk and enhance efficiency. Within this governance model, worker/producer responsibility and control in health and safety lie in their adherence to safety rules and procedures, and their reporting of perceived hazards through participation in audits, joint committee inspections, incident reports and safety talks; while management responsibility and control is expressed through the development and communication of rules and procedures, and by establishing the technologies and technical means through which hazards are to be identified and controlled. Through worker participation mechanisms, including things like quality circles and continuous improvement groups, workers are encouraged to recognize their interests and responsibility for rule conformity, while at the same time embrace revised notions of skill, control and responsibility based on their capacity to follow rules and avoid injuries (Belanger, Edwards and Wright, 2001; Gray, 2009; Rinehart, Huxley and Robertson, 1997). Management communications within and outside these group and team settings push the claim that workplace health and safety is a partnership grounded in common labour/capital interests in maximizing health and safety in concert with efficiency, productivity, profitability and employment security (Gray, 2009). In this neoliberal framing, one that is accepted by many workers, worker representatives and unions, state intervention and enforcement penalties are only required for a minority of bad actors; that is, individuals, managers or owners who have not realized the full benefits of best management practices in OHS (Vosko et al, 2020). The structural factors and interests behind management and corporate risk-taking are hidden in this discourse, although the contradictory practices of many firms and individual managers frequently expose some ugly truths about the value placed on worker health and safety (Gray, 2009; Walters et al., 2016).

These governance assumptions and features, to the extent they are accepted by different governments, unions and workers, form an important subjective foundation of workplace responses to hazards, underpinning labour consent while also shaping when and why resistance occurs. I have argued that the erosion of direct worker control, employment security and their associated power to negotiate risk exposures, both at the individual and collective level, are central to understanding why governments and corporations have increasingly focused their attention on formal OHS management systems along with strong elements of worker participation. More specifically, OHS management systems and the participative rights-based IRS are responses to the challenges of labour consent and resistance *and* the means of extending tighter management control over the production costs and disruptions of injury incidents and close calls within production regimes of lean production, flexible employment relations and shifting market demands.

With respect to labour consent, these new systems and the assumptions underlying them provide a foundation for reconstructing some sense of security for workers, both with respect to their body and mental health, and with respect to their employment or financial security. But, as the evidence also shows, whether the full political potential of neoliberal governance is realized, especially over the longer term, depends a great deal on whether employers and managers act in ways that provide workers with at least some substantive control and security (Brown et al., 2015; Hall, 2016; Mollering, 2006). However, employers and managers often implement their OHS systems or act personally in ways that undermine worker trust and the perceived integrity of management safety claims and controls. Some of these variations come from outside industry or market pressures, and the positions and power that firms exercise in those "fields," but others are understood as reflecting internal firm or occupational histories and dynamics, as well as the diverse histories of individual workers and managers. I've suggested several times that the resulting concerns and conflict are often addressed coercively especially in contexts where there are limited constraints on manager or supervisor power (e.g. weak or no union, low-skilled precarious employment, weak reprisal protections, strict management hierarchies and close direct supervision), and where firms failed to deploy meaningful worker participation and team approaches nesting more authority in workers (Tuohy and Simard, 1993; Walters et al., 2016). In turn, workers and worker reps. experiencing coercion are more likely to perceive and understand health and safety as a political outcome in the sense that they come to understand that their exposures are grounded explicitly in their lack of power and in conflicting interests (Basok, Hall and Rivas, 2014; Brown, 2006; King et al., 2019). However, where they perceive limited individual or collective power and limited options for employment, they often respond through various coping mechanisms including risk avoidance to the extent possible and denial of the risk (i.e. not thinking about it). Resistance in the form of refusals, slowdowns, union organizing etc. are more likely when workers construct the risks to their health as high, when they are less fearful of the consequences of challenging management and when they think resistance might make a difference.

Along with understanding the role of labour process changes in fueling and shaping OHS governance, I've argued that neoliberal ideas about governance became important social forces in their own right, shaping both management and worker habitus or dispositions towards risk. At the same time, the evidence demonstrated that managers and workers varied in their acceptance of neoliberal ideas, and indeed as seen in the farm case studies, some directly challenged neoliberalism or neoproductivism through alternative management and production practices (Vickers et al., 2005). However, in the farm context, we saw that the rejection of neoliberalism or neo-productivist methods did not mean safer farming practices.

The effects of neoliberal forms of management and OHS governance on worker responses to risk have also been shown to have their contradictions (Dean, 1999). Despite general reductions in worker security and power, some workers and worker representatives capitalize on aspects of the IRS and OHS management systems to mobilize around power resource opportunities within their particular work contexts and work relations and outside the workplace. In some cases, this involved the conscious political use of the auditing logic and associated inspection processes to collect and build effective evidence-based cases for change incorporating efficiency and productivity arguments, while in others it was more simply a recognition of management or supervisor vulnerability or needs around worker cooperation offering opportunities for negotiation (e.g. when auto firms need workers to put extra effort into rush factory orders in just-in-time contract arrangements).

As I've tried to show, Bourdieu's (1990) image of dominant and dominated parties engaged in a constant game of strategy and tactics seems apt in describing OHS relations and politics, contexts where workers and managers are seeking to obtain and mobilize the capital necessary to meet their respective safety, security and production interests within the "field's rules of the game" (Bourdieu, 1990, p. 60). However, just as some workers are more adept than others in resisting both intimidation and co-optation, some managers or firms are less likely than others to manage the contradictions well enough to gain worker trust, which is to acknowledge that there are variations in habitus, knowledge and skills among workers, managers and management teams within and between different workplaces. At the same time, of course, as we saw, some firms were in better positions within their industry and markets to offer concessions and cooperation.

While the challenge of labour consent within a restructuring contest is a consistent focus in this book in explaining why OHS governance and risk subjectivities have developed in certain ways, I also suggest that the dynamics underlying change in health and safety from the 1980s are less a response to overt worker resistance to risk than they are efforts, within the context of new labour processes, market developments and technologies, and the neoliberal emphasis on lean efficiency, to avoid the costs associated with incidents and close calls. Many of the management changes are reactive but those changes are often reacting as much to concerns about production disruptions, increased attention from regulators and damage to equipment than concerns about injured or sick workers or angry responses from workers and the public. While the owner in the auto case study was trying to prevent unionization in part through improved attention to OHS, worker resistance per se was not a primary motivation behind the move to formalize HR and OHS management within a participative framework. Moreover, the concern with both controlling and responsibilizing workers in all the workplaces was never just about preventing resistance, it was as much if not more about disciplining workers (or themselves as self-employed workers or managers) to meet the needs of flexible production, where workers were expected to work more for less, involved in more tasks with less control.

While suggesting that most workers in every industry and workplace continue to consent and comply to considerable OHS risks, even during the most politically volatile periods such as the 1980s, workers and worker health and safety representatives were shown to carve out areas of limited contestation and some increased influence in health and safety politics, which push both corporations and governments to temper or alter their reliance on neoliberal ideals. I have also demonstrated that the impact of restructuring and neoliberal forms of rule have been contradictory, often yielding management control failures, uncontrolled risks and spaces in which workers can contest and challenge corporate and state claims. While focusing principally on workplace politics, I also showed that it was these contradictory effects at the level of production which ultimately explained the variations not only in corporate applications of neoliberalism but also in the actions of the state and other key institutions. Here I was trying to build on the work of Tucker (2003) who argued that variations in neoliberal policy in health and safety are a function of power resources, in particular, the strength of labour unions. However, I have tried to demonstrate that along with the collective power resources emphasized by Tucker, it is important to understand the foundations of worker, union and corporate commitment to the rules and logic of health and safety as well as their conflicting interests, and the manner in which these tensions are managed both at the points of production and the public spheres.

Where do we go from here? What is to be done?

I've argued that neoliberalism underpins the current ways in which most workers are controlled by capitalist management but we should not make the mistake of thinking that the end of neoliberalism at least in terms of a greater emphasis on the regulation of OHS will mean the end of unsafe or unhealthy work (Dekker, 2020). A review of the OHS record during the heyday of Keynesian and social democratic policies in the 1950s and 1960s should also lead to some pause regarding any expectation that a "third way" or other so-called softer forms of capitalism will suddenly lead to real worker control and a workplace environment where safety priorities actually rule. Moreover, while we need much stronger enforcement with deterrence measures (King et al., 2019; Tucker, 2003), we should be in no hurry to go back to the command-and-control regimes of the early to mid-1900s where workers with essentially no rights to knowledge, representation or refusal were supposedly protected by a pluralist neutral state (Tucker, 1990). Those regimes didn't work through much of the 20th century (Ashford, 1976; Carson, 1982; Tucker, 1990) and are no more likely to work in the current context (Bittle and Snider, 2015; Quinlan, 1999). Even if neoliberalism is coming to an end or being transformed into some softer form with more state protections for workers, workers still require strategies for achieving meaningful change in these contexts, with a full understanding that the power dynamics, imbalances and contradictions that have undermined health and safety conditions since the early days of industrial capitalism are still going to be present in some form. It is not enough to simply state how we think OHS relations and practices should be conducted, we need to consider how we get there.

I began this book with a regulation LP theory perspective which argued that firms operating within capitalism seek to manage the inevitable political and economic costs of profit-based production without undermining the short- and long-term conditions necessary for maximum profit-making (Jessop, 1990, 1995; Thompson, 2010; Tucker, 1990). From this perspective, irrespective of the governance regime, workplace injuries and diseases within capitalism are variable costs that need to be controlled *to the extent possible and to the extent that they* disrupt production, damage or limit access to productive capital, increase compensation costs, raise wages and other labour costs, undermine labour relations, morale, and productivity, etc. Accordingly, two qualifications are understood as critical in shaping what firms will actually do to prevent injuries and diseases, again regardless of whether neoliberalism is dominant or not. First, what capitalist firms mean *by the extent possible* is not what is technologically or humanly possible, but rather what management and investors define as possible in terms of costs and in terms of the conditions and relations of production necessary to turn expected rates of profit. State regulations, and public and labour political pressure may alter the boundaries or calculus of acceptable versus unacceptable costs, and/or expected profit will also vary as it has under neoliberalism (Harvey, 2005); but ultimately, as the three industry case studies have hopefully shown, firms are constantly making prevention decisions based on implicit or explicit calculations of relative cost, profit benefit and production output. Cost and control over labour are critical elements underlying this calculus especially to the extent that capital is able to transfer or externalize the costs to labour in the form of deaths, injuries, wear and tear on workers' bodies and reduced life expectancies, and to the state and general public in terms of medical and other service costs. This logic also brings into play several other economic considerations including firm access to capital and market potential. Benefits almost invariably refer to efficiencies and productivity, but other consequences can be valued as economic interests by capital including public and consumer relations. In short, the steps taken by capital to protect workers *to the extent possible* are always constrained and buffeted by shifting and conflicting economic interests. While state policies can alter those interests in significant ways, the underlying dynamics that constrain attention to health and safety will still be operative (Navarro, 1982).

The second qualification is that capital has historically been able to reduce the costs of injuries and diseases without necessarily reducing their frequency or even severity. Firms do this in various ways including of course through the management of compensation costs which can involve everything from concealing injured workers from the compensation authorities to firing them for failing to conform to back to work regulations. Put more broadly, firms manage OHS costs and what they deem necessary or possible in terms of prevention by discouraging injury and disease reporting (Basok, Hall and Rivas, 2014; Breslin et al., 2003; 2007; Hall, 2016; Hall et al, 2018; MacEachen, 2000, 2005). Again, state policies and other shifts in governance can restrain these tendencies, especially through any measures that strengthen employment security and protect injured workers' rights, but there will always be limitations on the amount of restraint placed on employers' capacities to manage their employees and costs as they see fit within a capitalist economy.

Informed by this understanding, most of this book has sought to reveal the ways in which capital or management manages and externalizes its health and safety costs by internalizing responsibility and locating blame in workers. As has been demonstrated here and elsewhere (Basok, Hall and Rivas, 2014; Hall, 2016), workers (including self-employed workers), work with a host of injuries and health problems, often self-medicating or treating, in effect allowing capital to transfer the costs of production to workers' bodies and indeed to the extent that mental health is a critically under-examined problem in prevention practice, their minds. Workers accept these conditions and consequences for different reasons but most of those reasons revolve around the acceptance of injury as an occupational reality and employment necessity. As I've also tried to show, especially through the farm case studies, different managers and firms define and draw the boundaries around what they think is possible or reasonable, not just based on hard economics or immediate politics but also based on their particular histories, risk subjectivities and experiences. Certainly, neoliberalism accentuates these tensions by intensifying industry competition and the pressure on firms and actors within the firms to maximize profitability, which then fuels an ever-increasing maximum profit target, making it even harder for firms and managers to do the "right thing" for OHS even when inclined to do so (Boyd, 2003; Harvey, 2005; Thompson, 2010). Restraints on competition and movement away from trade liberalization would likely help to address this problem and in a post–COVID-19 virus world, who knows how far the pendulum might shift. But as we know from the aftermath of the 2008 financial crisis, it may be premature to announce the end of neoliberalism as a dominant form of corporate and state governance (Bittle and Snider, 2015).

Nested in this view of capitalism and the effects of capitalist dynamics on OHS decisions, this book has tried to narrow the focus somewhat by understanding why workers accept risk and when and why they don't, based on the view that ultimately it is workers' reactions and actions which are central to meaningful and sustained progress in OHS conditions. What the state and other institutions do, and how they operate in governing and regulating the market and the workplace, are demonstrably important, but the evidence also suggests that how much of a difference depends substantially on the public and production politics driven by workers, worker organizations and other social movements (Tucker, 2003). As such, rather than focusing on wishful proposals for a perfect OHS governance regime, such as equal worker/management power over health and safety decisions and conditions, or complete transparency in hazard knowledge and prevention information, I end this book by suggesting a strategy for re-politicizing health and safety so that we might limit corporate power and begin to move toward these ideals.

Where to go from here: building an OHS movement

First, rather than tearing down Ontario's neoliberal internal responsibility system, my view is that workers, activists and advocates need to exploit those aspects of the IRS which offer possibilities for more power and influence in the workplace. The core problem with the IRS is not that it is a rights-based governance system (Gray, 2009) but rather that these OHS rights have been cast in a discursive framework of responsibilization, which ignores the structural positions of workers relative to employers, and reinforces a risk acceptance and risk taking habitus operating in most workplaces. A right is not a substantive right if it is not being exercised, and the neoliberal casting of individual *and* collective worker responsibility for preventing injuries and disease in employment contexts where power is severely tilted to the employer's advantage, is ultimately a recipe for limited progress (Gray, 2002; Tucker, 2003; Weber et al., 2018). Most of what I have tried to reveal in this book are the various ways in which those rights are constrained and undermined and, of course, twisted and exploited to make worker reps. apolitical technocrats and workers self-regulating.

However, I've also pointed to the limitations, contradictions and political opportunities operating within neoliberal governance regimes and within capitalism more generally. Organizing around the expansion of the core IRS rights is one of those opportunities in as much as some workers and worker representatives (i.e. KAs) have demonstrated a capacity to push on the boundaries of information access, work refusals and participation in decisions. In effect, this is what KAs are doing when they use independently collected information to force managers to acknowledge hazards and risks, and over time, encourage managers to accept that they can't hide hazard information from workers or deny substantive worker input into risk decisions. This is also what workers are doing when they invoke their right to refuse. As rare as this is (Gray, 2002), some workers and worker representatives have learned to use this right as a significant political tool, not only individually but also collectively (Hall et al. 2006, 2016; Storey and Lewchuk, 2000; see also Chapter 5). The rights to representation and participation provide perhaps the most important spaces for change, and again, it is encouraging that there are variations in the extent to which workers and worker representative use these rights and refuse to be captured in a narrow technocratic box. As I've also argued, management is vulnerable in as much as some workers are able to access and successfully mobilize management language and knowledge as cultural and symbolic capital, which in field theory terms means that the dominant risk acceptance habitus and the subjective basis of management control have been disrupted at least for some workers. Thus, what the analysis suggests first and foremost is that the possibilities for change revolve around the identification of power resources within current workplaces and governance regimes (Tucker, 2003).

In nesting the study in LPT, I've emphasized that *labour power,* grounded in the relative skill and control that workers exercise in the labour process, remains a significant source of worker capacity to resist and yield concessions from managements. However, by drawing Bourdieu's theory into this analysis, I've also stressed that workers exercise influence through other power resources such as cultural, social and symbolic capital, which although structurally linked to the labour process and their employment security in important ways, are not determined by them. Grounded in governmentality theory, I have also argued that paradoxically neoliberal forms of management and OHS governance have offered channels of capital access and mobility for some workers.

The key message then is that when we say we need to identify sources of power that workers and reps. can use to extract changes from their employers, more attention needs to be focused on the cultural and political opportunities present within the workplace and within the dominant state and corporate discourses. In the current historical context, this means helping workers and worker reps. to use the knowledge, language and logic of neoliberal governance as cultural and symbolic capital to achieve more meaningful reductions in OHS risk, in effect flipping the neoliberal rules of the risk management game to worker advantage. If neoliberalism is dead or dying, which I doubt, other opportunities and constraints will need to be identified; but the general organizing principle is that worker organizations, activists and advocates should strategize around the power opportunities offered within the context of labour process transformations and dominant management and state governance systems.

As I have also tried to show, there are different opportunities and constraints in different workplaces and industries. For example, the extent to which workers and managers are captured within a technical-legal governance framework varies with the labour process and labour needs, industry market characteristics, and the degree and form of private and public regulation. Although not all small firms had poor health and safety practices, most small-scale auto parts suppliers operated with relatively limited health and safety management systems and/or worker participation programmes (Barrett, Mayson and Bahn, 2014; Eakin, 1992; Hasle et al., 2011; Walters, 2004). Most family farmers employing workers, for example, had very little in the way of HR or OHS management systems as ways of guiding their own work and management planning or as means of controlling their workers. Workers were mainly very transitory but some were longer-term with some history with the farm and the operator, albeit still with limited contract formality. In either case, the power resources that these farm workers were able to access were limited and constrained by their temporary and dependent status. Longer-term workers could rely to some extent on their personal relations with the owner and sometimes a lack of 'skilled' farm labour in the area gave them added power to negotiate better conditions with their employer. But, on the whole, and even though these workers often worked quite independently, they tended to largely accept what they were told and given.

As both the farm and the auto case studies suggest, firms tended to become more formal and systemic in their approaches to OHS and management more generally as they grew in size and as they embraced more mechanization and other technologies, developments which tended to be linked as well to neoliberal business ideas. In these contexts, firms also sought to develop a greater capacity to respond to market fluctuations and the dual demands for standardization and flexibility, which often placed new pressures undermining the possible health and safety benefits of risk assessment and management technologies. As noted, these ideas (i.e. OHS management systems and joint committees) were also being imposed or encouraged by outside institutions including government policy and law but also by a wide variety of private industry organizations such as accident prevention associations (OFSA, MAPAO, IAPA) and International Safety Organization (ISO), so there was some uptake even among some smaller firms as pointed out. This implies that despite the differences, there were similar opportunities and constraints occurring across industries and business sizes, offering possibilities of common political strategies.

External sources of power and influence also matter and have developed across industries in diverse ways with sometimes equivalent effects. While the absence of OHS legislation was likely significant in undermining the spread of safe farm practices, the threats of environment and OHS regulation were often enough to fuel efforts by farmers and the industry as a whole to present themselves as technically competent and voluntarily self-interested in controlling both environmental and OHS risks, which further encouraged the adoption of both technocratic and neoliberal governance ideas. Government agencies, most notably the federal and provincial agricultural agencies, motivated substantially by an interest in responding to competing political demands from food industry interests, consumers, environmentalists and labour, played along by pushing farmers to demonstrate their voluntary capacities to comply. But still, as shown, the threat of more restrictive environmental legislation also played its role in pushing a more technical and educational approach among some farmers which affected at least their risk-related practices, just as it happened in the auto sector with the increased imposition of ISO 14000 environmental standards. As also shown in the auto case study, even where unions are not successful in organizing plants, they can still exert an influence by pushing firms to adopt participative strategies and other measures which can help workers.

Understanding the dynamics of change is also important in identifying when and where a dominant risk habitus can be disrupted. As Bourdieu (1990) often pointed out, any dominant habitus operative within a given field, in this case the habitus underlying labour consent to risk, is resistant to change for various reasons. However, as we saw in all the industries, there are constantly occurring structural pressures which can shift the power relations and undermine the rules of the game at any time. Examples in auto parts and mining case study were the periodic shifts in the demands placed on workers and managers to take new risks or extend their risk-taking beyond the normative expectation in response to shifting market demands. However, as I've tried to show, it is not just a question of producing more or less risk, it is also a question of shifting the foundations of what workers know, perceive and understand about what hazards are in their workplace, who controls the risk and how. One of the more significant disruptive forces identified in this research is the restructuring of labour processes and the jobs within them. How managers and supervisor respond to these disruptions to their own habitus and the habitus of their workers is complex but I've shown that government and private agencies can impact firm management responses to these disruptions especially if they cause substantial conflict within or outside the workplace. Workers and managers may rely on the same old assumptions or understandings for some time even in contexts of transformation and conflict, again what Bourdieu (1977, p. 78) called "hysteresis," but over time it becomes harder to operate within the same assumptions and practices. Becoming cognizant of these changes and their effects on risk and risk habitus and recognizing them as political opportunities are critical steps in the development of successful change agents in the workplace. Certainly, the evidence from the two rep. studies as examined in Chapter 8 suggests that exploiting the cultural disruptions of production and management changes and structural shifts in interests and capital was what many KAs were doing.

A dominant habitus can also be disrupted by the introduction of new individuals (managers, supervisors, experts, workers, worker representatives, inspectors) and, to some extent, institutional players into a work field or context (e.g. workplaces unionize or de-unionize, governments increase or decrease enforcement), especially if they have different management orientations or habitus and are in positions of relative power to challenge the existing norms or rules of the game (see also Boyd, 2003; Walters, 1985; Lewchuk, Robb and Walters, 1996; O'Grady, 2000). The evidence from the joint health and safety committees suggests that the introduction of new managers and workers can introduce important dynamics in shifting worker representatives and committees towards or away from a TL, PA or KA orientation, although the strength of the effect is likely tied to the position that the new person has (e.g. supervisor vs. plant manager vs. OHS management; trades vs. operations vs. cleaning-job position). Although simply questioning existing assumptions or practices, or voicing alternatives can get people "thinking," what matters most is what people experience over time. When TL reps. or workers see major gains being made by a rep. using research and well-documented resolution proposals, they are more likely to expand their understanding of what they think is reasonable or possible, and recognize more effective ways of using their audits as tools to get those changes. This again suggests avenues and opportunities for change.

Along with recognizing opportunities for change which disrupt an existing risk habitus, I would argue there are several fairly practical ways in which worker advocates and organizations can help workers and worker representatives to exploit and subvert key aspects of neoliberal OHS governance. First, the existing knowledge activists need to be supported by labour unions and worker organizations as much as possible given their positioning within the workplace and health and safety field, which gives them, I would argue, the greatest potential for subverting the habitus of other workers in their workplaces. I argue further that these representatives also offer the best means of developing a broader OHS movement of KA representatives across regions, provinces and the country, which in turn is central to gaining structural and legal changes which can strengthen worker rights and power even if they follow the limited IRS pattern. By support, I'm suggesting more opportunities for both technical and political education, access to more research skills and resources and increased social contacts – that is, helping workers and reps. to access the cultural and social capital they can use in their relations with management and other workers. Ideally, existing knowledge activists should be recruited if numbers permit, to help in developing and delivering an alternative education regime for all workers and worker representatives whether through the WHSC, specific unions or other worker organizations. As noted, this may be happening already as we found in the second rep. study (Study 6) where many of the knowledge activists were also OHS certification instructors. Along with increased workplace-specific technical and health training aimed at giving workers greater capacity to challenge management claims, modified worker education programmes would encourage a critical political understanding of how and why OHS management systems and the IRS work in the way they do, and the different ways in which workers are made responsible for conditions over which they have limited control. At the same time, especially for worker and union representatives, political education would include instruction on basic research and networking skills, and political strategies consistent with those used effectively by knowledge activists. As the later implies, education would be focused on helping worker representatives develop the political and analytical skills necessary to identify political opportunities and weaknesses within specific firms and management, while also emphasizing the importance of connecting and building support with workers.

Representatives and workers in non-union and smaller workplaces with less-developed participative options are without a doubt disadvantaged (King et al., 2019). However, unions can offer be enormous help in pushing non-union firms in more "cooperative" and concessionary directions by continuing to mount organizing drives as we saw in the auto case study, although that evidence also showed that workers were not necessarily receptive to crude recruiting union messages about management coercion, deception and intransigence. Legislating networks of unionized OHS representatives who also represent non-union workplaces is also a model well worth exploring (Walters et al., 2016). However, failing that, the union impact could be expanded by providing or funding the same education, training and connection opportunities to non-union workers either directly or through other organizations such as worker centres (Fine and Gordon, 2010).

Also important in terms of support is to help knowledge activists make connections with their KA colleagues from other workplaces and unions. As noted, knowledge activists often develop their own networks which provide ideas and sources of OHS information, but some also talked about the importance of the social and emotional support they got from talking to other reps. about the stress and their accomplishments. Others without this support talked about feeling tired, isolated and underappreciated; and, there was certainly evidence in the KA interviews that some found the challenges of their role too difficult and either fell back into a more limited TL practice or quit altogether. However, increased connectedness was also a way of sharing stories about their union and other unions, and perhaps encouraging more political activism within their unions as they increasingly recognized that their union was not giving them enough support. Other support from unions that could help according to the reps. would be the collective bargaining negotiation of more training and on-the-job time for doing research, interacting with workers, and planning (Hall et al., 2014, 2016). As well, the mining study in particular showed that the more that union leaders and steward networks have the back of their reps. when they are pushing issues, the more secure they feel when challenging management and mobilizing workers. In practical terms, this means advocating on their behalf with senior management, supporting their requests for collective bargaining provisions, and using whatever organizational resources are available to defend reps. if they are threatened or penalized.

If there is some success in building networks of union reps. with regional and provincial ties, it would be ideal to recruit those people and networks to help with supporting and educating the development of knowledge activism within non-union contexts. This is also where community organizations such as worker centres and universities may be crucial in offering supports similar to the unions in terms of education and advocacy (Delp and Riley, 2015; Fine, 2015, 2017; Fine and Gordon, 2010), although funding support from unions and elsewhere are important in sustaining these organizations (see Vosko et al., 2020, Chapter 7).

As the director of the Labour Studies Program at the University of Windsor, I was involved in the Windsor Workers Action Center (WWAC) in Windsor, Ontario, for several years (2006–13) where we made some efforts in this direction, offering labour rights education programmes to high school students and young workers, and doing advocacy and organizing for workers in precarious employment situations. One of our projects was a collaboration with the local OHCOW clinic and a community organization called WOHIS (Windsor Occupational Health Information Services) to try to develop a local OHS rep. network involving union and non-union reps. Unfortunately, our efforts to connect union and non-union reps. in the Windsor area were largely unsuccessful in part because the union reps. reported being too busy in their workplaces to spend the time networking and helping non-union reps. We were also unsuccessful in getting the local union leadership to provide worker centre funding and after struggling with private donations and some lottery grant funding, WWAC effectively closed in 2016. However, other worker centres and programmes persist and offer at least some promise of what is possible (see Vosko et al., 2020, Chapter 7 for more on the challenges and activities of Worker Centres in Ontario; see also, Delp and Riley, 2015; Fine and Gordon, 2010; Workers Action Center, 2007; 2016). Certainly, I would argue that if properly supported by unions or even hands-off government grants, a network of worker centres across the province and the country with a significant focus on supporting non-union worker representatives and workers in small workplaces would have a major positive impact on health and safety.

While political as well as technical and medical education in OHS is certainly a critical aspect of this effort, education abstracted from the workplace and the structures governing relations in that workplace is not likely on its own to lead to widespread or sustained change in risk practices. If workers are still experiencing their work and the OHS risks in circumstances that are consistent with those that shaped the risk habitus in the first place, we would not expect education on its own to make a sufficient difference (Bourdieu, 1977, 1990). Having the abstract knowledge that alternative realities are possible is a start perhaps, but that's not the same thing as coming to believe they are reasonable or achievable. This is where structural change comes into play as forces which can disrupt the existing habitus and consent structures, whether the changes are in terms of the labour process, in management structures, in the firm's financial situation, in the diversity of workers, in the employment relationship or in the law.

While more research is always needed, I would argue that we know enough about how change happens in joint committees and in the habitus of workers and worker representatives to support the argument that *any structural change within or affecting the workplace,* even if it initially weakens worker control or security, has the potential to shift worker habitus and the subjective basis of consent, creating opportunities for resistance. Variations and inconsistencies in the management implementation of structural changes are other potential sources of tension and change which can subvert a dominant habitus. Legal reforms, even limited ones such as happened with Bill

208 in 1990, may have unexpected effects and paradoxically offer alternative opportunities or meanings to workers and worker representatives (e.g. the creation of OHCOW). Since change is a fairly constant feature within capitalism, I would argue further that the opportunities are always there. What's needed are the insights and supports to encourage workers and worker representatives to identify, organize and act on those opportunities, a central role which I think labour researchers and advocates can play.

I'll end with what I see as two positive messages. First, I want to point to a research collaboration in Ontario called LOARC which involves labour union activists, OHCOW researchers and practitioners and academics from several Ontario universities. Since around 2009, it has taken on the mandate of supporting health and safety representatives through research, education and other supports (see labourstudies.mcmaster.ca/loarc). Its impacts have been less than many of us hoped but based on its work (Hall et al., 2013, 2016; King et al., 2020), some unions such as the Ontario Nurses Association and the Ontario Public Service Employees Union have been moving to redevelop their training in line with knowledge activist principles and have been negotiating collective bargaining provisions to reflect their activism. One especially interesting project has been the development of a strategic approach for committee reps. to advocate major changes in mental health policies and programmes within workplaces, a process grounded in research and the application of the research information to the development of stress prevention programmes (Ramkissoon, Smith and Oudyk, 2019).

Second, while it may seem pie in the sky to be advocating an OHS representative–based movement, which is in effect what I'm suggesting, it is worth remembering that there were the beginnings of such a movement in Ontario in the 1980s, just when neoliberalism was getting its initial legs in the Canadian context. As reported in my interviews with activists from what was a very volatile period in Ontario health and safety history (Hall, 1991; see also Storey and Tucker, 2006), many health and safety representatives and union and community activists were organizing and advocating around demands for stronger power-sharing arrangements in workplaces with independent decision-making powers for workers, workers representatives and joint committees. This movement gained substantial political currency over a ten-year period within and outside workplaces. With more labour leadership support, it could well have led to much more substantial gains in workers' rights and changes in worker acceptance of risk (see also Storey and Tucker, 2006). Although this history also makes it clear that improvements for workers are never easily realized, the accomplishments of the past point to the need, the promise and the possibilities of a new stronger OHS movement in the future.

Bibliography and archival materials

Acosta-Leon, A., Grote, B., Salem, S., and Daraiseh, N. (2006) Risk factors associated with adverse health and safety outcomes in the Hispanic workforce. *Theoretical Issues in Ergonomic Science*, 7(3), 299–310.

Adler, P., Goldoftas, B., and Levine, D. (1997) Ergonomics, employee involvement, and the Toyota production system: A case study of Nummi's 1993 Model introduction. *International Labor Review*, 50(3), 416–437.

Adkin, L. (1992) Counter-hegemony and environmental politics in Canada. In W. Carroll (Ed.), *Organizing Dissent: Contemporary Social Movements in Theory and Practice* (pp. 247–263). Aurora: Garamond Press.

AgCare, Agricultural Groups Concerned About Resources (1992a) *Our Farm Environmental Agenda*. Guelph: AgCare.

AgCare, Agricultural Groups Concerned About Resources (1992b) *Peer Education through Environmental Farm Plans*. Guelph: AgCare.

AgCare, Agricultural Groups Concerned About Resources (2001) *Our Farm Environmental Agenda: Canada-Ontario Green Plan*. Guelph: AgCare.

Agriculture and Agri-Food Canada, News Release (2000) *Vanclief Announces $10 Million to Farm Environmental Program*. Leamington, Ontario, June 9.

Agriculture and Agri-Food Canada, News Release (2001a) *Vanclief Announces Funding to Help Organic Growers Seize New Market Opportunities*. Vancouver, B.C., June 8.

Agriculture and Agri-Food Canada, News Release (2001b) *Centre Set to Bolster Canada's Organic Expertise*. Baddeck, Nova Scotia, Agriculture and Agri-Food Canada, July 12.

Agriculture and Agri-Food Canada (2001c) *All about Canada's Organic Industry*. www.agr.ca/cb/factsheets/2industry_e.html. Ottawa: Accessed 09/27/20.

Agriculture Canada (1987) *Canadian Agricultural Research and Technology Transfer: Future Directions*. Working Paper. Ottawa: Agriculture Canada.

Agriculture Canada (1989) *Growing Together: A Vision for Canada's Agri-Food Industry*. Supply and Services Canada. Ottawa: Agriculture Canada.

Agriculture Canada's (1995) *The Health of Our Soils: Toward Sustainable Agriculture*. Supply and Services Canada. Ottawa: Agriculture Canada.

Agriculture Canada and OMAF (1992) *Canada/Ontario 1992 Agriculture Green Plan*. Queen's Printer. Toronto: Queen's Printer.

Agriculture Canada and OMAF (1992) *Horticultural Crops: Best Management Practices*. Environmental Sustainability Initiative. Toronto: Queen's Printer.

Agriculture Canada and OMAF (1994a) *Water Management*. Best Management Practices Series, Environmental Sustainability Initiative. Toronto: Queen's Printer

Agriculture Canada and OMAF (1994b) *A First Look: Practical Solutions for Soil and Water Problems*. Best Management Practices, Environmental Sustainability Initiative. Toronto: Queen's Printer.Alaszewski, A. (2015) Anthropology and risk: Insights into uncertainty, danger and blame from other cultures – A review essay. *Health, Risk & Society*, 17(3–4), 205–225.

Aldrich, M. (1997) *Safety First: Technology, Labor, and Business in the Building of American Work Safety, 1870–1939*. John Hopkins University Press, Studies in Industry and Society.

Allen, J. and Bernhardt, C. (1995) Farming practices and adherence to an alternative conventional agricultural paradigm. *Rural Sociology*, 60(2), 297–309.

Ames, D. (1987) *Bulk Mining and Worker safety: A Preliminary Study*. Ontario ministry of Labour, Occupational Health and Safety Branch.

Applebaum, E. (2002) The impact of new forms of work organization on workers. In G. Murray, J. Belanger, A. Giles and P. Lapointe (Eds.), *Work Employment Relations in the High Performance Workplace* (pp. 120–149). Continuum.

Archibald, H. (1999) Organic Farming Is Growing. *Canadian Agriculture at a Glance*. Ottawa: Agricultural Division, Statistics Canada.

Arcury, T. and Quandt, S. (1998) Occupational and environmental health risks in farm labor. *Human Organization*, 57(33), 331–334.

Arcury, T., Quandt, S., Rao, P., et al. (2005) Organophosphate pesticide exposure in farmworker families in western North Carolina and Virginia: Case comparisons. *Human Organization*, 64(1), 40–51.

Arksey, H. (1998) *RSI and the Experts: the Construction of Medical Knowledge*. London: UCL Press

Aronsson, G. (1999) Contingent workers and health and safety. *Work, Employment and Society*, 13(3), 439–459.

Ashcroft, J. and Taylor, T. (1982) Internal responsibility and the safety system Joint Audit Report of Inco Ontario and USWA Local 6500. Sudbury.

Asher, R. (1986) Industrial safety and labor relations in the United States, 1865–1917. In C. Stephenson and R. Asher (Eds.), *Life and Labor: Dimensions of American Working Class History*. SUNY Press, pp. 115–130.

Ashford, N. (1976) *Crisis in the Workplace: Occupational Disease and Injury*. MIT Press.

Ashmead, R. (1986) The myth and reality of the farm financial crisis. *Canadian Journal of Agricultural Economics*, 33(1), 63–73.

Atari, D., Yirideo, E., Smale, S., and Duinker, P.N. (2009) What motivates farmers to participate in the Nova Scotia environmental farm plan? Evidence and environmental policy implications. *Journal of Environmental Management*, 90, 1269–1279.

Aversa, T. and Hall, A. (Forthcoming 2021). PTSD and the limits of presumptive legislation. In R. Ricciardelli, S. Bornstein and A. Hall (Eds.), *Multidisciplinary Perspectives on Posttraumatic Stress: Causes, Consequences and Responses*. New York: Routledge Press.

AWCBC (2018) Association of workers Compensation Boards of Canada. *Detailed Key Statistical Measures (KSM) Report*. http://awcbc.org/

Babson, S. (1995) Lean or mean? The MIT model and lean production at Mazda. *Labour Studies Journal*, 18(3), 3–24.

Baker, B., Green, T., and Loker, A. (2020) Biological control and integrated pest management in organic and conventional systems. *Biological Control*, 140, 1–9.

Barkemeyer, R., Holt, D., Preuss, L., and Tsang, S. (2014) What happened to the 'development' in sustainable development? Business guidelines two decades after Brundtland. *Sustainable Development*, 22(1), 15–32.

Barnetson, B. (2010) *The Political Economy of Workplace Injury in Canada*. Athabasca University Press.

Barnetson, B., Foster, J., and Matsunaga-Turnbull, J. (2018) Estimating underclaiming of compensable workplace injuries in Alberta, Canada. *Canadian Public Policy*, 44(4), 400–410.

Barrett, R., Mayson, S., and Bahn, S. (2014) Small firms and health and safety harmonization: Potential regulatory effects of a dominant narrative. *Journal of Industrial Relations*, 56(1), 62–80.

Barth, P. and Hunt, H. (1982) *Worker's Compensation and Work-Related Illnesses and Diseases*. MIT Press.

Basok, Tanya. 2004. Post-National Citizenship, Social Exclusion, and Migrants' Rights: Mexican Seasonal Workers in Canada. *Citizenship Studies* 8.1: 47–64.

Basok, T., Hall, A., and Rivas, E. (2014) Claiming rights to workplace safety: Latin American immigrant workers in Southwestern Ontario. *Canadian Ethnic Studies*, 46(3), 35–53.

Basran, G. and Hay, D. (1988) *The Political Economy of Agriculture in Western Canada*. Garamond.Beardwood, B., Kirsh, B., and Clark, N. (2005) Victims twice over: Perceptions and experiences of injured workers. *Qualitative Health Research*, 15(1), 30–48.

Beaumont, P. (1983) *Safety at Work and the Unions*. Croom-Helm.

Beck, U. (1992) *Risk Society: Toward a New Modernity*. Sage.

Belanger, J., Edwards, P., and Wright, M. (2003) Commitment at work and independence from management: A study of advanced teamwork. *Work and Occupations*, 30(2), 234–252. doi:10.1177/0730888403251708Bennett, D. (2011) *Northern Exposures: A Canadian Perspective on Occupational Health and Environment*. Routledge.

Berdahl, T. (2006) *Occupational Injuries in the United States: A Longitudinal Analysis of Race-Gender Differences*. Dissertation Abstracts, 66(8), 3097-A. Ph.D. Thesis.

Berman, D. (1978) *Death on the Job*. Monthly Review Press.

Beus, C. and Dunlap, R. (1990) Conventional vs. alternative agriculture: The paradigmatic roots of the debate. *Rural Sociology*, 55(4), 590–616.

Birch, K. and S. Springer (2019) Peak neoliberalism: Revisiting and rethinking the concept of neoliberalism. *Theory and Politics in Organization*, 19(3), 467–485.

Bird, F.E. Jr. (1974) *Management Guide to Loss Control*. Toronto: Industrial Accident Prevention Association of Ontario (MAPAO).

Bittle, S. and Snider, L. (2015) Law, regulation, and safety crime: Exploring the boundaries of criminalizing powerful corporate actors. *Canadian Journal of Law and Society*, 30(3), 445–464.Bourdieu, P. (1977) *Outline of a Theory of Practice*. Cambridge University Press.

Bourdieu, P. (1990) *The Logic of Practice*. Stanford University Press.

Bourdieu, P. and Wacquant, J. (1992) *An Invitation to Reflexive Sociology*. Chicago: University of Chicago Press.

Boyd, C. (2003) *Human Resource Management and Occupational Health and Safety*. London: Routledge.

Braverman, H. (1974) *Labor and Monopoly Capital*. Monthly Review Press.

Breslin, C., Koehoorn, M., Smith, P., and Manno, M. (2003) Age related differences in work injuries and permanent impairment: A comparison of workers' compensation claims among adolescents, young adults and adults. *Occupational and Environmental Medicine*, 60(e10), 1–6.

Breslin, C., MacEachen, E., Polzer, J., Morrongiello B., and Shannon, H. (2007) Workplace injury or 'part of the job'? Towards a gendered understanding of injuries and complaints among young workers. *Social Science and Medicine*, 64(4), 782–793.

Breslin, F.C., Polzer, J., MacEachen, E., Morrongiello, B., and Shannon, H. (2006a) Workplace injury or "part of the job"?: Towards a gendered understanding of injuries and complaints among young workers. *Social Science and Medicine*, 64, 782–793.

Breslin, F.C. and Smith, P. (2005) Age-related differences in work injuries: A multivariate, population based study. *American Journal of Industrial Medicine*, 48, 50–56.

Breslin, F.C., Smith, P., Koehorn, M., and Lee, H. (2006b) Is the workplace becoming safer? *Perspectives on Labour and Income*, 18(3), 36–41.

Brown, M. (2006) Immigrant workers: Do they fear workplace injuries more than they fear their employers? In A.J. Schulz and L. Mulling (Eds.), *Gender, Race, Class and Health: Intersectional Approaches* (pp. 228–258). San Francisco, CA: Jossey Bass.

Brown, M.S. (1985) Disputed knowledge: Worker access to hazard information. In D. Nelkin (Ed.) *The Language of Risk: Conflicting Perspectives on Occupational Health*. Beverly Hills: Sage.

Brown, P. (2009) The phenomenology of trust: A Schutzian analysis of the social construction of knowledge by gynae-oncology patients. *Health, Risk and Society*, 11(5), 391–407.

Brown, P. and Calnan, M. (2012) *Trusting on the Edge: Managing Uncertainty and Vulnerability in the Midst of Severe Mental Health Problems*. Bristol: Policy Press.

Brown, P. and Calnan, M. (2013) Trust as a means of bridging the management of risk and the meeting of need: A case study in mental health service provision. *Social Policy and Administration*, 47(3), 242–261.

Brown, S., Gray, D., McHardy, J., and Taylor, K. (2015) Employee trust and workplace performance. *Journal of Economic Behavior and Organization*, 116, 361–378.

Brun, E. 2009. *Expert forecast on emerging chemical risks related to occupational safety and health*. Luxembourg, Office for Official Publications of the European Communities.

Burawoy, M. (1979) *Manufacturing Consent: Changes in the Labour Process under Monopoly Capitalism*. Chicago, IL: University of Chicago Press.

Burawoy, M. (1985) *The Politics of Production*. London: Verso.

Burkett, K., Riggin, R., and Rothney, K. (Burkett Report), (1981) *The Report of the Joint Federal-Provincial Inquiry Commission into Safety in Mines and Mining*

Plants in Ontario, Vol. 1 and 2. Joint Publication of Governments of Ontario and Canada.

Burnham, J. (2009) *Accident Prone: A History of Technology, Psychology and Misfits of the Machine Age*. University of Chicago Press.

Buttell, F. (1993) Socioeconomic impacts and social implications of reducing pesticide and agricultural chemical use in the United States. In D. Pimenthal and H. Lehman (Eds.), *The Pesticide Question: Environment, Economics and Ethics* (pp. 153–181). Chapman and Hall.

Calhoun, C. (1988) Coping with insidious injuries: The case of Johns-Manville Corporation and asbestos exposure. *Social Problems*, 35(2), 162–181.

Canada (1996) Canada's domestic and external reforms help create stronger base for economic expansion. Canadian Government Press Release, PRESS/TPRB/48, November 11. https://www.wto.org/english/tratop_e/tpr_e/tp50_e.htm#chair. Accessed 02/18/20.

Canada, House of Commons (1992) *Report of the House of Commons Standing Committee Report on Agriculture*, "The Path to Sustainable Agriculture". Supply and Services Canada.

Canada Safety Council (1987) *Smart Farming Puts Safety First*. https://canadasafety council.org/. Accessed 06/05/20.

Canadian Agricultural Safety Association (2000) *Stepping toward Health and Success in Your Farm Business: A Guide for a Healthy Farm Workplace*. Ottawa: Health Canada.

Canadian Centre for Occupational Health and Safety (CCOHS) (2020) Programs: Management systems. https://www.ccohs.ca/topics/programs/management/. Accessed 03/25/20.

Carson, W. (1979) The conventionalization of early factory crime. *International Journal for the Sociology of Law*, 7, 37–60.

Carson, W. (1982) *The Other Price of Britain's Oil: Safety and Control in the North Sea*. Oxford: Martin Robertson.

Castleman, B. and Ziem, G. (1988) Corporate influence on threshold limit values. *American Journal of Industrial Medicine*, 13, 55–559.

CAW, Canadian Auto Workers (1990) Submission to the Standing Committee on Resources Developments on Bill 208, An Act to Amend the Occupational Health and Safety Act.

CFIB, Canadian Federation of Independent Business (1990) *Partnership in the Workplace: An Effective Approach to Occupational Health and Safety*. Submission to the Standing Committee on Resources Development. Toronto: CFIB, January 15.

Clark, L. (2007) Business as usual: Corporatization and the changing role of social reproduction in the organic agrofood sector. *Studies in Political Economy*, 80(1), 55–74.

Clayson, Z. and Halpern, J. (1983) Changes in the workplace: Implications for occupational safety and health. *Journal of Public Health Policy*, 4(3), 279–297.

Clement, W. (1981) *Hardrock Mining: Industrial Relations and Technological Changes at Inco*. McClelland Stewart.

Clunies Ross, T. and Cox, G. (1994) Challenging the productivist paradigm: Organic farming and the politics of agricultural change In P. Lowe, T. Marsden, and S. Whatmore (Eds.), *Regulating Agriculture* (pp. 53–74). David Fulton Publishers.

Crawford R. (1977) You are dangerous to your health: the ideology and politics of victim blaming. *International Journal of Health Services* 7, 663–680.

Creswell, J. and V. Clark (2011), Designing and Conducting Mixed Methods Research Thousand Oaks, CA: Sage).

Cunningham, N. and D. Sinclair (2014) The impact of safety culture on systemic risk management. *European Journal of Risk Regulation*, 5(4), 505–516.

Dean, M. (1999) *Governmentality: Power and Rule in Modern Society.* Sage.

Dean, M. (2007) *Governing Societies: Political Perspectives on Domestic and International Rule.* Open University Press.

Dekker, S. (2020) Safety after neoliberalism. *Safety Science*, 125. doi:10.1016/j.ssci. 2020.104630

Delbridge, R. and Ezzamel, M. (2005) The strength of difference: Contemporary conceptions of control. *Organization*, 12(5), 603–618.

Delp, L. and Riley, K. (2015) Worker engagement in the health and safety regulatory arena under changing models of worker representation. *Labor Studies Journal*, 40(1), 54–83.

Denis, W. (1988) Causes of health and safety hazards in Canadian agriculture. *International Journal of Health Services*, 18(3), 419–436.

Derickson, A. (2018) Naphtha drunks, lead colic, and the smelter shakes: The inordinate exposure of immigrant workers to occupational health hazards at the turn of the twentieth century. *Journal of American Ethnic History*, 37(2), 37–61.

Dick, P. and Hyde, R. (2006) Consent as resistance, resistance as consent: Rereading part-time professionals' acceptance of their marginal positions. *Gender, Work and Organization*, 13(6), 543–564.

Doern, G. (1977) The political economy of regulating occupational health: The Ham and Beaudry reports. *Canadian Public Administration*, 21(Spring), 1–35.

Dorman, P. (1996) *Markets and Mortality: Dangerous Work, and the Value of Human Life.* Cambridge University Press.

Dorman, P. (2006) Regulating risk at work: Is expert paternalism the answer to worker irrationality? In V. Mogensen (Ed.), *Worker Safety under Siege: Labor, Capital, and the Politics of Workplace Safety in a Deregulated World* (pp. 34–57). M.E. Sharpe.

Dosman, J., Graham, B., Hall, D., Van Loon, P., Bhasin, P., and Froh, F. (1987) Respiratory symptoms and pulmonary function in farmers. *Journal of Occupational Medicine*, 29(1), 38–43.

Doyal, L. and Epstein, S. (1983) *Cancer in Britain: The Politics of Prevention.* Pluto Press.

Dudgeon, M. (2016) In a split second. OHS, Canada's occupational health and safety magazine. https://www.ohscanada.com/features/in-a-split-second/. Accessed 05/24/2020.

Dwyer, T. (1981) The production of industrial accidents: A sociological approach. *Australian and New Zealand Journal of Sociology*, 6(4), 12–28.

Dwyer, T. (1992) The industrial safety professionals: A comparative analysis from World War I until the 1980s. *International Journal of Health Services*, 22(4), 705–727.

Dyck, D. (2015) *Occupational Health and Safety: Theory, Strategy and Industrial Practice* (3rd ed.). LexisNexix Canada.

Eakin, J. (1992) Leaving it up to the workers: Sociological perspectives on the management of health and safety in small workplaces. *International Journal of Health Services*, 22(4), 689–704.

Eakin, J., Champoux, D., and MacEachen, E. (2010) Health and safety in small workplaces: refocusing upstream. *Canadian Journal of Public Health*, 101(Suppl 1), 29–33. doi:10.1007/BF03403843

Eakin, J.M. and MacEachen, E. (1998) Health and the social relations of work: A study of the health-related experiences of employees in small workplaces. *Sociology of Health and Illness*, 20, 896–914.

Eaton, A. and Nocerino, T. (2000) The effectiveness of health and safety committees: Results of a survey of public-sector workplaces. *Industrial Relations*, 39(20), 265–290.

Edwards, R. (1979) *Contested Terrain*. New York: Basic Books.

Edwards, R., Geary, J., and Sisson, K. (2002) New forms of work: Transformative, exploitive, or limited and controlled. In G. Murray, J. Belanger, A. Giles, and P. Lapointe (Eds.), *Work Employment Relations in the High Performance Workplace* (pp. 72–119). Continuum.

Elefterie, L. (2012) Risk assessment audit versus work accidents prevention. *Contemporary Readings in Law and Social Justice*, 4(2), 552–561. ISSN 1948-9137.

Elling, R. (1986) *The Struggle for Workers' Health: A Study of Six Industrialized Countries*. Baywood Publishing.

Elliot, C. and Long, G. (2016) Manufacturing rate busters: Computer control and social relations in the labour process. *Work, Employment and Society*, 30(1), 135–151.

Emirbayer, M. and Johnson, V. (2008) Bringing Bourdieu into the organizational field: A symposium. *Theory and Society*, 37(1), 1–44.

Emslie, C., Hunt, K., and Macintyre, S. (1999) 'Gender' or 'job' differences? Working Conditions amongst men and women in white-collar occupations. *Work, Employment and Society*, 13(4), 711–729.

Epstein, S. (1979) *The Politics of Cancer*. Doubleday.

Facey, M. (2016) 'Maintaining talk' among taxi drivers: Accomplishing health-protective behaviour in precarious workplaces. *Health and Place*, 16, 1259–1267.

Facey, M., MacEachen, E., Vermi, A., and Morales, K. (2017) The everyday function of joint health and safety committees. *Policy and Practice in Health and Safety*, 15(2), 160–173.

Farndale, E., Hope-Hailey, V., and Kelliher, C., (2011) High commitment performance management: The roles of justice and trust. *Personnel Review*, 40(1), 5–23.

Figlio, K. (1985) What is an accident? In P. Weindling (Ed.), *The Social History of Occupational Health* (pp. 181–206). London: Croon-Helm.

Fine, J. (2015) Alternative labour protection movements in the United States: Reshaping industrial relations. *International Labour Review*, 154(1), 15–26.

Fine, J. (2017) Enforcing labor standards in partnership with civil society: Can co-enforcement succeed where state along has failed. *Politics and Society*, 45(3), 359–388.

Fine, J. and Gordon, J. (2010) Strengthening labor standards enforcement through partnerships with workers organizations. *Politics & Society*, 38(4), 552–585. doi:10.1177/0032329210381240.

Fingnet, A. and Smith, A. (2013) *Occupational Health: A Practical Guide for Managers*. Routledge.

Fitzpatrick, J. (1980) Adapting to danger: A participant observation study of an underground mine. *Sociology of Work and Occupations*, 7(2), 131–158.

Fontana, A. and Frey, J. (2003) The interview: From structured questions to negotiated text. In N. Denzin and Y. Lincoln (Eds.), *Collecting and Interpreting Qualitative Materials* (2nd ed., pp. 61–106). Sage.

Foucault, M. (1977) *Discipline and Punishment*. Pantheon Books.

Fowke, V. (1947) Canadian agriculture in the postwar world. *Annals of the American Academy of Political and Social Science*, 253(1), 44–51.

Frederiksen, M. (2014) Trust in the face of uncertainty: A qualitative study of intersubjective trust and risk. *International Review of Sociology*, 24(1), 130–144.

Frick, K. (1997) Can managers see any profit in health and safety? *New Solutions: A Journal of Environmental and Occupational Health Policy*, 7(2), 32–40.

Frick, K. (2000) *Systematic Occupational Health and Safety Management: Perspectives on an International Development*. Pergamon.

Frick, K. (2011) Worker influence on voluntary OHS management systems – A review of its ends. *Safety Science*, 49, 974–987.

Frick, K. and Walters, D. (1998) Worker representation on health and safety in small enterprises: Lessons from a Swedish approach. *International Labour Review*, 137(3), 367–387.

Friedmann, A. (1977) *Industry and Labor*. MacMillan.

Fudge, J. and Glasbeek, H. (1992) The politics of rights: A politics with little class. *Social and Legal Studies*, 1, 45–70.

Fudge, J. and Tucker, E. (2001) *Labour before the Law: The Regulation of Workers' Collective Action in Canada, 1900–1948*. Toronto: University of Toronto Press.

Galizzi, M., Miesmaa, P., Punnett, L., Slatin, C., and the Phase in Healthcare Research Team (2009) Injured workers underreporting in the health care industry: An analysis using quantitative, qualitative and observational data. *Industrial Relations*, 49(1), 22–43.

Geldart, S., Shannon, H., and Lohfeld, L. (2005) Have companies improved their health and safety approaches over the last decade? A longitudinal study. *American Journal of Industrial Medicine*, 47, 227–236.

Geldart, S., Smith, C., Shannon, H., and Lohfield, L. (2010) Organizational practices and workplace health and safety: A cross-sectional study in manufacturing companies. *Safety Science*, 48, 562–569.

Geller, E.S. (2000) *The Psychology of Safety Handbook*. CRC Press.

Gertler, M. and Murphy, T. (1987) The social economy of Canadian agriculture: Family farming and alternative futures. In B. Galeski and E. Wilening (Eds.), *Family Farming in Europe and America* (pp. 239–270). Westview Press.

Giddens, A. (1991) *Modernity and Self-Identity: Self and Society in the Late Modern Age*. Polity Press.

Gindin, S. (1995) *The Canadian Auto Workers: The Birth and Transformation of a Union*. J. Lorimer.

Glasbeek, H. and Tucker, E. (1999) Death by consensus at Westray. In C. McCormick (Ed.), *The Westray Chronicles: A Case Study in Corporate Crime* (pp. 71–96). Fernwood Publishing.

Goodrum, P. and Dai, J. (2005) Differences in occupational injuries, illnesses and fatalities among Hispanic and non-Hispanic construction workers. *Journal of Construction Engineering and Management*, 131(9), 1021–1028.

Guse, M. and Gunn, T. (1980) Internal safety audit: Inco Metals Co. Paper presented to the MAPAO Annual Meeting, Toronto.

Graham, L. (1995) *On the Line at Subaru-Isuzu: The Japanese Model and the American Worker*. Ithaca, NY: ILR Press.

Gramsci, A. (1971) *Selections from the Prison Notebooks of Antonio Gramsci*. New York: International Publishers.

Gravel, S., Fournier, M., Boucheron, L., and Patry, L. (2005) Barriers in access to compensation of immigrant workers who have suffered work injuries. *Canadian Issues* 32, Spring, 94–98.

Gray, G. (2002) A sociolegal ethnography of the right to refuse unsafe work. *Studies in Law, Politics and Society*, 24, 133–169.

Gray, G. (2009) The responsibilization strategy of health and safety: Neo-liberalism and the reconfiguration of individual responsibility for risk. *British Journal of Criminology*, 49(3), 326–342.

Gray, W. (1993) Does regulatory enforcement work: A panel analysis of OSHS enforcement. *Law and Society Review*, 27(1), 177.

Greenwood M. and Woods, H. (1919) The incidence of industrial accidents upon individuals with special reference to multiple accidents. *Industrial Fatigue Research Board, Medical Research Committee*, Report No. 4. Her Majesty's Stationery Office, London.

Grieshop, J., Stiles, M., and Villanueva, N. (1996) Prevention and resiliency: A cross-cultural view of farm-workers and farmers' beliefs about work safety. *Human Organization*, 55(1), 25–32.

Grunberg, L. (1985) The effects of the social relations of production on productivity and workers' safety. An ignored set of relationships. *International Journal of Health Services*, 13(4), 621–634.

Grunberg, L., Everard, J., and O'Toole, M. (1984) Productivity and safety in worker cooperatives and conventional firms. *International Journal of Health Services*, 14(3), 413–432.

Guaimet, J. (2019) Workplace theory: An anthropological approach to emotional labour, subjectivity and control/resistance processes. *Anthropological Theory*, 19(4), 470–488.

Guldenmund, F. (2000) The nature of safety culture: A review of theory and research. *Safety Science*, 34(1–3), 215–257.

Guldenmund, F. (2018) Understanding safety culture through models and metaphors. In C. Gilbert et al. (Eds.), *Safety Cultures, Safety Models* (pp. 21–34). Springer.

Gunderson, M. and Hyatt, D. (2000) Workforce and workplace change: Implications for injuries and compensation. In T. Sullivan (Ed.), *Injury and the New World of Work*. Vancouver: UBC Press, 46–68.

Gunningham, N. (2008) Occupational health and safety: Worker participation and the mining industry in a changing world. *Economic and Industrial Democracy*, 29(3), 336–361.

Gunningham, N. and Sinclair, D. (2014) The impact of safety culture on systemic risk management. *European Journal of Risk Regulation*, 4, 505–516.

Guse, M. and Gunn, T. (1980) Internal safety audit: Inco Metals Company, Ontario Division. Paper presented at MAPAO annual meeting, May 22.

Guthman, J. (1998) Regulating meaning, appropriating nature: The codification of California organic agriculture. *Antipode*, 30(2), 135–154.

Guthman, J. (2004) *Agrarian dreams: The paradox of organic farming in California*. Berkeley: University of California.

Haas, J. (1977) Learning real findings: A study of high steel ironworkers reactions to fear and danger. *Sociology of Work and Occupations*, 4(2), 147–168.

Hadler, N. (1993) *Occupational Musculoskeletal Disorders*. New York: Raven Press.

Halfacre-Hitchcock, A., D. McCarthy, T. Burkett, and A. Cavajal (2006) Latino migrant farmworkers in low country South Carolina: A demographic profile and an examination of pesticide risk perception and protection in two pilot case studies. *Human Organization*, 65(1), 55–72.

Hakkinen, K. (2015) Safety management: From basic understanding towards excellence. In S. Vayrynen, K. Hakkinen, and T. Niskanen (Eds.), *Integrated Occupational Safety and health Management: Solutions and Industrial Cases* Springer, 7–15.

Halfacre-Hitchcock, A., McCarthy, T., Burkett, T., and Carvajal, A. (2004) Latino migrant farmworkers in lowcountry South Carolina: A demographic profile and an examination of pesticide risk perception and protection in two pilot case studies. *Human Organization*, 65(1), 55–72.

Hall, A. (1989) *Production Politics and the Construction of Consent: A Case Study of Health and Safety in Mining*. Doctoral Thesis, University of Toronto.

Hall, A. (1991) *Organized labour and its approach to reform in occupational health and safety: 1980–1990*. Paper presented to the Annual Meeting of the Canadian Sociology and Anthropology Association. Toronto.

Hall, A. (1993) The corporate construction of occupational health and safety: A labour process analysis. *Canadian Journal of Sociology*, 18(1), 1–20.

Hall, A. (1996) The ideological construction of risk in mining. *Critical Sociology*, 22(1), 93–116.

Hall, A. (1998a) Pesticide reforms and globalization: Making the farmers responsible. *Canadian Journal of Law and Society*, 13(1), 187–213.

Hall, A. (1998b) Sustainable agriculture and conservation tillage: Managing the contradictions. *Canadian Review of Sociology and Anthropology*, 35(2), 1–31.

Hall, A. (1999) Understanding the impact of mine health and safety programs. *Labor Studies Journal*, 23(4), 51–76.

Hall, A. (2003) Canadian Agricultural Policy: Liberal, Global and Sustainable. In Jane Adams (ed.) *Fight for the Farm: Rural America Transformed*. Philadelphia: University of Pennsylvania Press, 209–229.

Hall, A. (2007) Restructuring, environmentalism and the problem of farm safety. *Sociologia Ruralis*, 47(4), 342–368.

Hall, A. (2016) Trust, Uncertainty and the Reporting of Workplace Hazards and Injuries. *Health, Risk and Society*, 18, 427–448.

Hall, A., Basok, T., Essex, J., Kwantes, K., and Soni-Sinha, U. (2012) Worker responses to workplace hazards and injuries: the influences of immigration status and visible minority identity. Final Report to WSIB.

Hall, A., Forrest, A., Sears, A., and Carlan, N. (2006) Making a difference: Knowledge activism and worker representation in joint OHS committees. *Relations*

Industrielles/Industrial Relations, 61(3), 408–436.Hall, A., King, A., Lewchuk, W., Oudyk, J., and Naqvi, S. (2013) *Making participation work in the new economy: Final report to OWSIB*. Workplace Safety Insurance Board, August 23.

Hall, A. and Mogyorody, V. (2001) Organic farmers in Ontario: An examination of the conventionalization argument. *Sociologia Ruralis*, 42(1), 399–422.

Hall, A. and Mogyorody, V. (2002) The marketing practices of Ontario's organic farmers: Local or global? *Capitalism, Nature, Socialism*, 13(2), 3–34.

Hall, A., Oudyk, J., King, A., Lewchuk, W., and Naqvi, S. (2016) Knowledge activism and effective worker representation in health and safety. *American Journal of Industrial Medicine*, 59, 42–56.

Hall, A., R. Ricciardelli, K. Sitter, D. Simas Medeiros, C. de Boer, and S. Small. (2018) Occupational Stress Injuries in Two Atlantic Provinces: A Policy Analysis. *Canadian Public Policy*. 44(4), 384–399.

Hall, A., Toldo, J., and Gerard, Z. (2012) *Responding to employment standards and OHS violations*: The case of young workers. Paper presented at the Canadian Sociology Association Conference, Waterloo, Ontario, May 30.

Harvey, D. (2005) *A Brief History of Neoliberalism*. Oxford University Press.

Hasle, P., Jorgen Limborg, H., Kallehave, T., Klitgaard, C., and Rye Andersen, T. (2011) The working environment in small firms: Responses from owner-managers. *International Small Business Journal*, 30(6), 622–639.

Hendrickson, M., Howard, P., Constance, D., and Houston, S. (2017) Power, food and agriculture: Implications for farmers, consumers and communities. College of Food, Agriculture and Natural Resources, University of Missouri. https://www.researchgate.net/publication/320837077_Power_Food_and_Agriculture_Implications_for_Farmers_Consumers_and_Communities. Accessed 05/28/20.

Herman, A. (1982) Conceptualizing control: Domination and hegemony in the capitalist labour process. *Insurgent Sociologist*, 11, 7–22.

Heron, C. and Storey, R. (1986) Work and struggle in the Canadian Steel Industry, 1900–1950. In C. Heron and R. Storey (Eds.), *On the Job: Confronting the Labour Process in Canada* (pp. 210–244). McGill-Queen's Press.

Hilgartner, S. (1985) The political language of risk: Defining occupational health. In D. Nelkin (Ed.), *The Language of Risk*. Sage.

Hilgers, M. and Mangez, E. (2015) Introduction to Bourdieu's theory of social fields. In M. Hilgers and E. Mangez (Eds.), *Bourdieu' Theory of Social Fields* (pp. 1–35). New York: Routledge.

Hodson, R. (2002) Worker participation and teams: New Evidence from analyzing organizational ethnographies. *Economic and Industrial Democracy*, 23(4), 491–528.

Horiguchi, R. (2011) *Birth of Safety First: A Social History of Safety Campaigns*. Fuji Shuppan.

Hutter, B. (2001) *Regulation and Risk: Occupational Health and Safety on the Railways*. Oxford University Press.

Hynes T. and Prasad, P. (1999) The Normal Violation of Safety Rules. In C. McCormick (ed.) *The Westray Chronicles: A Case Study in Corporate Crime*. Halifax, Fernwood, 117–134.

ILO Encyclopedia of Occupational Health and Safety (2016) *Mining and Quarrying*. https://www.iloencyclopaedia.org/part-xi-36283/mining-and-quarrying/item/605-lighting-in-underground-mines.

Inco (ND) Inco Training Manual, Neil George Safety System.

Inco (1980a) Inco announces new safety programs. Inco Press Release, September 16.

Inco (1980b) Direct Line: News of Interest to Employees of the Ontario Division of Inco Ltd. December 17.

Inco (1980c) Inco Submission to the Burkett Inquiry. Sudbury, Ontario.

Inco (1980–86) *Annual Reports*. Toronto: Inco Metals Ltd.

Inco (1981) *All Mines Standard Practices Manual*. Sudbury: Ontario Division Mines. Inco (1982–86) OSHE Minutes (Levack, McCreedy and South Mine). Sudbury.

Inco (1983) *Inco Innovation: A Report to Inco Employees*. President Chuck Baird.

Inco (1985a) Direct Line: News of Interest to Employees of the Ontario Division of Inco Ltd., April 24 and November 6 and 29. Sudbury: INCO Metals Inc.

Inco (1985b) *Safety Program: Mining*. Ontario Division, Inco Metals Inc.

Inco (1985c) Annual Report. Toronto: Inco Metals Inc.

Inco (1985d) *Brief Submitted to the Provincial Inquiry into Ground Control and Emergency Preparedness* (Stevenson Commission). Sudbury.

Inco (1986) Direct Line: News of Interest to Employees of the Ontario Division of Inco Ltd., March 7.

Inco/Local 6500 (1982–84). Gravimetric Dust Test Results for Mines. Compiled from Levack, McGreedy and South Joint Health and Safety Committee Minutes.

Inco Mines Equipment Files, Sudbury Operations, 1980–82.

Inco, Ontario Division, Mine Injury Statistics, Provided on Request, August 11, 1986.

Inco, Ontario Division, Ground Fall Occurrence Reports, 1983–86.

Inco Triangle (1981) Excerpts from the Company Submission to the Joint Federal/Provincial Inquiry into Safety in mines and mining plants (Burkett). The Triangle: Inco Employee Newsletter. 41(1), February, p. 10.

Inco Triangle (1982) Five star safety program. The Triangle: Inco Employee Newsletter. 42, p. 1

Inco Triangle (1983) Vertical retreat mining leads to increased productivity. The Triangle: *Inco Employee Newsletter*, 43(4), p. 4.

Innovative Farmers of Ontario (ND) *The E-Plus Efficiency Program*, Pamphlet being distributed in Essex County, 1992. https://www.ifao.com/

ISO, International Standards Organization (2017) ISO and agriculture. https://www.iso.org/publication/PUB100412.html

Ison, T. (1983) *Workers' Compensation in Canada*. Toronto: Butterworths.

Ison, T. (2015) Reflections on the state of workers' compensation and occupational health & safety in the United States and Canada. *Compensation & Benefits Review*, 47(1), 27–38.

Jamieson, D. (ND) The life and death of an Amazon Warehouse Temp: What the future of low wage work really looks like. The Huffington Post, https://highline.huffingtonpost.com/articles/en/life-and-death-amazon-temp/. Accessed 05/31/20.

Jeffcott, S., Pidgeon, N., Weyman, H., and Walls, J. (2006) Risk, trust and safety culture in train operating companies. *Risk Analysis*, 26(5), 1105–1121.

Jessop, B. (1990) Regulation theories in retrospect and prospect. *Economy and Society*, 19(2), 153–216.

Jessop, B. (1995) The regulation approach, governance, and post-fordism: Alternative perspectives on economic and political change. *Economy and Society*, 24(3), 307–333.

Johnson, D. (2002) Empirical study of second tier automotive suppliers achieving QS 9000. *International Journal of Operations and Product Management*, 22(8), 902–928.

Johnstone, R., Quinlan, M., and Walters, D. (2005) Statutory occupational health and safety workplace arrangements for the modern labour market. *The Journal of Industrial Relations*, 47(1), 93–116.

Kalleberg, A. (2012) Job quality and precarious work: Clarifications, controversies, and challenges. *Work and Occupations*, 39(4), 427–448.

Kath, L., Magley, V., and Marnet, M. (2010) The role of organizational trust in safety climate's influence on organizational outcomes. *Accident Analysis and Prevention*, 42, 1488–1497.

Keating, N. (1987) Reducing stress of farm men and women. *Family Relations*, 36(4), 358–363.

Kelloway, K., Francis, L., and Gatien, B. (2016) *Management of Occupational Health and Safety* (5th ed.). Nelson.

King, A., Lewchuk, W., MacEachen, E., and Goyal, J. (2019) Making worker voice a reality under the Internal Responsibility System: The limits of Section 50 protections for workers experiencing OHS reprisals. Report provided to the Ontario Ministry of Labour Occupational Health and Safety Research Program. Toronto, January.

Kitchener, L. (1979) Safety and Health aspects of recent developments in mining equipment. Inco Paper presented to the Mining Accident Prevention Association Annual Meeting, May 24.

Knights, D. and Willmott, H. (1999) *Management Lives: Power and Identity in Work Organizations*. Sage.

Kochan, T., Lansbury, R., and MacDuffie, J.P. (1997) *After Lean Production: Evolving Employment Practices in the World Auto Industry*. Cornell University Press.

Kosla, M. (2015) The safety dance: Men without [hard]hats. *Health, Risk and Society*, 17(5–6), 388–403.

Kosny, A., MacEachen, E., Lifshen, M., Smith, P., Joya Jafri, G., Neilson, C., Pugliese, D., and Shields, J. (2012) Delicate dances: Immigrant worker's experiences of injury reporting and claim filing. *Ethnicity and Health*, 17(3), 267–290.

Landbergis, P., Cahill, J., and Schnall, P. (1999) The impact of lean production and related new systems of work organization on worker health. *The Journal of Occupational Health Psychology*, 4(2), 108–130.

Lane, C. (2011) *A Company of One: Insecurity, Independence and the New World of White-Collar Employment*. Ithaca, NY: ILR Press.

Law Commission of Ontario (2012) *Interim Report on Vulnerable Workers and Precarious Work*. Toronto.

Lawson, J. (2009/10) Explaining workplace injuries among BC loggers: Cultures of risk and of desperation. *BC Studies*, 164(Winter), 51–74.

Lee, H. (2001) Paternalistic human resource practices: Their emergence and characteristics. *Journal of Economic Issues*, 35(4), 841–865.

Leeth, J. and Ruser, J. (2006) Safety segregation: The importance of gender, races and ethnicity on workplace risk. *Journal of Economic Inequality*, 4, 123–152.

Legendre, C. (1987) When your time has come...Gold miners' views on occupational hazards. In R. Argue, C. Gannage, and D. Livingstone (Eds.), *Working People and Hard Times*. Garamond Press, 46–59.

Lerner, J. and D. Keltner (2001) Fear, anger and risk. *Journal of Personality and Social Psychology*, 81(1), 146–159

Leslie, D. and Butz, D. (1998) 'GM suicide': Flexibility, space and the injured body. *Economic Geography*, 74(4), 360–378.

Lewchuk, W., Clarke, M., and de Wolfe, A. (2009) Precarious employment and the internal responsibility system: Some Canadian experiences. In D. Waltersand T. Nichols (Editor) (Ed.), *Workplace Health and Safety: International Perspectives on Worker Representation*. London: Palgrave-McMillan. 109–133.

Lewchuk, W., Clarke, M., and de Wolff, A. (2011) *Working without Commitments: The Health Effects of Precarious Employment*. Montreal and Kingston: McGill-Queen's Press.

Lewchuk, W. and Dassinger, J. (2016) Precarious employment and precarious resistance: "We are people still". *Studies in Political Economy*, 97(2), 143–158. doi:10.10 80/07078552.2016.1211397

Lewchuk, W., de Wolff, A., King, A., and Polanyi, M. (2005) The Hidden cost of precarious employment. In L. Vosko (Ed.), *Precarious Employment: Understanding Labour Market Insecurity in Canada* (pp. 141–162). Montreal and Kingston: McGill-Queen's University Press.

Lewchuk, W., Robb, A., and Walters, V. (1996) The effectiveness of Bill 70 and joint health and safety committees in reducing injuries in the workplace. *Canadian Public Policy*, 22, 225–243.

Lewchuk, W. and Robertson, D. (1997) Production without empowerment: Work reorganization from the perspective of motor vehicle workers. *Capital and Class*, 21(3), 37–64.Lewchuk, W., Stewart, P., and Yates, C. (2001) Quality of working life in the automobile industry: A Canada-UK comparative study. *New Technology, Work and Employment*, 16(2), 72–87.

Lewchuk, W. and Wells, D. (2006) When corporations substitute for adversarial unions: Labour markets and human resource management at Magna. *Relations Industrielle/Industrial Relations*, 61(4), 639–664.

Leyton, E. (1975) *Dying Hard: The Ravages of Industrial Carnage*. McLelland & Stewart.

Lighthall, D. (1995) Farm structure and chemical use in the corn belt. *Rural Sociology*, 60(3), 505–520.

Lind, C. (1995) *Something's Wrong Somewhere: Globalization, Community and Moral Economy of the Farm Crisis*. Fernwood.

Lippel, K. (2005) Precarious employment and occupational health and safety regulation in Quebec. In L. Vosko (Ed.), *Precarious Employment: Understanding Labour Market Insecurity in Canada* (pp. 241–255). McGill-Queen's University Press.

Littler, C. and Salaman, G. (1982) Bravermania and beyond: Recent theories of the labour process. *Sociology*, 16(2), 251–269.

Littler, C. and Salaman, G. (1984) *Class at Work: The Design, Allocation and Control of Jobs*. London: Batsford Academic.

Lofland, J. and Lofland, L. (1995) *Analyzing Social Settings: A Guide to Qualitative Observation and Analysis*. Wadsworth.

Loh, K. and Richardson, S. (2004) Foreign born workers: Trends in fatal occupational injuries. *Monthly Labor Review*, 127(6), 42–53.

Luhmann, N. (2005) *Risk: A Sociological Theory*. Aldine.

Lupton, D. (1999) *Risk*. Routledge.

Luria, G. (2010) The social aspects of safety management: Trust and safety climate. *Accident Analysis and Prevention*, 42, 1288–1295.

Lynggaard, K. (2001) The farmer within the institutional environment: Comparing Danish and Belgian organic farming. *Sociologia Ruralis*, 41(1), 85–111.

MacDowell, L. Sefton (2012) The Elliot Lake uranium miners' battle to gain occupational health and safety improvements, 1950–1980. *Labour/Le Travail*, 69(Spring), 91–118.

MacEachen, E. (2000) The mundane administration of worker bodies: From welfarism to neoliberalism. *Health, Risk and Society*, 2(3), 315–327.

MacEachen, E. (2005) The demise of repetitive strain injury in skeptical governing rationalities of workplace managers. *Sociology of Health and Illness*, 27(4), 490–514.

Macey, A. (2000) Canadian organic statistics update. *Eco-Farm and Garden*, (Winter), 26.

Macrae, R., Henning, J., and Hill, S. (1993) Strategies to overcome barriers to the development of sustainable agriculture in Canada: The role of agribusiness. *Journal of Agricultural and Environmental Ethics*, 6(1), 21–51.

Mae-Hitchcock, A., McCarthy, D., Burkett, T., and Carvajal A. (2006) Latino migrant farmworkers in lowcountry South Carolina: A Demographic profile and an examination of pesticide risk perception and protection in two pilot case studies. *Human Organization*, 65(1), 55–72.

Mandell, E. (1977) *Marxist Economic Theory*. Merlin Press.

Manning, E. (1986) Planning Canada's resource base for sustainable production. In I. Knell and J. English (Eds.), *Canadian Agriculture in a Global Context*. Centre on Foreign Policy and Federalism, Waterloo: University of Waterloo Press.

MAPAO, Mining Accident Prevention Association (1985a) Ground control bulletin: Multi-channel seismic monitoring. Bulletin No. 2 September.

MAPAO, Mining Accident Prevention Association of Ontario (1985b) *The Five Star Loss Management Audit*. Toronto.

MAPAO, Mining Accident Prevention Association of Ontario (1983–86). Individual Mine Statistics Request: Medical and Lost Time Accident Rates per Mine Job Classification.

Maluske, M. (2020). Windsor company fined $175K after death of worker. https://windsor.ctvnews.ca/windsor-company-fined-175k-after-death-of-worker-1.4775402

Marr, E., Howley, P., and Burns, C. (2016) Sparing or sharing? Differing approaches to managing agricultural and environmental spaces in England and Ontario. *Journal of Rural Studies*, 48, 77–91.

Marx, K. (1977) *Capital Vol 1*. New York: Vintage Books.

McCormick, C. (1999) *The Westray Chronicles: A Case Study in Corporate Crime*. Halifax, Fernwood.

McIntyre, D. (2005) 'My way or the highway': Managerial prerogative, the labour process and worker health. *Health Sociology Review*, 14(1), 59–68.

McKenzie, G. and Laskin, J. (1987) *Report on the Administration of the Occupational Health and Safety Act of Ontario*. Government of Ontario.

Meredith, R. (2013) *The Meredith Report: Royal Commission to Investigate Worker Compensation*. Government of Ontario.

Meyer, S. and Ward, P. (2013) Differentiating between trust and dependence of patients with coronary heart disease: Furthering the sociology of trust. *Health, Risk and Society*, 15(3), 279–293.

Middlesworth, M. (ND) Workplace safety – Does you CEO get it? https://www.ergo-plus.com/workplace-safety-culture-ceo/

Milgate, N, Innes, E., and O'Loughlin, K. (2002) Examining the effectiveness of health and safety committees and representatives: A review. *Work: Journal of Prevention, Assessment and Rehabilitation*, 19(3): 281–290.

Mining Accident Prevention Association of Ontario (MAPAO). (1985) *The Five Star Management Audit*. Toronto: MAPAO.

Mogensen, V. (Ed.) (2006) *Worker Safety under Siege: Labor, Capital, and the Politics of Workplace Safety in a Deregulated World*. M.E. Sharpe.

Mojtehedzadeh, S. (2019) Another worker dies at Fiera Foods. The Labour Ministry is investigating. Toronto Star, September 25. https://www.thestar.com/news/canada/2019/09/25/labour-ministry-investigating-fatal-industrial-accident-at-fiera-foods.html

Mojtehedzadeh, S. (2019) The life and death of Fiera Foods temp worker Enrico Miranda. Toronto Star, October 4.

Mollering, G. (2006) *Trust, Reason, Routine and Reflexivity*. Oxford: Elsevier.

Morse, T., Bracker, A., Warren, N., Goyzuetta, J., and Cook, M. (2012) Characteristics of effective health and safety committees: Survey results. *American Journal of Industrial Medicine*, 56, 163–179.

NAFTA (1992) North American Free Trade Agreement: Canada, US, Mexico. Signed December 17, 1992, enacted 1994.

National Center for Productivity and Quality of Working Life (1976) *Alternatives in the World of Work*. Washington, DC: Committee on Alternative Work Patterns.

National Center for Productivity and Quality of Working Life (1978) Annual Report: 1977–78. Washington.

National Farmers Union (NFU) and Catholic Rural Life Conference (CRLC) (1990) Vision 2000: A sustainable Canadian agricultural policy. Stratford, Ontario.

Navarro, V. (1982) The labor process and health: An historical materialist interpretation. *International Journal of Health Services*, 12, 5–29.

NDP (1983) *Not yet healthy, not yet safe*. A report of the Ontario NDP Task Force on Occupational Health and Safety. Submitted to the Ontario Legislature.

NDP (1986) *Still not healthy, still not safe*. Report of the Ontario New Democrat Caucus. Second Task Force Report on Occupational Health and Safety.

Nelkin, D. and Brown, M. (1984) *Workers at Risk: Voices from the Workplace*. Chicago, IL: The University of Chicago Press.

Nichols, T. (1986) Industrial injuries in British manufacturing in the 1980s — A commentary on Wright's article. *Sociological Review*, 34(2), 290–306.

Nichols, T., Walters, D., and Tasiran, A. (2007) Trade unions, institutional mediation and industrial safety. Evidence from the UK. *Journal of Industrial Relations*, 49(2), 211–225.

Novek, J., Yassi, A., and Speigel, J. (1991) Mechanization, the labor process and injury risks in the meat-packing industry. *International Journal of Health Services*, 21, 281–296.

O'Doherty, D. and Willmott, H. (2001) Debating labour process theory: The issue of subjectivity and the relevance of poststructuralism. *Sociology*, 35(2), 457–476.

O'Doherty, D. and Willmott, H. (2009) The decline of labour process analysis and the future sociology of work. *Sociology*, 43(5), 931–951.

OFL, Ontario Federation of Labour (1990) *Submission to Standing Committee Resources*. Toronto: OFL.

O'Grady, J. (2000) Joint health and safety committees: Finding a balance. In T. Sullivan (Ed.), *Injury and the New World of Work* (pp. 162–197). University of British Columbia Press.

OHSA, Occupational Health and Safety Act (1990) Heat Stress Section 25(2)(h).

OHSA, Occupational Health and Safety Act (1979/1990) Ontario Queen's Printer.

OHSA, Occupational Health and Safety Act, Control of Exposure to Biological or Chemical Agents, RRO 1990, Reg. 833, http://canlii.ca/t/54c5d retrieved on 2020-08-23

Olle-Espluga, L., Menendez-Fuster, M., Muntaner, C., Benach, J., Vergara-Duarte, M., and Luisa Vaquez, M. (2014) Safety representatives' views on their interaction with workers in a context of unequal power relations: An exploratory qualitative study in Barcelona (Spain). *American Journal of Industrial Medicine*, 57, 338–350.

Olle-Espluga, L., Vergara-Duarte, M., Belvis, F., Mendandez-Fuster, M., Jodar, P., and Benach, J. (2015) What is the impact on occupational health and safety when workers know they have safety representatives. *Safety Science*, 74, 55–58.

OMAF, Ontario Ministry of Agriculture and Food (1989) *Conservation Tillage Handbook*. Produced by Soil and Water Environmental Enhancement Program (SWEEP). Toronto.

OMAF, Ontario Ministry of Agriculture and Food (1990) *Soybean Production*. Publication 173. Toronto.

OMAF, Ontario Ministry of Agriculture and Food (1991) *Common Ground Update: The Strategic Plan for the Ontario Ministry of Agriculture and Food*. Toronto: Queen's Printer.

OMAF, Ontario Ministry of Agriculture and Food (1992) *Final Reports: Technical Evaluations of Technological Development and Subprograms of SWEEP*. Produced by Soil and Water Environmental Enhancement Program (SWEEP). Toronto.

OMAF, Ontario Ministry of Agriculture and Food (1993) *No-Till: The Basics*. Produced by Soil and Water Environmental Enhancement Program (SWEEP). Toronto: Queen's Printer.

OMAF and OMEE, Ontario Ministry of Agriculture and Food and Ontario Ministry of Environment and Energy (1992) *Food Systems 2002: A Program to Reduce Pesticides in Food Production*. Toronto: Queen's Printer.

OMAF and OMEE, Ontario Ministry of Agriculture and Food and Ontario Ministry of Environment and Energy (1993) *Grower Pesticide Safety Course: Learner Manual*. Ridgetown College of Agricultural Technology.

O'Malley, P. (2004) *Risk, Uncertainty and Government*. London: Glass House Press.

OMOL, Ontario Ministry of Labour (1989a) Occupational Health and Safety Reform, 1989. Toronto: Queen's Printer.

OMOL, Ontario Ministry of Labour (1989b) *Labour Minister Sorbara Announces Major Reform of Workplace Health and Safety Law*. News Release, January 24.

OMOL, Ontario Ministry of Labour (Dean, T., Panel Chair) (2010) *Final Report of the Advisory Expert Panel on Occupational Health and Safety*. Toronto: Ontario Ministry of Labour.

OMOL, Ontario Ministry of Labour (2014) Machine guarding. https://www.labour.gov.on.ca/english/hs/sawo/pubs/fs_machineguarding.php

OMOL, Ontario Ministry of Labour (2016) Guidelines for pre-start health and safety reviews: How to apply Section 7 of the Industrial Establishments Regulation. Government of Ontario.

OMOL, Ontario Ministry of Labour (2019) Guide to Occupational Health and Safety Act. Health and Safety Branch, November 4. https://www.ontario.ca/document/guide-occupational-health-and-safety-act?

OMOL, Ontario Ministry of Labour (1983) *Control of Diesel Emissions: An Overview*. Mining Health and Safety Branch, Toronto: Queen's Printer.

OMOL, Ontario Ministry of Labour (1985a/2014/2019) *Guide to the Occupational Health and Safety Act*. Toronto: Queen's Printer.

OMOL (1985b). Ministry of Labour Annual Report. Toronto: Queen's Printer.

OMOL, Ontario Ministry of Labour (1985c) Report of the Ontario Task Force on Health and Safety in Agriculture. Toronto: Queen's Printer.

OMOL, Ontario Ministry of Labour (1991) Farming operations in Ontario: The need for health and safety legislation. Ministry Discussion Paper. Toronto: Queen's Printer.

OMOL, Ontario Ministry of Labour (1995) *Report of the Ontario Task Force on Health and Safety in Agriculture*. Toronto: Queen's Printer.

OMOL, Ontario Ministry of Labour (2006) Occupational health and safety policy and programmes for farming operations. http://www.labour.gov.on.ca/English/farming/policy.html. Accessed 05/23/07.

Ontario (2005) O. Reg. 414/05: FARMING OPERATIONS under Occupational Health and Safety Act, R.S.O. 1990, c. O.1.

Ontario, Advisory Council on Occupational Health and Safety (ACOHOS) (1986) An evaluation of Joint Health and Safety Committees in Ontario. Published in ACOHOS Annual Report. Vol. 2. Toronto: Queen's Printer.

Ontario, AG (2019) *Health and Safety in the Workplace*. Office of Auditor General Annual Report, Chapter 3 Section 3.07, pp. 395–451. Toronto: Queen's Printer.

Ontario, Bill 160 (2011) Occupational Health and Safety Statute Law Amendment Act. https://www.ola.org/en/legislative-business/bills/parliament-39/session-2/bill-160. Toronto: Queen's Printer.

Ontario, Bill 208 (1990) *Act to Reform the Occupational Health and Safety Act*. Toronto: Queen's Printer.

Ontario/Canada (Burkett Report) (1981) *Report of the Joint Federal/Provincial Inquiry into Safety in Mine and Mining Plant in Ontario*, Vols. 1 and 2. Toronto: Queen's Printer.

Ontario/Canada Task Force on Inco Limited and Falconbridge Limited (1982) Final Report. Ontario Ministry of Environment and Ministry of Consumer and Commercial Relations. Toronto: Queen's Printer.

Ontario Farmer's Western Edition (1993) Conservation Farming. Ontario Farmer's Western Edition, July 6, p. 42.

Ontario Farm Safety Association (OFSA) (1985a) Noise – Sound without Value. Fact Sheet, September. Guelph: Farm Safety Association Inc.

Ontario Farm Safety Association (OFSA) (1985b) Silo gas – A swift and silent killer. Fact Sheet, September. Guelph: Farm Safety Association Inc.

Ontario Farm Safety Association (OFSA) (1985c) Tractor safety. Fact Sheet, September. Guelph: Farm Safety Association Inc.

Ontario Farm Safety Association (OFSA) (1985d) Dealing with stress. Fact Sheet, September. Guelph: Farm Safety Association Inc.

Ontario Farm Safety Association (OFSA) (1985e) Flowing grain entrapment. Fact Sheet, September. Guelph: Farm Safety Association Inc.

Ontario Farm Safety Association (OFSA) (1990) Safety Film Catalogue. *NH3: The Cold Facts.* Guelph: Farm Safety Association Inc.

Ontario Farm Safety Association (OFSA) (1992) PTO safety. Fact Sheet, December. Guelph: Farm Safety Association Inc.

Ontario Farm Safety Association (OFSA) (1993a) Harvest safety reminders. *Farmsafe*, 18(4), 2. Guelph: Farm Safety Association Inc.

Ontario Farm Safety Association (OFSA) (1993b) *Ontario Farm Fatality Reports – 1987–1992.* Guelph: Farm Safety Association Inc.

Ontario Farm Safety Association (OFSA) (1993c) Various articles. *Farmsafe.* 18(2), Spring. Guelph: Farm Safety Association Inc.

Ontario Farm Safety Association (OFSA) (1993d) Farm safety checklist. *Farmsafe* 18(4), Fall. Guelph: Farm Safety Association Inc.

Ontario Federation of Agriculture (OFA) (1994) *Brief to Ontario Cabinet.* February 9. Guelph: OFA.

Ontario Federation of Agriculture (1995) OFA Policy Handbook. Guelph: OFA.

Ontario Federation of Agriculture (2020) Farm Labour and Safety: OFA Position https://ofa.on.ca/issues/famr-labour-safety/. Accessed 04/27/20.

Ontario Legislature (1914) *Workmen's Compensation Act*, S.O. 1914, c. 25.

Ontario, Legislature Standing Committee on Resources (1998) *Inquiry Report into Mine Safety.* Toronto: Queen's Printer.

Ontario, Regulation 751 of the Ontario Pesticides Act, R.S.O, 1990. Toronto: Queen's Printer.

Ontario Roundtable on the Economy and Environment (1991) *Ontario Food Sector Task Force Report.* Toronto: Queen's Printer.

Ontario (1976) *Royal Commission Report on the Health and Safety of Workers in Mines* (Ham Report). Toronto: Ministry of Attorney General Ontario, Queen's Printer.

Ontario R.R.O. OHSA (1979, 1990) Reg. 851: Industrial Establishments. Toronto: Queen's Printer.

Ontario, R.R.O. OHSA (1979, 1990) Reg. 854: Mines and Mining PLANTS. Toronto: Queen's Printer.

Ontario, R.R.O. OHSA (1990) Regulation 833, Control of Exposure to Biological or Chemical Agents, OHSA. Toronto: Queen's Printer.Ontario QWL Centre (1981) *Annotated Bibliography, 1974–80.* Ontario Ministry of Labour.

Ontario QWL Centre (1984) A labour perspective on QWL. *QWL Focus: Journal of the Ontario Quality of Work Life Centre*, 4(1), 1–4.

Ontario QWL Centre, Annotated Bibliography (1974–80). Toronto: Ontario Ministry of Labour.Ontario

Workplace Health and Safety Agency (WHSA)(1994) *The impact of joint health and safety committees on health and safety trends in Ontario.* Toronto: WHSA. Ontario

WSIB (2018) By the Numbers: WSIB Statistical Report. http://www.wsibstatistics.ca/

Orica (2013) *AMEX, Safety data sheet.* Issued August 6. Federal Register Vol. 77, No. 58, Rules and Regulations.

Paap, K. (2006) *Working Construction: Why Working Class Men Put Themselves – And the Labor Movement – In Harm's Way.* Cornell University Press.

Pelucha, M. and Kveton, V. (2017) The role of EU rural development policy in neo-productivist agricultural paradigm. *Regional Studies,* 51(12), 2860–2870.

Pender, T. (1986a) Union blames Inco for increase in rockbursts. Globe and Mail, June 2, p.A11.

Pender, T. (1986) Rescuers seek buried worker in Inco mine. Globe and Mail, August 8, p.A1.

Pender, T. Minister is assailed over cave-in of Inco mine. August 9, p.A1.

Perrow, C. (1984) *Normal Accidents.* New York: Basic Books.

Peters, S., Reid, A., Fritschi, L., de Klerk, N., and Musk, A. (2013) Long terms effects of aluminum dust inhalation. *Occupational and Environmental Medicine,* 70(12), 864–868.

Piazza, J. (2005) Globalizing quiescence: Globalization, unions density and strikes in 15 industrialized countries. *Economic and Industrial Democracy,* 26(2), 289–314.

Pickett, W., Hartling, L., Brison, R., and Guernsey, (1999) Fatal work-related farm injuries in Canada, 1991–1995. Canadian Agricultural Injury Surveillance Program. *Canadian Medical Association Journal,* 160(13), 1843–1848.

Piore, M. and Sabel, C. (1984) *The Second Industrial Divide: Possibilities for Prosperity.* Basic Books.

Pransky, G., Snyder, T., Dembe, A., and Himmelstein, J. (1999) Under-reporting of work-related disorders in the workplace: A case study and review of the literature. *Ergonomics,* 42, 171–182.

Probst, T., Barbaranelli, C., and Petitta, L. (2013) The relationship between job insecurity and accident under-reporting: A test in two countries. *Work and Stress,* 27(40), 383–402.

Probst, T. and Estrada, A. (2009) Accident under-reporting among employees: Testing the moderating influence of psychological safety climate and supervisor enforcement of safety practices. *Accident Analysis and Prevention,* 42, 1438–1444.

Probst, T. and Graso, M. (2013) Pressure to produce = Pressure to reduce accident reporting? *Accident Analysis and Prevention,* 59, 580–587.

Quandt, S., Grzywacz, J., Marín, A., Carrillo, L., Coates, M., Bless, M. Burke and Arcury, T. (2006) Illnesses and injuries reported by Latino Poultry workers in western North Carolina. *American Journal of Industrial Medicine,* 49, 343–351.

Quinlan, M. (1999) The implications of labour market restructuring in industrialized societies for occupational health and safety. *Economic and Industrial Democracy,* 20(3), 427–460.

Quinlan, M. (2005) The hidden epidemic of injuries and illness associated with the global expansion of precarious employment. In C. Peterson and C. Mayhew (Eds.), *Occupational Health and Safety: International Influences and the "New Epidemics"* (pp. 53–74). Baywood.

Quinlan, M., Lippell, K., Johnstone, R., and Walters, D. (2016) Governance, change and the work environment. *Policy and Practice in Health and Safety,* 13(2), 1–5.

Quinlan, M. and Mayhew, C. (1999) Precarious employment and workers' compensation. *International Journal of Law and Psychiatry,* 22(5–6), 491–520.

Quinlan, M., Mayhew, C., and Bohle, P. (2001) The global expansion of precarious employment, work disorganization, and consequences for occupational health: A review of recent research. *International Journal of Health Services*, 31(2), 335–414.

Rae, A., Provan, D., Weber, E., and Dekker, W. (2018) Safety clutter: The accumulation and persistence of safety work that does not contribute to operational safety. *Policy and Practice in Health and Safety*, 16(2), 194–211.

Radecki, H. (1981) *One Year Later: The 1978–79 Strike at Inco*. SIS analysis.

Ramkissoon, A., P. Smith and J. Oudyk (2019) Dissecting the effect of workplace exposures on workers' rating of psychological health and safety. *American Journal of Industrial Medicine*, 62(2), 412–421.

Ramsay, H. (1977) 'Cycles of control': Workers' participation in sociological and historical perspective. *Sociology*, 11(3), 481–506.

Rawson, Arnold J. (1944) Accident proneness. *Psychosomatic Medicine*, 6(1), 88–94.

Reed, J. (1990) Safety programs of the Farm Safety Association in Ontario, Canada. *American Journal of Industrial Medicine*, 18(4), 409–411.

Reese, C. (2017) *Occupational Health and Safety Management: A Practical Approach* (3rd ed.). CRC Press.

Reid, D. (1981) The role of mine safety in the development of working class consciousness and organization: The case of the Aubin coal basin. *French Historical Studies*, 12(1), 98–119.

Reilly, B., Paci, P., and Holl, P. (1995) Unions, safety committees and workplace injuries. *British Journal of Industrial Relations*, 33(2), 275–288.

Rennie, R. (2005) The historical origins of an industrial disaster: Occupational health and labour relations at the Fluorspar Mines, St. Lawrence, Newfoundland, 1933–45. *Labour/Le Travail*, 55(Spring), 107–142.

Richardson, D., Loomis, D., Bena, J., and Bailer, A. (2004) Fatal occupational injury rates in southern and non-southern states, by race and ethnicity. *American Journal of Public Health*, 94(10), 1756–1761.

Rinehart, J. (1975) *The Tyranny of Work*. Toronto: Academic Press.

Rinehart, J., Huxley, C., and Roberston, D. (1997) *Just Another Car Factory: Lean Production and Its Discontents*. Ithaca, NY: ILR Press.

Robens, A., et al. (1972) Safety and health at work: Report of the Committee 1970–72. Presented to Parliament by the Secretary of State for Employment by Command of Her Majesty, July.

Robinson, G. (2006) Ontario's environmental farm plan. *Geoforum*, 37, 859–873.

Robson, L., Clarke, J., Cullen, K., Bieleckya, A., Severina, C., Bigelow, L., Irvina, E., Culerac, A., and Mahooda, Q. (2007) The effectiveness of occupational health and safety management system interventions: A systematic review. *Safety Science*, 45(3), 329–353.

Rolston, J. (2010) Risky business: Neoliberalism and workplace safety in Wyoming coal mines. *Human Organization*, 69(4), 331–342.

Rose, N. (1993) Government, authority and expertise in advanced liberalism. *Economy and Society*, 22, 283–299.

Ross, G. (1984) Perspectives on rock mechanics, ground control procedures, and safety preparedness in the mines of the Ontario division of Inco Ltd. Sudbury, Ontario.

Russell, B. (1999) *More with Less: Work Reorganization in the Canadian Mining Industry*. Toronto: University of Toronto Press.

Russell, E. and Dufour, M. (2016) Why the rising tide doesn't lift all boats: wages and bargaining power in neoliberal Canada. *Studies in Political Economy*, 97(1), 37–55. doi:10.1080/07078552.2016.1174462

Saar, P.E., Dimich-Ward, H., Kelly, D.K., and Voaklander, C.D. (2006) Farm Injuries and fatalaties in British Columbia, 1990–2000. *Canadian Journal of Public Health*, 97(2), 100–104.

Saksvik, P.O. and Quinlan, M. (2003) Regulating systematic occupational health and safety management. *Relations Industrielles/Industrial Relations*, 58(1), 33–59.

Saloniemi, A., Virtanen, P., and Vahtera, J. (2004) The work environment in fixed term jobs: Are poor psychosocial conditions inevitable? *Work, Employment and Society*, 18(1), 193–208.

Sass, R. (1979) The underdevelopment of occupational health and safety in Canada. In W. Leiss (Ed.), *Ecology versus Politics in Canada*. Toronto: University of Toronto Press.

Sass, R. (1986) Workplace health and safety: Report from Canada. *International Journal of Health Services*, 16(4), 565–582.

Sass, R. and Grook, G. (1981) Accident proneness: Science or non-science. *International Journal of Health Services*, 11(2), 175–190.

Schacter, H. (2010) The role played by Frederick Taylor in the rise of the academic management fields. *Journal of Management History*, 16(4), 437–448.

Schofield, T. (2005) The impact of neoliberal policy on workplace health. *Health Sociology Review*, 14(1), 5–7.

Savino, D.M. (2016) Frederick Winslow Taylor and his lasting legacy of functional leadership competence. *Journal of Leadership, Accountability and Ethics*, 13(1). doi:10.33423/jlae.v13i1.1926

Shannon, H. (2000) Firm level organizational practices and work injury. In T. Sullivan (Ed.), *Injury and the New World of Work* (pp. 140–161). UBC Press.

Shannon, H. and G. Lowe. 2002. How many injured workers do not file claims for worker compensation benefits. *American Journal of Industrial Medicine*, 42, 467–473.

Shannon, H., Walters, V., Lewchuk, W., Richardson, R., Verma, D., Haines, T., and Moran, L. (1991) *Health and Safety Approaches in the Workplace*. Toronto: Industrial Accident Prevention Association.

Shortall, S., McKee, A., and Sutherland, L. (2019) Why do farm accidents persist? Normalising danger on the farm within the farm family. *Sociology of Health and Illness*, 41(3), 470–483.

Sirrs, C. (2015) Accidents and apathy: The construction of the 'Robens philosophy' of occupational safety and health regulation in Britain, 1961–1974. *Social History of Medicine*, 29(1), 66–88.

Sirs, C. (2016) Risk, responsibility and Robens: The transformation of the British system of OHS regulation, 1961–1974. In T. Crook and M. Ebester (Eds.), *Governing Risks in Modern Britain: Danger, Safety and Accidents, c. 1800–2000* (pp. 249–276). Palgrave Macmillan.

Smith, B. (1987) *Digging Our Own Graves: Coal Miners and the Struggle Over Black Lung Disease*. Temple University Press.

Snider, L. (2009) Accommodating power: The 'common sense' of regulators. *Social and Legal Studies*, 18(2), 179–197.

Sniders, T. and Bosker, R. (1999) *Multilevel Analysis*. London: Sage.

Sopko, M. (1985) Interview with Mike Sopko, INCO Ontario President, *Northern Life*, October 9.

Sopko, M. President of Inco, Ontario (1985) Letter to Employees, December 16.

Sopko, M. President of Inco Ontario (1986) Letter to William Wrye, Ontario Minister of Labour, April 17.

Stake, R. (2003) Case studies. In N. Denzin and Y. Lincoln (Eds.), *Strategies of Qualitative Inquiry* (2nd ed., pp. 134–164). Sage.

Stanford, J. (2008) *Economics for Everyone: A Short Guide to the Economics of Capitalism*. Pluto Press.

Stanford, J. (2014) *Canada's Auto Industry and the New Free Trade Agreements: Sorting through the Impacts*. Canadian Center for Policy Alternatives.

Stelman, J. and Daum, S. (1971) *Work Is Dangerous to Your Health*. New York: Vintage.

Stevenson, T. (1986) Improving Ground Stability and Mine Rescue: The Report of the Provincial Inquiry into Ground Control and Emergency Preparedness. Toronto: Queen's Printer.

Stonehouse, D. and Bohl, M. (1993) Selected government policies for encouraging soil conservation in Ontario cash-cropping farms. *Journal of Soil and Water Conservation*, 48, 343–349.

Storey, R. (2004) From the environment to the workplace…and back again? Occupational health and safety activism in Ontario, 1970s–2000. *Canadian Review of Sociology and Anthropology*, 41(4), 419–447.

Storey, R. (2005) Activism and the making of occupational health and safety law in Ontario, 1960s–1980. *Policy and Practice in Occupational Health and Safety*, 1, 41–68.

Storey, R. (2006) Social assistance or a worker's right: Workmen's compensation and the struggle of injured workers in Ontario, 1970–85. *Studies in Political Economy*, 78(Autumn), 67–91.

Storey, R. and Lewchuk, W. (2000) From dust to dust: Asbestos and the struggle for worker health and safety at Bendix Automotive. *Labour/Le Travail*, 45(Spring), 103–140.

Storey, R. and Tucker, E. (2006) All that is solid melts into air: Worker participation and occupational health and safety regulation in Ontario, 1970–2000. In V. Mogenson (Ed.), *Worker Safety under Siege: Labor, Capital and the Politics of Workplace Safety in a Deregulated World* (pp. 157–186). M.E. Sharpe.

Strutt, G. (1986) *Improving safety at (Inco) Frood Mine*. Paper presented at the MAPAO Annual Meeting. Toronto, May 28–30.

Surgeoner, G. and Roberts, W. (1993) Reducing pesticide use by 50% in the province of Ontario. In D. Pimenthal and H. Lehman (Eds.), *The Pesticide Question* London: Chapman and Hall, 206–220.

Suruda, A., Philips, P., Lillquist, D., and Sesek, R. (2003) Fatal injuries to teenage construction workers in the US. *American Journal of Industrial Medicine*, 44, 510–514.

Sustainable Farming (1994) *Quebec Moves to Develop Organic Farming*. Fall.

Suthers, J., et al. (1986) *Health Aspects of Worker Exposure to Oil Mists*. CONCAWE, Report No. 86/69, October.

Sweeney, B. and Mordue, G. (2017) The restructuring of Canada's automotive industry. *Canadian Public Policy*, 43(Supplement 1), S1–S15.

Swift, J. (1977) *The Big Nickel: Inco at Home and Abroad.* Between the Lines.

Taksa, L. (2009) Intended or unintended consequences? A critical reappraisal of the safety first movement and its non-union safety committees. *Economic and Industrial Democracy,* 30(1), 9–36. doi:10.1177/0143831X08099432

Tarasuk, V. and Eakin, J. (1995) The problem of legitimacy in the experience of work-related back injury. *Qualitative Health Research,* 5(2), 204–221.

Tataryn, L. (1979) *Dying for a living: The politics of industrial death.* Deneau and Greenberg Publishers.

Thompson, P. (1983) *The Nature of Work: An Introduction to Debates on the Labour Process.* London: MacMillan Press.

Thompson, P. (2010) The capitalist labour process: Concepts and connections. *Capital and Class,* 34(1), 7–14.

Thompson, P. and Smith, C. (2009) Labour power and labour process: Contesting the marginality of the of the sociology of work. *Sociology,* 43(5), 1–18.

Thu, K. (1998) The health consequences of industrialized agriculture in the United States. *Human Organization,* 57(3), 335–341.

Tilzey, M. and Potter, C. (2008) Productivism versus post-productivism? Modes of agri-environmental governance in post-fordist agricultural transitions. In G.M. Robinson (Ed.), *Sustainable Rural Systems – Sustainable Agriculture and Rural Communities* (pp. 41–66). Hampshire: Ashgate.

Tombs, S. (1996) Injury, death and the deregulation fetish: The politics of occupational safety regulation in UK manufacturing industries. *International Journal of Health Services,* 26(2), 309–329.

Tombs, S. and Whyte, D. (2007) *Safety Crimes.* Devon: Willan Publishing.

Trenoweth, S. 2003–05 Perceiving risk in dangerous situations: risk of violence among mental health inpatients. Journal of Advanced Nursing, 42 (3), 278–287.

Tucker, E. (1990) *Administering Danger in the Workplace: The Law and Politics of Occupational Health and Safety Regulation in Ontario, 1850–1914.* Toronto: University of Toronto Press.

Tucker, E. (1995) And the defeat goes on: An assessment of third wave health and safety regulation. In F. Pearce and L. Snider (Eds.), *Corporate Crime* (pp. 245–267). University of Toronto Press.

Tucker, E. (2003) Diverging trends in worker health and safety protection and participation in Canada, 1985–2000. *Relations Industrielles/Industrial Relations,* 58(3), 395–426.

Tucker, E. (2005) Will the vicious circle of precariousness be unbroken? The exclusion of Ontario farm workers from the occupational health and safety act. In L. Vosko (Ed.), *Precarious Employment: Understanding Labour Market Insecurity in Canada* (pp. 256–276). McGill-Queen's University Press.

Tucker, S., Diekrager, D., Turner, N., and Kelloway, E. (2014) Work related injury underreporting among young workers: Prevalence, gender differences and explanations for underreporting. *Journal of Safety Research,* 50, 67–53.

Tuohy, C. and Simard, M. (1993) *The Impact of Joint Health and Safety Committees in Ontario and Quebec.* Toronto: Report to the Canadian Association of Administrators of Labour Law.

U.S. Department of Labor (1978) *Occupational Health Guideline for Mineral Oil Mist.* Occupational Safety and Health Administration, September.

USWA, Local 6500 (ND) Job Classification Files, Local Union Office. Sudbury.

USWA, Local 6500 (1975) *Submissions to the Royal Commission on the Health and Safety of Workers in Mines*. Sudbury and Toronto. USWA, Local 6500 (1980) Submission to the Joint Federal Provincial Inquiry Commission into Safety in Mines and Mining Plants in Ontario (Burkett Commission). Submission by Mr. R. Ramsay, September 17.

USWA, Local 6500 (1981) Better lighting needed underground. *The Searcher* (Newsletter), 19(4), 1.

USWA, Local 6500 (1984) Submission to the Provincial Inquiry into Ground Control and Emergency Preparedness (Stevenson Commission).

USWA, Local 6500 (1985) Letter from D. Sweezy (Local 6500 OHS Chairman) to Norm Carriere (USWA, National Staff Representative).

USWA, Local 6500 (1987) *Technological Change at Inco and Its Impact on Workers.* Study sponsored by USWA Local 6500. Sudbury, Ontario.

USWA, Local 6500 (1982–86) Grievance Committee Records. Sudbury.

USWA, Local 6500/Inco (1984) Annual Report of the Joint Manpower Adjustment Committee.

USWA, Local 6500/Inco Ontario (1979/1982/1985) Collective Bargaining Agreements.

USWA, Local 6500/Inco Sudbury Operations (1982–85) Minutes of joint committee for cooperative wage support. Sudbury.

USWA, United Steelworkers of America, District 6 (1990) Submission to the Standing Committee on Resources Development on Bill 208, Act to Amend the Occupational Health and Safety Act. Toronto.

Vallas, S. (2003) Why teamwork fails: Obstacles to Workplace change in four manufacturing plants. *American Sociological Review*, 68(2), 223–250.

Vickers, I., Philip, J., Smallbone, D., and Baldock, R. (2005) Understanding small firm responses to regulation: The case of workplace health and safety. *Policy Studies*, 26(2), 149–169.

Villegas, P. (2014) 'I can't even buy a bed because I don't know if I'll leave tomorrow:' Temporal orientations among Mexican precarious status migrants in Toronto. *Citizenship Studies*, 18(3), 277–291.

Voice of the Essex Farmer (1992) Look for technology and innovation for economic prosperity. *Voice of the Essex Farmer*, March 31, p. 4.

Vosko, L. (2006) Precarious employment: Towards an improved understanding of labour market insecurity. In L. Vosko (Ed.), *Precarious Employment: Understanding Labour Market Insecurity in Canada* (pp. 3–42). Montreal and Kingston: McGill-Queen's University Press.

Vosko, L. (2010) *Managing the Margins: Gender, Citizenship and the International Regulation of Precarious Employment*. Oxford University Press.

Vosko, L., Akkaymak, G., Casey, R., Condratto, S., Grundy, J., Hall, A., et al. (2020) *Closing the Enforcement Gap: Improving employment Standards Protections for People in Precarious Jobs*. University of Toronto Press.

Wall, E. (1996) Unions in the field. *Canadian Journal of Agricultural Economics*, 44, 515–526.

Walls, J., Pidgeon, N., Weyman, A., and Horlick-Jones, T. (2004) Critical trust: Understanding lay perceptions of health and safety risk regulation. *Health, Risk and Society*, 6(2), 133–150.

Walters, D. (1996) Trade unions and the effectiveness of worker representation in health and safety in Britain. *International Journal of Health Services*, 26(4), 625–641.

Walters, D. (1997) Trade unions and the training of health and safety representatives: Challenges of the 1990s. *Personnel Review*, 26(5), 357–376.

Walters, D. (2003) *Workplace Arrangements for OHS in the 21st Century*. Working Paper 10, The Australian National University.

Walters, D. (2004) Worker representation and health and safety in small enterprises in Europe. *Industrial Relations Journal*, 35(2), 169–186.

Walters, D. (2006) One step forward, two steps back: Worker representation in Health and Safety in the United Kingdom. *International Journal of Health Services*, 36(2), 87–111.

Walters, D., Johnstone, R., Quinlan, M., and Wadsworth, E. (2016) Safeguarding workers: A study of health and safety representatives in the Queensland coalmining industry, 1990–2013. *Relations Industrielle/Industrial Relations*, 71–73, 418–441.

Walters, V. (1983) Occupational health and safety legislation in Ontario: An analysis of its origins and content. *Review of Sociology and Anthropology*, 20, 413–434.

Walters V. (1985) The politics of occupational health and safety: Interviews with workers, health and safety representatives and company doctors. *Canadian Review of Sociology and Anthropology*, 22(1), 56–79.

Walters, V. and Haines, T. (1990) Workers' use and knowledge of the internal responsibility system: Limits to participation in occupational health and safety. *Canadian Public Policy*, 14, 411–432.

Walters, V. and Haines, T. (1988) Workers' perceptions, knowledge and responses regarding occupational health and safety: A report on a Canadian study. *Social Science and Medicine*, 27(11), 1189–1196.

Weber, D., MacGregor, S., Provan, D., and Rae, A. (2018) We can stop work, but then nothing gets done: Factors that support and hinder a workforce to discontinue work for safety. *Safety Science*, 108, 149–160.

Weil, D. (2012) "Broken windows", vulnerable workers and the future of worker representation. *Labor in American Politics*, 10(1), 1–21.

Wells, D. (1986) *Soft Sell: Quality of Working Life Programs and the Productivity Race*. Ottawa: Canadian Centre for Policy Alternatives.

Westaby, J. and Lowe, J. (2005) Risk-taking orientation and injuries among youth workers: Examining the social influence of supervisors, coworkers, and parents. *Journal of Applied Psychology*, 90(25), 1027–1035.

White, B. (1987) *Hard Bargains: My Life on the Line*. McClelland and Stewart.

White, G. (2014) The Toyota way: How the automotive giant manages health and safety. *Global Manufacturing Newsletter*. www.manufacturingglobal.com, November 10 issue.

Wicks, D. (2002) Institutional bases of identity construction and reproduction: The case of undergound coal mining. *Gender, Work and Organization*, 9(3).

Wilde, G. (1999) The awareness and acceptance of risk at Westray. In C. McCormick (Ed.), *The Westray Chronicles: A Case Study in Corporate Crime* (pp. 97–116). Halifax: Fernwood Publishing.

Williams, K. (1995) Deregulating Occupational Health and Safety: Deregulation and Contracting Cut Act 1994, section 37. *Industrial Law Journal*, 24(2), 133–140.

Wilpert, B. and Ruiz Quintanilla, A. (1984) The German humanization of work programme: Review of its first twenty publications. *Journal of Occupational Psychology*, 57(3), 185–195.

Wilson, G.A. (2001) From productivism to post-productivism … and back again? Exploring the (un)changed natural and mental landscapes of European agriculture. *Transactions of the Institute of British Geographers*, 26(1), 77–102. doi:10.1111/1475-

Wilson, G. and Burton, R. (2015) Neo-productivist' agriculture: Spatio-temporal versus structuralist perspectives. *Journal of Rural Studies*, 38, 52–64.

Winson, A. (1992) *The Intimate Commodity*. Garamond Press.

Winson, A. and Leach, B. (2002) *Contingent Work, Disrupted Lives: Labour and Community in the New Rural Economy*. University of Toronto Press.

Womack, J., Jones, D., and Roos, D. (1990) *The Machine That Changed the World*. Rawson and Associates.

Workers Action Centre (WAC) (2007) Working on the Edge, [Online], Available: http://www.workersactioncentre.org/Documents/pdfs%20policy/Working%20 on%20the%20Edge-May%202007.pdf.

Workers Action Center (WAC) (2016) *Building Decent Jobs from the Ground Up*. Toronto. https://workersactioncentre.org/wp-content/uploads/2017/12/Building-Decent-Jobs-from-the-Ground-Up_eng.pdf.

Workplace Safety Insurance Board (WSIB), Ontario (2018) By the Numbers: 2018 WSIB Statistical Report. http://www.wsibstatistics.ca/

Wright, C. (1986) Routine deaths: Fatal accidents in the oil industry. *Sociological Review*, 34(2), 265–289.

Xia, N., X. Wang, M. Griffin, C. Wu and B. Liu (2017) Do we see how they perceive risk: An integrated analysis of risk perception and its effect on workplace safety behaviour. *Accident Analysis and Prevention*, 106, 234–242.

Xinhui, J., Chen, C., and Shi, K. (2013) Favor in exchange for trust. The role of subordinate attribution of supervisory favors. *Asian Pacific Journal of Management*, 30, 513–536.

Yassi, A., Lockhart, K., Sykes, M., Buck, B., Stime, B., and Spiegel, J.M. (2012a) Effectiveness of joint health and safety committees: A realist review. *American Journal of Industrial Medicine*, 56(4), 424–438.

Yassi, A., Spiegel, J., Buck, B., Sykes, M., Lockhart, K., and Stime, B. (2012b) *The Effectiveness of Joint Health and Safety Committees: A Systematic Review*. Vancouver, BC: WorkSafeBC.

Yates, C. (1993) *From Plant to Politics: The Autoworkers Union in Postwar Canada*. Temple University Press.

Yates, C., Lewchuk, W., and Stewart, P. (2001) Empowerment as a Trojan Horse: New systems of work organization in the North American automobile industry. *Economic and Industrial Democracy*, 22(4), 517–541.

Zejda, J., Semchuk, K., McDuffie, H., and Dosman, J. (1991) A need for population-based studies of health and safety risks in Canadian agriculture. *Canadian Medical Association Journal*, 145(7), 773–775.

Zinn, J. (2004) Health, risk and uncertainty in the life course: A typology of biographical certainty constructions. *Social Theory and Health*, 2, 199–221.

Zinn, J. (2008a) *Social Theories of Risk and Uncertainty: An Introduction*. Malden, MA: Blackwell Publishing.

Zinn, J. (2008b) Heading into the unknown: Everyday strategies for managing risk and uncertainty. *Health, Risk and Society*, 10(5), 439–450.

Zinn, J. (2015) Towards a better understanding of risk-taking: Key concepts, dimensions and perspectives. *Health, Risk and Society*, 17(2), 99–114.

Zuckerman, A. (1997) International standards desk reference: Your passport to world markets, ISO 9000, CE Mark, QS9000, SSM, ISO 14000, Q 9000 American European and global standards systems. New York: AMACOM.

Index